SILICON GERMANIUM

SILICON GERMANIUM

Technology, Modeling, and Design

RAMINDERPAL SINGH
DAVID L. HARAME
MODEST M. OPRYSKO

IEEE PRESS

WILEY-INTERSCIENCE

A JOHN WILEY & SONS, INC., PUBLICATION

Published simultaneously in Canada.

Library of Congress Cataloging-in-Publication Data:

Singh, Raminderpal.
Silicon germanium : technology, modeling, and design / Raminderpal Singh, David L. Harame, Modest M. Oprysko.
p. cm.
ISBN 0-471-44653-X (cloth)
1. Silicon. 2. Germanium. I. Harame, David Louis. II. Oprysko, Modest Michael, 1957– III. Title.

TK7871.9.S56 2003
621.39'732—dc22 2003057676

Printed in the United States of America.

10 9 8 7 6 5 4 3 2 1

To our wives and kids,
Thank you.

CONTENTS

Contributors ix

Foreword xiii

Preface xvii

Acknowledgments xxi

Acronyms xxiii

Introduction 1

A Historical Perspective at IBM 21

1 Technology Development 47

Overview 47
1.1 Active Devices 48
1.2 Technology Development: Advanced Passives and ESD Protection 58
1.3 Process Development 77
1.4 Technology Implications in SiGe Design 84

2 Modeling and Characterization 103

Overview 103
2.1 Predictive Modeling 104
2.2 Characterization 116

2.3 Compact-Model Development: Active Devices 127
2.4 Compact-Model Development: Advanced Passives 148

3 Design Automation and Signal Integrity 163

Overview 163
3.1 Design Automation Overview 164
3.2 ESD: Best-Practice CAD Implementation 180
3.3 Interconnect Extraction and Modeling 194
3.4 Substrate Noise Isolation and Modeling 217

4 Leading-Edge Applications 233

Overview 233
4.1 Wired Communications: SONET Design 234
4.2 Wireless Design: A Direct Conversion Receiver IC for WCDMA Mobile Systems 271
4.3 Wireless Design: Ericsson Power-Amplifier Design 296
4.4 Memory Design: A 32-Word by 32-Bit Three-Port Bipolar Register File 304

Appendix. Summary of IBM Foundry Offerings 319

Index 335

About the Authors 338

CONTRIBUTORS

David Ahlgren, IBM, Research Division, East Fishkill, New York

Herschel Ainspan, IBM, Research Division, East Fishkill, New York

Brent Anderson, IBM, Microelectronics Division, Essex Junction, Vermont

William E. Ausley, IBM, Microelectronics Division, East Fishkill, New York

P.-O. Brandt, Rensselaer Polytechnic Institute, New York

Troy Beukema, IBM, Research Division, Yorktown Heights, New York

Tony Bonaccio, IBM, Microelectronics Division, Essex Junction, Vermont

John Boquet, IBM, Microelectronics Division, Essex Junction, Vermont

Douglas Coolbaugh, IBM, Microelectronics Division, Essex Junction, Vermont

John D. Cressler, Georgia Institute of Technology, Georgia

Carl E. Dickey, RF Micro Devices, North Carolina

Jim Dunn, IBM, Microelectronics Division, Essex Junction, Vermont

Mete Erturk, IBM, Microelectronics Division, Essex Junction, Vermont

Natalie Feilchenfeld, IBM, Microelectronics Division, Essex Junction, Vermont

Brian Floyd, IBM, Research Division, Yorktown Heights, New York

Greg Freeman, IBM, Microelectronics Division, East Fishkill, New York

Daniel Friedman, IBM, Research Division, Yorktown Heights, New York

Matthew D. Gallagher, IBM, Microelectronics Division, Essex Junction, Vermont

Usha Gogineni, IBM, Microelectronics Division, Essex Junction, Vermont

David Greenberg, IBM, Research Division, East Fishkill, New York

Robert Groves, IBM, Microelectronics Division, East Fishkill, New York

Fernando Guarin, IBM, Microelectronics Division, East Fishkill, New York

Dean A. Herman, IBM, Microelectronics Division, East Fishkill, New York

Donald Jordan, IBM, Microelectronics Division, Essex Junction, Vermont

Jeffrey Johnson, IBM, Microelectronics Division, Essex Junction, Vermont

Alvin Joseph, IBM, Microelectronics Division, Essex Junction, Vermont

Michael P. Keene, IBM, Microelectronics Division, Essex Junction, Vermont

Rajendran Krishnasamy, IBM, Microelectronics Division, Essex Junction, Vermont

Peter Krusius, Cornell University, Ithaca, New York

Mukesh Kumar, IBM, Microelectronics Division, Essex Junction, Vermont

Young Kwark, IBM, Research Division, Yorktown Heights, New York

Lousi Lanzerotti, IBM, Microelectronics Division, Essex Junction, Vermont

Lawrence E. Larson, University of California San Diego, California

MICHAEL LIEHR, IBM, Microelectronics Division, Essex Junction, Vermont

JOHN F. MCDONALD, Rensselaer Polytechnic Institute, New York

MOUNIR MEGHELLI, IBM, Research Division, Yorktown Heights, New York

BERNARD MEYERSON,, IBM, Research Division, Yorktown Heights, New York

KIM M. NEWTON, IBM, Microelectronics Division, Essex Junction, Vermont

BRAD ORNER, IBM, Microelectronics Division, Essex Junction, Vermont

BENJAMIN PARKER, IBM, Research Division, Yorktown Heights, New York

SCOTT M. PARKER, IBM, Microelectronics Division, Essex Junction, Vermont

ULLRICH PFEIFFER, IBM, Research Division, Yorktown Heights, New York

VIDHYA RAMACHANDRAN, IBM, Microelectronics Division, Essex Junction, Vermont

JAE-SUNG RIEH, IBM, Microelectronics Division, East Fishkill, New York

MARK RITTER, IBM, Research Division, Yorktown Heights, New York

SCOTT REYNOLDS, IBM, Research Division, Yorktown Heights, New York

PARKER ROBINSON, Insyte Corporation, Florida

ALEXANDER PYLYAKOV, IBM, Research Division, Yorktown Heights, New York

LEI SHAN, IBM, Research Division, Yorktown Heights, New York

DAVID SHERIDAN, IBM, Microelectronics Division, Essex Junction, Vermont

STEPHEN ST. ONGE, IBM, Microelectronics Division, Essex Junction, Vermont

MICHAEL SORNA, IBM, Microelectronics Division, East Fishkill, New York

MEHMET SOYNER, IBM, Research Division, Yorktown Heights, New York

SAMUEL A. STEIDL, Sierra Monolithics, California

KENNETH STEIN, IBM, Microelectronics Division, East Fishkill, New York

Sue Strang, IBM, Microelectronics Division, Essex Junction, Vermont

Seshadri Subbhanna, IBM, Server Division, Poughkeepsie, New York

Susan Sweeney, IBM, Microelectronics Division, Essex Junction, Vermont

Youri Tretiakov, IBM, Microelectronics Division, Essex Junction, Vermont

Lars Tilly, Ericsson Mobile Platforms, Lund, Sweden

Ping-Chuan Wang, IBM, Microelectronics Division, East Fishkill, New York

Steven Voldman, IBM, Microelectronics Division, Essex Junction, Vermont

Dennis Whittaker, Insyte Corporation, Florida

Wayne H. Woods, IBM, Microelectronics Division, Essex Junction, Vermont

Jeffrey Yang, IBM, Microelectronics Division, East Fishkill, New York

Steven Zier, IBM, Microelectronics Division, East Fishkill, New York

Michael Zierak, IBM, Microelectronics Division, Essex Junction, Vermont

Thomas Zwick, IBM, Research Division, Yorktown Heights, New York

FOREWORD

IBM's silicon germanium (SiGe) program has been the focus of a great deal of external analysis, performed by both technical and business experts. This close and ongoing inspection has led to the generalized realization that SiGe-derived technology will find application in a great preponderance of high-performance chipsets produced worldwide within the coming few years. With the widespread adoption of bandgap-tailored technology [as in SiGe heterojunction bipolar transistors (HBTs)] and enhanced mobility device technology, revenues derived from such devices will likely exceed $30B in the 2005–2006 time frame. Although IBM led the way to the demonstration and commercialization of SiGe-based bandgap engineered device technology through the 1990s, the key to success was a constant search to ensure that the program moved swiftly by leveraging rather than reproducing the achievements of those who came before.

The genesis of this effort was the realization in the early 1980s that device scaling in homojunction silicon bipolar technology was in its terminal stages, requiring the identification and reduction to practice of a different strategy for performance enhancement. In addressing this challenge, early work by the now Nobel laureate, Herbert Kroemer, provided the foundation for the design of graded-base SiGe HBTs, notable for their efficient use of a given level of germanium to produce performance benefits. Similarly, great attention was paid to early work in silicon and germanium materials science and film growth. This rich background in research identifying roadblocks, ultimately led to the breakthroughs in SiGe materials quality required for reduction to practice of heretofore theoretical devices. This inclusive approach created a team spirit in the conduct of this endeavor that crossed technical, organizational, and company boundaries, greatly enriching the invention process through the diversity of experiences and learning that were brought to bear on the numerous challenges encountered.

The team, assembled early in the program, made seminal choices that have shaped the results presented herein, and ultimately the global industry. However, most fundamental to success was an adherence to absolute rather than relative standards. Competing with mainstream silicon technology, the team's focus was not to produce a better version of an inferior device, but rather the achievement of a best-of-breed device. This often required reaching out across and then beyond IBM, taking an inclusive approach to leadership.

I am fortunate to have had the privilege of leading this effort, but this program would have failed utterly were it not for teamwork at every juncture. In developing the required understanding of silicon surface chemistry for low temperature SiGe epitaxy, an entire community of chemists within and external to IBM rose up to the address the problem. In challenging the very nature of silicon surface chemistry, ultimately disproving widely accepted notions of native oxide formation on silicon, an entire field of low-temperature epitaxy was enabled. The "mainstream" silicon community within IBM contributed its best and brightest individuals to drive device design and technology integration in a field then thought by many to be a legacy of the past, i.e., silicon bipolar technology. Numerous alliances with commercial leaders in the application of HBT technology brought commercial insights and drove ever higher quality and performance standards. To this day, alliances begun almost a decade ago continue.

In a continuing demonstration of the worth of interdisciplinary teams, early 1990s work in the field of mobility-enhanced silicon technology has now transitioned to center stage. With the discovery of a class of mobility-limiting defects in strained alloy structures of silicon and germanium, and the means to suppress such defects, ultimate low-temperature mobilities possible in silicon were shown to be hundreds of times greater than thought feasible as late as the early 1990s. An explosion of study followed this revelation, and continues to today. Room temperature electron and hole mobilities in strained structures [high-electron mobility transistors (HEMTs)] were similarly improved by an order of magnitude by the mid-1990s, the device and circuit results of that period leading to commercial exploitation now well underway. The headroom this early work still provides to the industry is remarkable and valuable, in that the scaling of complementary metal-oxide transistor (CMOS) technology finds itself in the same endgame position today as bipolar technology occupied in the late 1980s.

This text cannot possibly present the entire history of this endeavor, nor reveal all achievements along the way, nor properly credit all who contributed, but the breadth and depth of what was accomplished should be noted as much for how it was done as for what was done. Having chosen to take a radical innovation from first concept to a multibillion dollar industry—while at the same time remaining deeply engaged with the broadest possible scientific and technical community—there is a source of immense pride within this team, which is arguably a major factor in its success. I hope this book serves to provide technical clarity to the reader as to underpinnings of SiGe technology, as well as a sense of the continued excitement and growth in this field.

Bernard Meyerson
IBM Fellow,
Chief Technology Officer
Technology Group,
International Business Machines

PREFACE

This book has the distinction of being the first of its kind presenting IBM's extremely successful endeavor into the silicon germanium (SiGe) market. Over the last 10–20 years, literally hundreds of technical experts have contributed to this effort, and it is with great pride that we are now able to publish the details of IBM's work in this area for the benefit of our readers.

GOALS OF THE BOOK

This book is aimed at radio frequency (RF)/analog and mixed-signal integrated-circuit (IC) designers, computer-aided design (CAD) engineers, semiconductor students, and foundry process engineers worldwide. The goals of this book can be summarized as:

1. To give the reader a thorough introduction to SiGe bipolar complementary metal-oxide transistors (BiCMOS) as a technology, including the history of its development at IBM.
2. To provide a detailed insight into the modeling and design automation requirements to enable leading-edge RF/analog and mixed-signal IC products.
3. To illustrate in-depth applications, implemented using IBM's advanced SiGe process technologies and design kits.

BOOK STRUCTURE

To this end, we have structured this book to cover all key aspects of technology and enablement. We begin with a brief introduction to and historical perspective of IBM's SiGe technology. Following this, the book is divided into four main chapters, as shown in Fig. 1:

- **Chapter 1.** Details of the many IBM SiGe technology development programs;
- **Chapter 2.** IBM's approach to device modeling and characterization, including predictive technology CAD (TCAD) modeling;
- **Chapter 3.** IBM's design automation and signal integrity knowledge and implementation methodologies, including best-practice implementation of CAD solutions for electrostatic discharge (ESD);
- **Chapter 4.** Design applications in a variety of IBM's SiGe technologies, including implemented wired, wireless transceiver, power amp, and high-speed memory designs.

In each section, we provide detailed coverage of the key issues together with a full set of measurement and simulation data, as well as references. In addition, an overview of IBM's SiGe and RF-CMOS offerings, in 2002, is provided in the Appendix.

Note that, as with any technical area, SiGe performance and enablement is an ever-moving target. The performance and data presented in this book can only be as up-to-date as the book's publishing schedule allows. As such, the data and discussions in this book are presented as accurate and current as of December 2002.

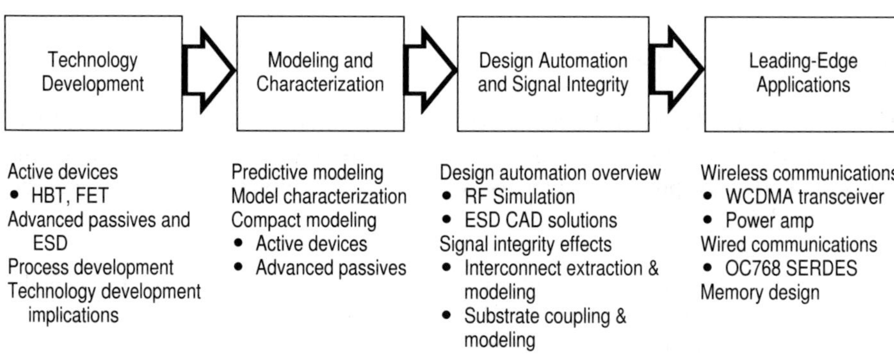

Figure 1 IBM provides a front–back enablement of the SiGe process technology family. This flow diagram shows the key points described in this book, and will be used to highlight the topic at hand, in chapter overviews throughout the book.

Please note that we (the authors) have made a determined effort to check that the content of this book is accurate, but do not guarantee the accuracy of the presented data.

RAMINDERPAL SINGH
DAVID L. HARAME
MODEST M. OPRYSKO

International Business Machines
October 2003

ACKNOWLEDGMENTS

Many technical experts have contributed to the content in this book. A complete list of all contributors to the content in this book is shown on page ix. The authors would like to thank these individuals for their contributions to this book and dedication to this technical field. We also extend our thanks to many technical and business line professionals, at IBM and in this industry as a whole, who have contributed to the success of *Silicon Germanium.*

Special thanks go to Dr. Bernard Meyerson, Chief Technology Officer of IBM's Technology Group. With his technical vision, determination, and leadership, the multitude of silicon germanium research and development projects in IBM have been very successful.

We also offer our thanks to the IEEE Press and John Wiley & Sons teams, who have done a professional and quality job in getting this book published. We would also like to thank Larry Cooke for his time and feedback, in reviewing this book. Finally, thanks go to the IBM team of Editors for the *Journal of Research and Development.*

ACRONYMS

3G	third generation cellular telephone protocol
ABB	analog baseband
AC	alternating current
ACLR1	adjacent-channel leakage ratio
ACPR	adjacent-channel power rejection
AD	drain area
A/D	analog to digital
ADC	analog-to-digital Converter
ADS	advanced-design system
AIM	adaptive integral methods
ALU	arithmetic logic unit
AMS	analog and mixed signal
AS	source area
ASIC	application-specific integrated circuit
ASTC	Advanced Semiconductor Technology Center
BANANA	boron artifact nonartifact nefarious anomaly
BBVGA	baseband variable-gain amplifier
BEOL	back end of the line (interconnects)
BER	bit error rate
BiCMOS	bipolar CMOS
BiFET	bipolar FET
BIST	built-in self-test

BJT	bipolar junction transistor
BLER	block-error rate
BSIM	Berkeley short-channel IGFET model
BV_{CEO}	emitter-collector junction breakdown voltage
CAD	computer-aided design
CB	collector base
CDF	component description form
CDMA	code division multiplexing access
CDR	clock and data recovery
CDS	Cadence Design Systems
CISP	circuit implementation of skin and frequency effects
CLM	channel-length modulation
CML	common-mode logic
CMOS	complementary metal-oxide transistor
CMP	chemical-mechanical polishing
CMU	clock multiplier unit
CPU	central processing unit
CPW	coplanar waveguide
CSIM	compact short-channel IGFET model
CV	capacitance–voltage
CVD	chemical vapor deposition
CW	continuous wave
2D	two-dimensional
DA	design automation
DAC	digital-to-analog converter
DARPA	Defense Advanced Research Projects Administration
DC	direct current
DIBL	drain-induced barrier lowering
divclk	divided clock
DJ	deterministic jitter
DMACS	device measurement and characterization system
DPSA	double-polysilicon self-aligned
DRAM	dynamic random-access memory
DRC	design rules checking
DR_f	free dynamic range
DSP	digital-signal processor
DT	deep trench
DUT	device under test
EAM	electroabsorption modulator
ECL	emitter-coupled logic
EDA	electronic design automation
EM	electroMagnetics
epi	epitaxial
ESD	electrostatic discharge
ETX	epitaxial transistor structure

eV	electron volt
FD	fully depleted
FDD	frequency-domain duplex
FDTD	finite difference time domain
FEC	forward error correction
FEM	focus exposure matrix
FEOL	front end of the line
FET	field-effect transistor
FIFO	first-in, first-out
FMM	fast multipole methods
FPGA	field-programmable gate array
GA	genetic algorithm
GaAs	gallium arsenide
GICCR	general ICCR
GMSK	Gaussian minimum shift key
GPIB	general-purpose interface bus
GPRS	General Packet Radio Services
GPS	Global Positioning System
GR	guard ring
GSG	ground–signal–ground
GSM	global system for mobile communication
GSPS	gigasamples per second
HA	hyperabrupt
HB	harmonic balance
HBM	human-body model
HBT	heterojunction bipolar transistor
HEMT	high-electron mobility transistor
HF	high frequency
HFSS	High-Frequency Structure Simulator
HiCUM	high current model
HiPOX	high-pressure oxidation
H-parameter	hybrid parameter
HPSK	hybrid phase-shift keying
HTO	hot thermal oxide
I/O	input/output
I/Q	in-phase quadrature phase
IC	integrated circuit
ICCR	integral charge-control relation
IEDM	International Electron Devices meeting
IF	intermediate frequency
IGFET	insulated-gate FET
IIPx	*x*th-order intermodulation intercept pont
ILD	interLayer dielectric
IM3	third-order intermodulation
IM5	fifth-order intermodulation

IMxD	*x*th-order intermodulation distortion
inP	indium phosphide
IP3	third-order intercept point
IP5	fifth-order intercept point
IsoNFET	isolated NFET
I-V	current–voltage
LAN	Local Area Network
LC-VCO	low-conduction VCO
LEFF	effective gate length
LFSR	linear feedback shift-register
LNA	low-noise amplifier
LO	local oscillator
LOCOS	LOCal oxidation silicon
LRP	limited reaction processing
LTCC	low-temperature cured ceramic
LTE	low-temperature epitaxy
LVS	layout versus schematic
MAG	maximum-available power gain
MBE	molecular-beam epitaxy
MEXTRAM	most exquisite transistor model
MIM	metal insulator metal
MIMCAP	metal insulator metal capacitor
MM	machine model
MOD	MIM over dielectric
MOM	method of moments
MOS	metal-oxide semiconductor
MOSCAP	metal-oxide semiconductor capacitor
MOSFET	metal-oxide semiconductor FET
MR	magneto resistive
MSST	mesa shallow trench isolated transistor
NBTI	negative-bias temperature instability
NF	noise figure
NPN	bipolar transistor with N-type emitter, P-type base, N-type collector
NS	n-subcollector
NSA	non-self-aligned
NRZ	nonreturn to zero
NTX	nitride self-aligned transistor structure
OEM	original equipment manufacturer
PA	power amplifier
PAE	power-added efficiency
PC	personal computer
PCB	printed circuit board
PCELL	parameterized cell
PCI	peripheral component interconnect
PCS	personal communications service

PD	drain perimeter
PDC	personal digital communications
PDK	process design kit
PEC	perfect electrical conductor
PECVD	plasma-enhanced chemical vapor deposition
PFD	phase-frequency detector
PIP	polysilicon–insulator–polysilicon
PLL	phase lock loop
POH	power-on hours
POR	Plan of Record
PPG	pulse pattern generator
PRBS	pseudorandom binary sequence
PRML	partial response maximum likelihood
PS	source perimeter
PSRO	performance sort ring oscillator
PSS	periodic steady state
PTAT	proportional-to-absolute-temperature
Q	quality factor = store power / dissipated power
RAM	random-access memory
RC	resistance–capacitance
R&D	research and development
refclk	reference clock
RF	radio frequency
RF-CMOS	radio frequency CMOS
RLC	resistance–inductance–capacitance
RMS	root-mean-square
ROO	region of operation
ROX	recessed field oxide
RTA	rapid thermal anneals
Rx	receive path
RxP	parallel Rx
RxS	serial Rx
RXB	raised extrinsic base
S/D	source/drain
SAW	surface acoustic wave
SCBE	substrate current-induced body effect
SEEW	selective epitaxial-emitter window
SEM	scanning electron microscope
SERDES	serializer–deserializer
SFDR	spurious free dynamic range
SGP	spice Gummel poon
SHF	super high frequency
Si	silicon
SiGe	silicon (Si) germanium (Ge)
SIMS	secondary ion mass spectroscopy

SLC	surface laminar circuit
SNR	signal-to-noise ratio
SOA	safe operating area
SOC	system on a chip
SOI	silicon on insulator
SONET	synchronous optical network
S-parameter	scattering parameter
SPC	statistical process control
SPICE	simulation program with integrated-circuit emphasis
SRAM	static random-access memory
STI	shallow-trench isolation
SX GR	simple p^+ guard ring
TaN	tantalium nitride
TAS	trans-admittance stage
TCAD	technology CAD
TCC	thermal coefficient of capacitance
TCR	low-temperature coefficient
TDDB	time-dependent dielectric breakdown
TEM	transmission electron micrograph
TIS	transimpedance stage
TLine	transmission line
TLP	transmission-line pulse
TSEP	temperature-sensitive electrical parameters
TW	shaped wire compact conductor
Tx	transmit path
TxP	parallel Tx
TxS	serial Tx
UHV	ultrahigh vacuum
ULSI	ultra large scale integration
UMTS	universal mobile telecommunications system
UTRA	UMTS terrestrial radio access
VBIC	vertical bipolar intercompany
VCC	voltage coefficient of capacitance
VCO	voltage controlled oscillator
VLSI	very large-scale integration
VSWR	voltage standing-wave ratio
WAN	wireless area network
WCDMA	wideband CDMA
XNOR	exclusive-NOR
Y-parameter	admittance parameter

INTRODUCTION

Today, silicon germanium bipolar complementary metal-oxide (SiGe BiCMOS) transistor is a well-established pervasive technology in the marketplace, and continues to be used in an ever-expanding number of commercial integrated-circuit (IC) products. SiGe BiCMOS technology has not only displaced III–V compound semiconductors [gallium arsenide (GaAs) and indium phosphide (InP)] in many communications applications, but more importantly has made many new applications and functions possible. Where did the idea of SiGe come from? How does it work in simple terms? What are its advantages? What are the latest advances in SiGe and SiGe designs? What has been and currently is IBM's role in this industry?

THE SILICON GERMANIUM HETEROJUNCTION BIPOLAR TRANSISTOR

The idea of using silicon and germanium together in IC transistors is as old as the transistor invention itself. In fact, in the first transistor patent filed by Shockley, there is a drawing that depicts a wide-band gap emitter on a narrow-band gap base, as shown in Fig. 1 [1]. Given that in 1947 there were no III–Vs, this had to be a combination of silicon and germanium.

In 1954, Herb Kroemer published "Zur theorie des diffusions und des drift transistors part III," in [2] in a little-known journal outside of Germany. This article contained the first hint of using "alloys" to guide carrier transport in seminconductors. Kroemer continued working on these ideas, and in 1957 published a landmark article, "Quasi-Electric and Quasi-Magnetic Fields in Non-Uniform Semiconductors" [3]. This was the first paper in which the concept of quasi-electric fields in semiconductors from alloy grading was described. This is the basic concept for today's SiGe heterojunction bipolar transistors (HBTs). The quasi-electric field idea can be explained with the aid of the band diagrams shown in Fig. 2. When a nega-

Silicon Germanium: Technology, Modeling, and Design. By Singh, Harame, and Oprysko
ISBN 0-471-44653-X

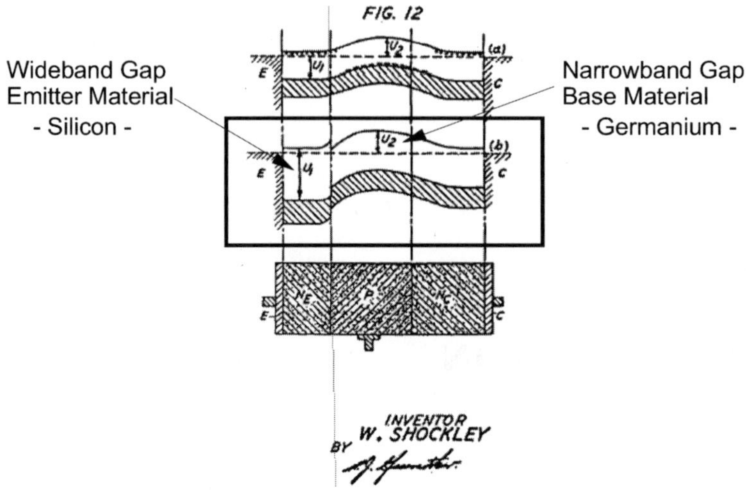

Figure 1 This figure is from Shockley's 1948 original transistor patent. It shows a homojunction band diagram and a heterojunction (wide bandgap emitter) band diagram for an NPN bipolar transistor. It shows the connections to the transistor regions. This was patented more than 50 years ago!!!

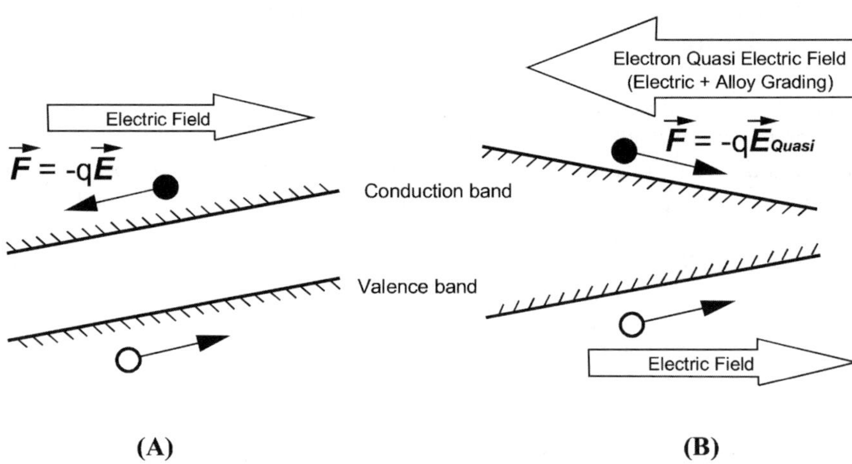

Figure 2 The principle of quasi-electric fields from alloy grading.

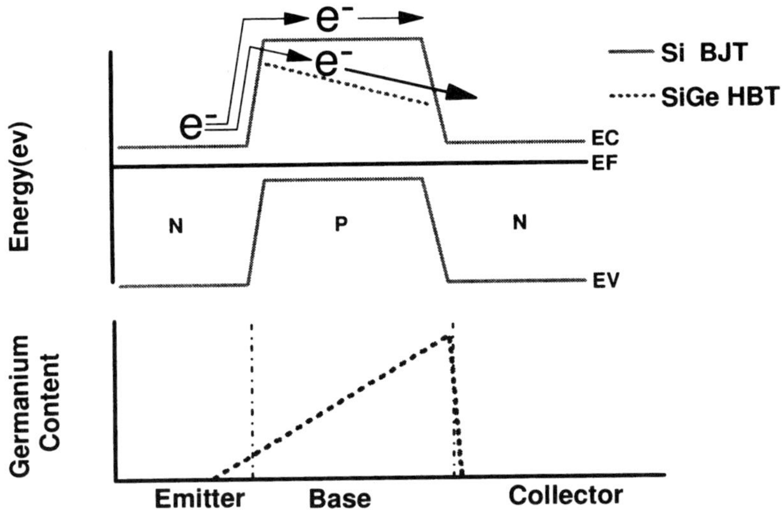

Figure 3 Bandgap diagram (top) showing reduction of conduction band resulting from graded doping of germanium (bottom) across the base region of the Si/SiGe HBT (dashed) in comparison to a conventional silicon-only bipolar junction transistor (BJT, solid). The grading of the germanium in the active device also creates a "drift field," further accelerating electrons across the base.

tive potential is applied to a uniformly doped semiconductor, the electron energy is increased and the hole energy is decreased. This will introduce an electric field across the semiconductor region. Electrons and holes will move in opposite directions in response to the field. The field is shown in the band diagram as a sloping conduction and valence band in Fig. 2A. Kroemer postulated that with alloy grading, the bandgap could be altered such that the electrostatic force could be overcome by a "quasi-electric" field that could change the direction of carrier transport, as is shown in Fig. 2B.

In a graded-base SiGe HBT, the Ge is graded across the base with the higher Ge content at the collector side. A profile and band diagram for a silicon homojunction and graded-base SiGe HBT is shown in Fig. 3. Note the valence and conduction band for the silicon homojunction transistor is flat, implying that there is no electrostatic field in the base. However, alloy grading of the Ge changes the band structure and introduces a quasi-electric field that drives the electrons across the base.

In the late 1980s—primarily through research-and-development (R&D) activities at IBM's T.J. Watson Research Center in Yorktown Heights, NY, USA—scientists and engineers began to focus on a new class of silicon devices. The goal of this work, the Si/SiGe heterojunction HBT, in effect mimicked the bandgap-engineered attributes of compound semiconductors in a silicon device. The addition of germanium to silicon-only technology to form a SiGe layer and heterojunction band structure has since created a revolution in the semiconductor industry.

THE IMPACT OF SiGe BiCMOS

Based on the preceding discussions, the primary advantages of using the SiGe process should be apparent, and can be summarized as:

- The SiGe epitaxial base layer seamlessly incorporates into the silicon fabrication process with the addition of one tool, a \$2.5M SiGe deposition tool.
- The result is 10× increase in speed over silicon bipolar transistor's speed at a fractional investment of the lithography-scaling approach.
- Modern CMOS technology enhances the performance generation to generation by scaling the device, a process that takes a major financial investment in new lithography tools.

Compared to InP, IBM's advanced 8HP SiGe process technology offering demonstrates clear performance advantages, as shown in Fig. 4. The figure also shows the 5HP, 7HP, and 8HP to compare to each other.

Some of the early payoff in using the Si/SiGe HBT was its ability to perform at very high speeds: e.g., 65-GHz maximum oscillation frequency (f_{MAX}) in IBM's earliest production technology, BiCMOS 5HP. Since device switching at these speeds is not necessary for the bulk of wireless circuits operating at frequencies from 900 MHz to 2.4 GHz, the usefulness of the SiGe HBT comes at being able to trade this excess speed for improvement in other device figures of merit; most notably operation at lower power levels (see Fig. 5).

The Si/SiGe HBT has also demonstrated the ability to provide excellent high-performance characteristics with very low noise, at high power gain, and with excellent linearity, all allowing designers wide latitude in solving challenges for specific circuit requirements. Additionally, due to SiGe's proven ability to achieve power-added efficiencies reaching 70% [4], the use of SiGe HBTs for power amplification is a very rich area of design activity.

Another essential requirement for successful development of advanced analog circuits is the availability of high-quality passive elements. IBM's SiGe technology has developed and integrated excellent passives, including high-Q inductors [5] and high-value metal insulator metal (MIM) capacitors [6]. The performance margin achieved from this combination of active and passive devices has given SiGe technology the leverage required to meet the stringent specifications for circuits designed for use in a wide range of wireless protocols: personal communications service (PCS) (digital), personal digital communications (PDC) (Japan), global system for mobile communications (GSM) (European standard), and code division multiple access (CDMA) as well as next generation WCDMA (3G cellular).

Table 1 summarizes some key aspects of SiGe that make it superior over other Si processes for radio-frequency (RF) IC design. Notably, SiGe BiCMOS offers the best solution for a *full* complement of key characteristics for RF design. In addition, another key point is that older, more mature SiGe BiCMOS offerings can often offer better yield and lower end costs than advanced CMOS processes. The additional cost associated with the SiGe bipolar devices and analog metal layers, is superseded by

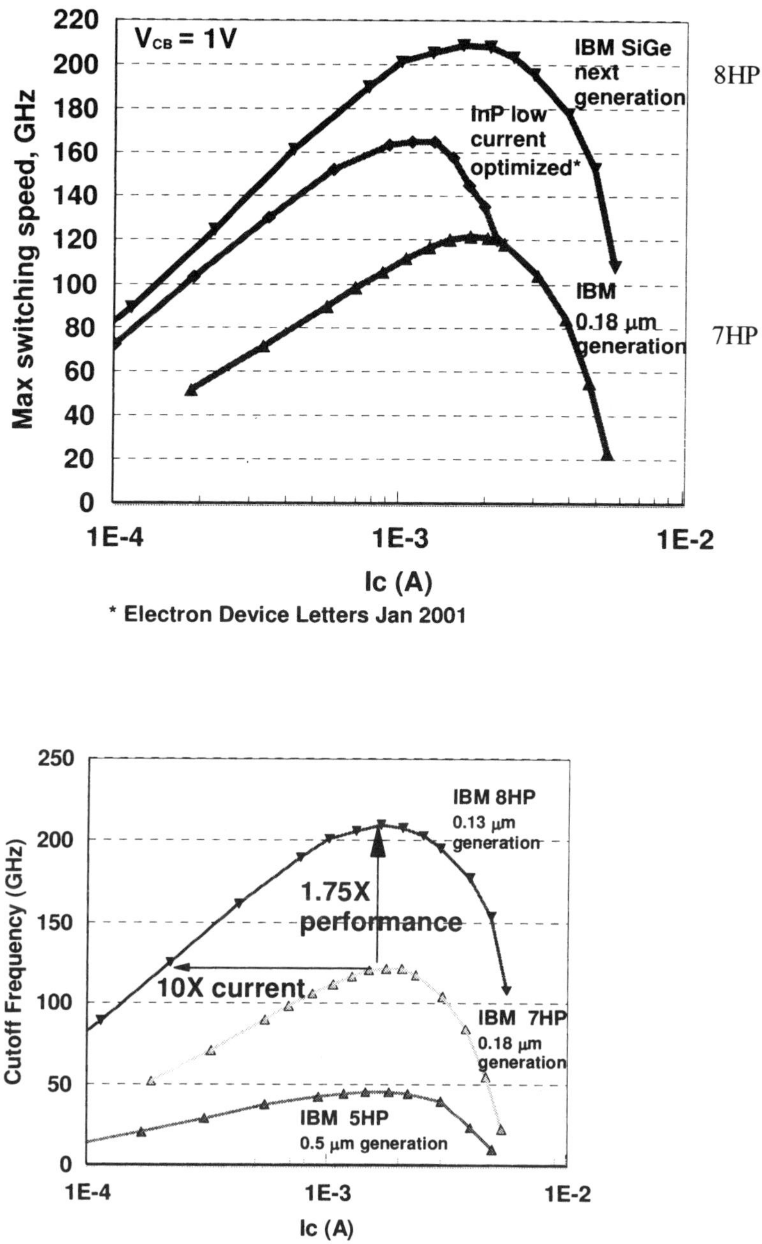

Figure 4 (Top) A look at how IBM's advanced SiGe offering, called 8HP, compares to 2002 InP offerings, as well as (Bottom) IBM's 5HP 0.5-μm and 7HP 0.18-μm process technologies.

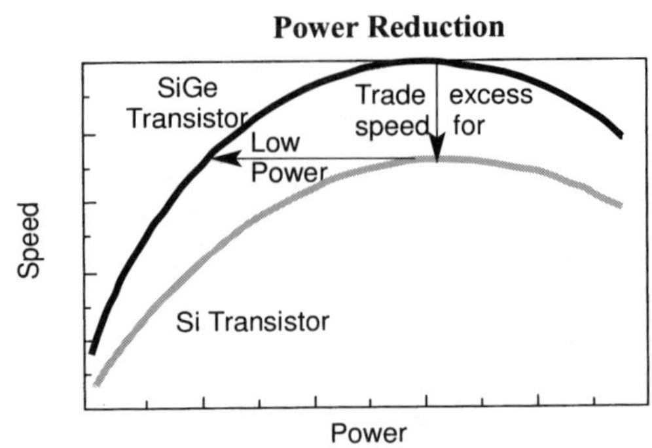

Figure 5 SiGe offers IC designers significant opportunities for circuit optimization. For example, by reducing operating currents, a designer can trade excess speed for substantially reduced power consumption in applications where power consumption rather than high-frequency operation is critical, such as wireless handsets.

the complexity and mask costs of advanced CMOS processes. For example, industry estimates for the 130-nm CMOS node put early mask costs in excess of $1M!!

Fundamentally why is the HBT a superior device for RF/analog design? Compared to CMOS the SiGe HBT has higher transconductance, lower $1/f$ noise, higher voltage capability, better matching, earlier availability of higher f_T, etc. Transport in the SiGe HBT is primarily vertical, and it is these transport properties that fundamentally determine the speed of the device, the cutoff frequency. Power gain (f_{MAX}) and parasitics in the device are primarily determined by lateral dimensions. Lateral and vertical scaling can proceed somewhat independently to optimize device performance. Many RF/analog figures of merit are weakly dependent on layout. In CMOS the speed of the device is strongly dependent on one lateral parameter, the gate length. Scaling difficulties are limiting how rapidly we scale the device vertically. The parasitics are also largely determined by lateral dimensions, particularly the gate width and number of fingers. Therefore power gain (f_{MAX}) and RF/analog characteristics are very strongly dependent on the device layout. This makes it

Table 1 Overall, SiGe Offers Many Key Performance Advantages over Other Silicon-Based Processes, for RF Wireless and Wired Communications Design

	Foundry CMOS	RF CMOS	SOI CMOS	SiGe BiCMOS
Low $1/f$ noise	×	×	×	✓
Low noise figure	●	●	●	✓
High breakdown	●	●	●	✓
High f_{max}	●	●	✓	✓
High Q inductors/MIMs	×	✓	✓	✓
Linear varacter with wide tuning	●	✓	●	✓

× Poor ● OK ✓ Advantage

much more difficult to design RF/analog circuits with CMOS than with the SiGe HBT, giving a significant time to market advantage for the bipolar.

Perhaps the biggest use for SiGe in the marketplace is its capability for integration with CMOS, providing unsurpassed value as a BiCMOS technology. For effective combination of the high-performance attributes of the HBT in analog function and the advantage that CMOS holds in design of very large digital designs, compatibility with ASIC design methodologies is crucial. The ability to produce these Si/SiGe devices (rivaling the performance of commercial compound semiconductors) in a standard CMOS silicon facility leverages the billions of dollars of development and capital investment already made by the industry. Thus by using the existing CMOS infrastructure, SiGe chips can be fabricated on 200-mm (8-in.) silicon substrates, with very low defect densities, at CMOS economies of scale.

Production of SiGe in the CMOS environment leverages tools and process maturity driven by low-cost CMOS requirements. In addition, the ability to integrate the SiGe HBT with a standard, application-specific integrated-circuit compatible (ASIC) CMOS technology makes possible the production of BiCMOS with unprecedented levels of integration. This paves the way for high-performance analog and RF circuits with dense CMOS logic and the production of a wide range of new products for wireless and wired communications.

This compatibility has been demonstrated [7] and is an essential element of all IBM BiCMOS technologies. In fact, SiGe BiCMOS chips have been fabricated with as many as 80,000 HBTs, with yields much higher than currently possible with III–V technologies. These dense analog designs may be combined with very high levels of CMOS gates, resulting in true system-on-a-chip levels of integration and with yields and associated chip costs that make production feasible. Additionally, the ability to eliminate large numbers of off-chip drivers and passive components also results in substantial power savings and a reduction of packaging costs.

SiGe BiCMOS DESIGN APPLICATIONS

The use of SiGe BiCMOS across the semiconductor marketplace is largely due to the flexibility that SiGe BiCMOS brings to a given designer's product requirements. There are basically three ways to use SiGe BiCMOS:

- At the device level, exploit the raw speed of the intrinsic SiGe HBT to design at frequencies unattainable by other silicon technologies, e.g., >20 GHz and 40 Gb/s.
- At the circuit level, trade-off excess gain-bandwidth of SiGe to achieve whatever figure of merit is of material significance for a given application: low power, high linearity, low noise, high dynamic range.
- At the system level, exploit the rich feature set of SiGe BiCMOS to rearchitect systems from the top down.

This design flexibility of SiGe BiCMOS is observed in a wide range of applications that use SiGe BiCMOS. SiGe has found an obvious home in microelectronic RF/analog products, with uses in low-noise amplifiers, voltage-controlled oscillators, mixers, and transceivers. High-performance analog designs include analog-to-digital converters, digital-to-analog converters, frequency synthesizers, IF filters and Global Positioning System (GPS) receivers. Other product design activities focus on SiGe power amplifiers. SiGe BiCMOS is also used for storage applications including high-speed partial-response maximum-likelihood (PRML) read channels, as discussed later in the chapter. In addition, the same process techniques that allow SiGe BiCMOS to be adapted for use in power amplifiers are being used to design state-of-the-art magnetoresistive (MR) preamplifiers for use in hard drives. In addition, SiGe BiCMOS has been able to meet the performance requirements for current cellular protocols and new standards for 3G cell phones, wideband code division multiplexing access (WCDMA), home wireless IEEE 802.11, and Bluetooth. SiGe also promises to be a major factor in emerging standards for new communications technologies. With the rapid expansion of the Internet and local- and wide-area networking, applications for data transport are in the forefront of the information system attention. SiGe has demonstrated that it is the cost/performance leader for 10-Gb/s synchronous optical network (SONET) (OC192) data rates, and has established its ability to achieve the demanding jitter performance requirements for the 40-Gb/s (OC768) market with IBM's latest production technology, SiGe 7HP featuring a 120-GHz HBT.

A sample of the wide variety of SiGe BiCMOS circuits, in order to illustrate wide-ranging applicability of these technologies, is shown in Table 2. This list demonstrates the advanced usage of IBM's SiGe processes.

Communications Applications

From the trickle of early circuits, there is now a flood of new SiGe products in almost every wired and wireless application area. The wireless applications continue to leverage the 0.5- and 0.25-μm generations where cost and time to market drive the technology choice. Wired applications such as SONET OC-768 (40-Gb/s transport) currently require 0.18- and 0.13-μm technology nodes where large-scale integration is possible and the high speeds are available with SiGe HBT 120 and 210-GHz f_T performance. In order to expand the application space for SiGe, derivative technologies with cost-reduced device offerings or application-specific device optimized processes have been developed.

Figure 6 shows how the telecommunications fiber bandwidth has grown over the years, and projects out into the future. The demands on the IC process technologies are shown to grow rapidly, as higher levels of digital circuit integration occurs, and a matrix of possible technology solutions arises in this application space. This matrix is shown in Fig. 7, which describes appropriate solution points for using RF-CMOS and SiGe. A similar matrix is shown for wireless telecommunications designs in Fig. 8. Both figures show how RF-CMOS and SiGe BiCMOS have a continuing role to play as the design requirements; i.e., RF-CMOS being more applicable for very high levels of integration (millions of gates), and SiGe playing a

Table 2 Circuits Demonstrated Using IBM's SiGe Technologies, Showing the Wide-Ranging Applicability and Utility of this Technology

Application/ Circuit	Comments	Figure of Merit	Reference
Model-hardware correlation			
Ring oscillators	ECL differential, typ. 250–300 mV swing	5HP–16 ps 7HP–9 ps, 8HP–4.2 ps	Nortel[a] unpublished (7HP, 8HP)]
Storage			
PRML read channel	Highly integrated BICMOS design	>75 MB/s (600 Mbit/s), product	IBM[b]
RF/WLAN (2–2.5 GHz)			
Integrated VCO	0.5 μm SiGe,tuning using MOS cap	2.5 V, –95 dBc/Hz @ 25 kHz, 9 mA core	IBM[c]
TDMA power amplifier	IS–54 compliant at 800 MHz		IBM [www.chips.ibm.com]
CDMA power amplifier	IS–95 compliant at 1800 or 1900 MHz		IBM [www.chips.ibm.com]
Wireless LAN chipset	3 SiGe chips and 1 CMOS chip to replace 8 GaAs chips	Commercial production part in PCM-CIA cards	Intersil[d]
Wireless downconverter	1V design, integrated transformer coupling and feedback	Mixer 2.5 mA, LNA 2.5 mA @ 1V, LNA 10.5 dB gain, 0.9-dB noise figure	Nortel[a]
2.5-GHz frequency synthesizer	3 bipolar, 4 CMOS blocks (200 HBTs, 2500 FETs, ~150 passives)	–91 dBc/Hz at 100 kHz offset, 2375–2550 MHz, 1-MHz spacing, 44-mW core	IBM[e]
GPS chipset	SiGe 0.5 μm BICMOS	Direct-conversion front end	IBM and SMI[f]
Microwave/WLAN (5+ GHz)			
Integrated VCO	Fully integrated *L*,*C*, varactor tank	To 26 GHz, 3 V, 22 mW core power, 3.6% tune, –84 dBc/Hz at 100-kHz offset	IBM[g]
Frequency divider	1.9 V, 0.5 μm SiGe BICMOS	1.9 V, 220 uA, 2.3–5.9 GHz	Nortel[a]
K-band static frequency divider	1/128, inductively peaked input buffer, 0.5 μm SiGe	23-GHz operation demonstrated	HRL[h]
Base station			
Digital, DAC chips	8-GHz clock, highly integrated	–140-dBc/Hz dynamic range	Siemens[i]
Networking			
Integrated VCO - 40 G	0.18 μmSiGe, fully integrated	To 25-GHz, digital coarse tuning w/ MOS cap.	IBM[j]
Broadband amplifier	0.5 μm SiGe BICMOS design	9-dB gain, 22 GHz BW, 6 dB NF	Nortel[k]
Broadband amplifier	For optical networking receiver	Upto 50-GHz bandwidth in 7 HP	AMCC[j]
High-gain amp	0.5 μm SiGe BICMOS design	Integrated 60-dB stable gain for 12.5 G	Nortel[l]

(continued)

Table 2 Circuits Demonstrated Using IBM's SiGe Technologies, Showing the Wide-Ranging Applicability and Utility of this Technology *(continued)*

Application/ Circuit	Comments	Figure of Merit	Reference
Networking (*cont.*)			
Dynamic frequency dividers	Building block, divide by 2	5 HP–50 GHz, 7 HP–up to 98 GHz	[unpublished], Anonymous
Multiplexer	For SONET applications	5 HP–12.5 Gbit/s, 7 HP–56 Gbit/s	IBM[j,m]
Demultiplexer	For SONET	Up to 45 Gbit/s	
SERDES	0.5 μm single-chip solution	12.5 Gbaud	IBM[n]
Modulator driver	Distributed large-signal amplifier	Up to 48 Gbit/s, 3.5 V pk–pk	AMCC [data-sheet]
Network switch	68×69, 150,000 SiGe HBTs	2.5 Gb/s, >200 Gb/s throughput	AMCC [data-sheet]
10 Gb/s chipset	Complete chipset for STM64/ OC-192 designed by Alcatel	Mux & demux, laser driver, preamp, limiting amp., CDR (data recovery)	Alcatel[o]
Data conversion			
D-to-A converter	Analog devices design	12 bits, >1-G sample/sec	ADI[p]
A-to-D converter	IBM research	4 bit, 8-G sample/sec	IBM[q]
$\Delta\Sigma$ modulator	Fourth-order, 0.5-μm SiGe, *LC* resonators w/Q enhancement	5 V, 350 mW @ 4 GHz, max SNR 53 dB, SFDR 69 dB (11 bits)	Carleton U[r]
Memory			
Bipolar Cache	RPI design	0.3-ns access time	RPI[s]
Ultrawide-Band			
Timing Generator Chip	0.5 μm SiGe HBT design	5 V,0.5 W, up to 2.5 GHz—2 ps accuracy, 10 ps jitter in a 100-ns window	TDSI/SMI[t]
Instrumentation			
Pin electronics driver			IBM[u]
Digital			
RISC engine	Simulation/ analysis of methods to achieve >16 GHz RISC engine	SiGe higher HBT-count, and CMOS integration has major benefits	RPI[v]
Highly integrated			
OC48 Mapper	0.5 μm SiGe BiCMOS	2.5 Gbps highly integrated	AMCC[d]
ASIC Test-Site	1.8 M CMOS, ASIC qual. vehicle	Equivalent to base 0.5 μm CMOS	IBM[w]
Radar			
X-band phase shifters	PIN diode ccts w/ thick metal add-on module (Hughes/ Raytheon)	2- and 3-bit fully integrated phase shifters at 6–10 GHz	Hughes/ IBM[d]

[a]L. Larson *et al., Tech. Digest IEEE ISSC,* pp. 80–81, 1996.

[b]S. St. Onge *et al.,* "A 0.24 μm SiGe BiCMOS Mixed-Signal RF Production Technology Featuring a 47 GHz ft HBT and 0.18 μm Leff CMOS," *IEEE Proc. BCTM,* p. 117, 1999.

(continued)

Table 2 *Continued*

[c]M. Mourant *et al., 2000 IEEE Radio Frequency Integrated Circuits Symp. Dig.*, pp. 65–68, June 2000.
[d]S. Subbanna *et al., Tech. Dig. Int. Electron Devices Meeting (IEDM),* pp. 845–848, 1999.
[e]M. Soyuer, H. A. Ainspan, M. Meghelli, and J.-O. Plouchart, "Low-Power Multi-GHz and Multi-Gb/s SiGe BiCMOS Circuits," *Proc. IEEE,* Vol. 85(10), pp. 1572–1582, Oct. 2000.
[f]J. Ceccherelli, *IBM MicroNews,* Vol. 1, pp. 38–40, Mar.2000.
[g]J.-O. Plouchart, B.-U. Klepser, H. A. Ainspan, and M. Soyuer, "Fully-monolithic 3-V SiGe Differential Voltage-controlled Oscillators for 5-GHz and 17 GHz Applications," *Proc. Eur. Solid-State Circuits Conf.,* pp. 332–335, Sept. 1998.
[h]M. Case *et al., Microwave Journal,* pp. 264–276, May 1997.
[i]A. Splett, H.-J. Dressler, A. Fuchs, R. Hofmann, B. Jelonnek, H. Kling, E. Koenig, and A Schultheiss, "Solutions for Highly Integrated Future Generation Software Radio Basestation Transceivers," *Proc. IEEE Custom Integrated Circuits Conf.,* pp. 511–518, May 2001.
[j]G. Freeman *et al., Tech. Digest IEEE GaAs Integrated Circuits Conf.,* pp. 89–92, 2001
[k]S. P. Voinigescu *et al., Tech. Digest IEEE Int. Electron Devices Meeting,* pp. 307–310, 1998.
[l]Y. M. Greshishchev *et al., IEEE ISSCC Dig. Tech. Papers,* pp. 382–383, Feb. 1999.
[m]D. Friedman *et al., Tech. Digest IEEE VLSI Circuits Symp.,* pp. 132–135, 2000.
[n]D. Friedman, M. Meghelli, B. Parker, J. Yang, H. Ainspan, and M. Soyuer, "A Single-Chip 12.5Gbaud Transceiver for Serial Data Communication," *IEEE VLSI Symp. Dig. Tech. Papers,* pp. 145–148, June 2001.
[o]T. Brenner, B. Wedding, and B. Coene, "Alcatel's Revolutionary 10 Gbps Transmission System Enabled by IBM's SiGe High-Speed Technology," *IBM MicroNews,* Vol. 5(1), pp.1–4, Mar. 1999.
[p]D. Harame *et al., Tech. Dig. IEEE Int. Electron Devices Meeting,* pp. 437–440, 1994.
[q]P. Xiao, K. Jenkins, M. Soyuer, H. Ainspan, J. Burghartz, H. Shin, M. Dolan and D. Harame, "A 4-b 8-Gsample/s A/D converter in SiGe bipolar technology," *IEEE ISSCC Dig. Tech. Papers,* pp. 124–125, Feb. 1997.
[r]W. Gao, J. A. Cherry, and W. M. Snelgrove, "A 4GHz Fourth-Order SiGe HBT Band Pass ΔΣ Modulator," Symp. VLSI Circuits Dig. Tech. Papers, pp. 174–175, June 1998.
[s]S. Steidl *et al., IEEE ISSCC Dig. Tech. Papers,* pp. 194–195, Feb. 1999.
[t]D. Rowe, B. Pollack, J. Pulver, W. Chon, P. Jett, L. Fullerton, and L. Larson, "A Si/SiGe HBT Timing Generator IC for High-Bandwidth Impulse Radio Applications," *Proc. IEEE Custom Integrated Circuits Conf.,* pp. 221–224, May 1999.
[u]Subbanna *et al.,* Slide Suppl. to *IEEE ISSCC Dig. Tech. Papers,* p. 387, Feb. 1999.
[v]S. Steidl, S. Carlough, M. Ernest, A. Garg, R. Kraft, and J. F. McDonald, "A 16GHz Fast RISC Engine Using GaAs/AlGaAs and SiGe HBT Technology," *Proc. IEEE Int. Conf. Innovative Systems in Silicon,* pp. 72–81, May 1997.
[w]R. Johnson *et al.,* "A 1.8 Million Transistor CMOS ASIC Fabricated in a SiGe BiCMOS Technology," *Tech. Dig. IEEE Int. Electron Devices Meeting (IEDM),* pp. 217–220, 1998.

key role in leading-edge high-speed applications. Notably, the most demanding space (up to 2002) in which SiGe products have found a home is wired telecommunications, specifically, OC-768 serialize–deserialize (SERDES) design. However, there are also active R&D programs in place, investigating the effectiveness of high-end SiGe processes for millimeter-wave applications, such as 60 GHz for high-speed wireless data, and 77 GHz for radar applications. For this latter application space, the 8HP and 9HP processes are well placed as effective solutions. Demonstrating the importance of SiGe for telecommunications, in Chapter 4, we present comprehensive details of a selection of wired and wireless IC designs at the leading edge of today's telecommunications' requirements.

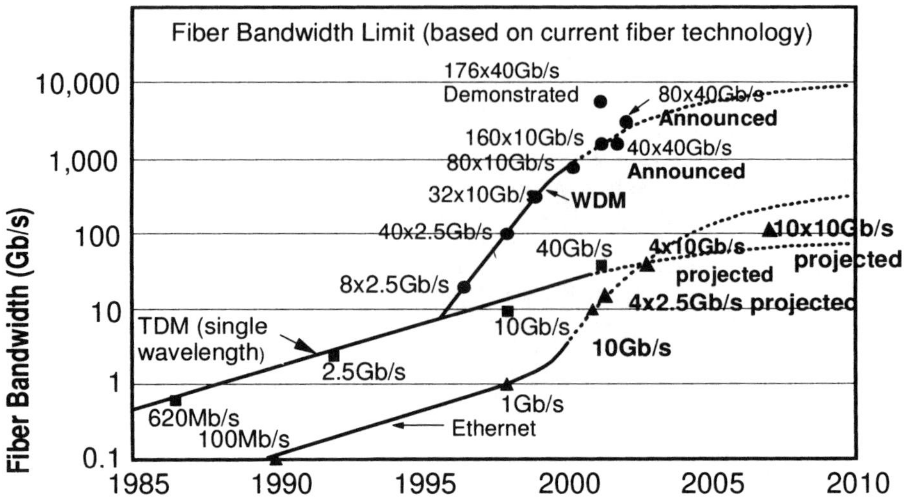

Figure 6 The timeline shown here depicts the historical and anticipated exploitation evolution of fiber bandwidth.

— **Frequency Rang e**—

CMOS Content	50-120 Gb/s	16-50 Gb/s	5-16 Gb/s	2-6 Gb/s	1-2 Gb/s	<1Gb/s
1M - 20M gates			100's channel SERDES Prizma	Network Processor		
100k - 1M gates			Switch Adapter			
20k - 100k gates		OC-768, SERDES, Mux-demux	OC-192, 10G Ethernet	XAUI Infiniband	OC-48	
negligible CMOS	laser driver, transimped. amplifier					

90 nm ↑ **CMOS** 0.25μm

0.13μm ← 0.5μm

SiGe BiCMOS

Figure 7 Wired communications market. Optimizing technology to the application space: Silicon, germanium, BiCMOS, and CMOS (bulk and SOI) have their own characteristics that make them fit well in certain areas.

--Frequency Range--

CMOS Content	5-10 GHz	2.5-5 GHz	2.0-2.5 GHz	0.9-2.0 GHz	< 0.9 GHz
100K - 1M+ gates	Highly Integrated Wireless Access SOC	Wireless MAC Solution Platform	Bluetooth SOC	FPGA Basestation Transciever Platform	baseband processor all protocols
20k - 100K gates	10GHz 802.11, 3G	UWB	ISM, 802.11	Integrated GSM Chipset	IF Digital Down Converters
1K - 20K gates	3G	5GHz 802.11 Integrated 2.5G	Bluetooth Front end	GSM, DECT, Chipsets GPS SOC	Handset Chipsets
Negligible CMOS	Basestation Front End	WCDMA 3G, UWB RF Front End	WCDMA, GSM Dual Band RF Front End	CDMA, GSM, DCS, PCS, Dual Band Front End	TDMA, CDMA, GSM Dual Band Front End

CMOS: 0.25μm → 90nm

0.13μm ← 0.5μm

SiGe BiCMOS

Figure 8 Wireless communications market. Optimizing technology to the application space. Silicon, germanium, BiCMOS, and CMOS (bulk and SOI) have their own characteristics that make them fit well in certain areas.

Storage Applications

SiGe HBTs play a dominant role in wired and wireless applications areas, but there are also many other important application areas. One important application is that of hard disk drive storage applications where the storage capacity continues to increase at a compounded annual growth rate of 60% [8]. BiCMOS technology has traditionally been used for read/write ICs or MR preamps [9], and PRML read channels [10].

There are both stringent cost and technical requirements for the read/write IC electronics. The write driver must pass current through an inductive write head generating magnetic fields for writing data on the disk surface. The required write currents can be as high as 120 mA_{PP}, and nanosecond current reversal times are called for [11]. A SiGe HBT is an ideal choice to meet the safe operating voltage requirements of 10 V, high speed operation, and low noise. For data retrieval, a high-speed disk drive requires a wide bandwidth, low-noise amplifier with typical input load conditions of $Z_0 = 50\ \Omega$, $\tau_{pd} = 175\text{-}250$ ps, and 25 Ω to 100 Ω for the read-transducer source impedance. The measured results, as seen in Fig. 9, for a differential low-noise amplifier using SiGe BiCMOS technology are well beyond the requirements for today's applications.

Storage density is achieved by increasing the linear density (bits per inch) on the disk media, which translates to faster data rates that have lower signal-to-noise ratio. PRML systems have a lower bit error rate than other detection systems, but re-

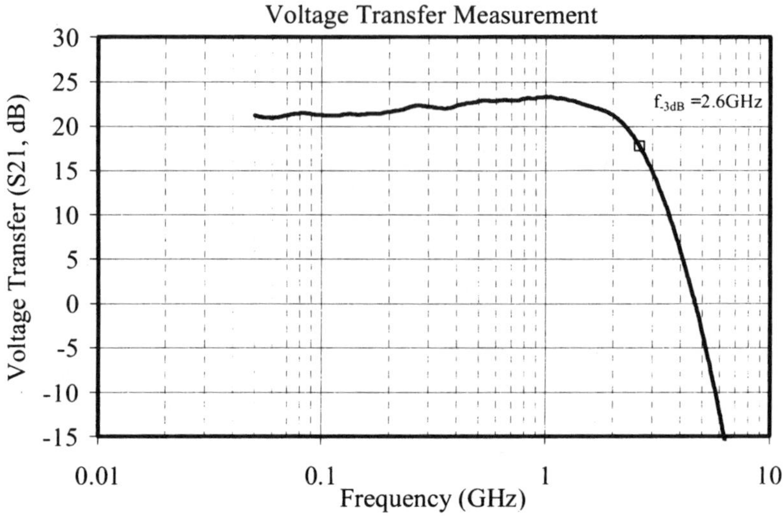

Figure 9 Measured voltage transfer for a wide bandwidth differential amplifier using a typical read transducer impedance value.

quire a complex analog and mixed-signal chip design. Such a chip was implemented in a 0.24-μm SiGe BiCMOS process [12]. The SiGe 6HP BiCMOS process boasts a 47-GHz f_T, 65-GHz f_{MAX}, and BV_{CEO} of 9.6 V. In addition, there are high-voltage 0.24-μm, 5-V FETs (5-nm gate oxide), 0.18-μm (LEFF) n field-effect transistor (nFET) and 0.18-μm LEFF pFETs, spiral inductors with a 4.0-μm thick aluminum metal, and a full complement of resistors and capacitors. A microphotograph of a successful IBM PRML read-channel product (SiGe 6HP), with functional blocks indicated, is shown in Fig. 10. The figure shows the significant digital integration, and the highly integrated analog blocks.

IBM'S FOUNDRY OFFERING

IBM has a comprehensive silicon foundry offering, which has been expanding rapidly in the last 2 to 3 years. Figure 11 shows a snapshot of IBM's silicon foundry offerings, as of December 2002. The figure shows how IBM provides a complete offering of SiGe BiCMOS and CMOS offerings, for both RF/mixed-signal and digital designers. The figure shows many supported IBM SiGe technologies and derivatives offered to foundry customers, demonstrating the aggressive development and productization program that IBM is committed to. In the Appendix, we present an overview of the key technical data in most of these technologies, as well as key RF-CMOS technologies offered by IBM in 2002.

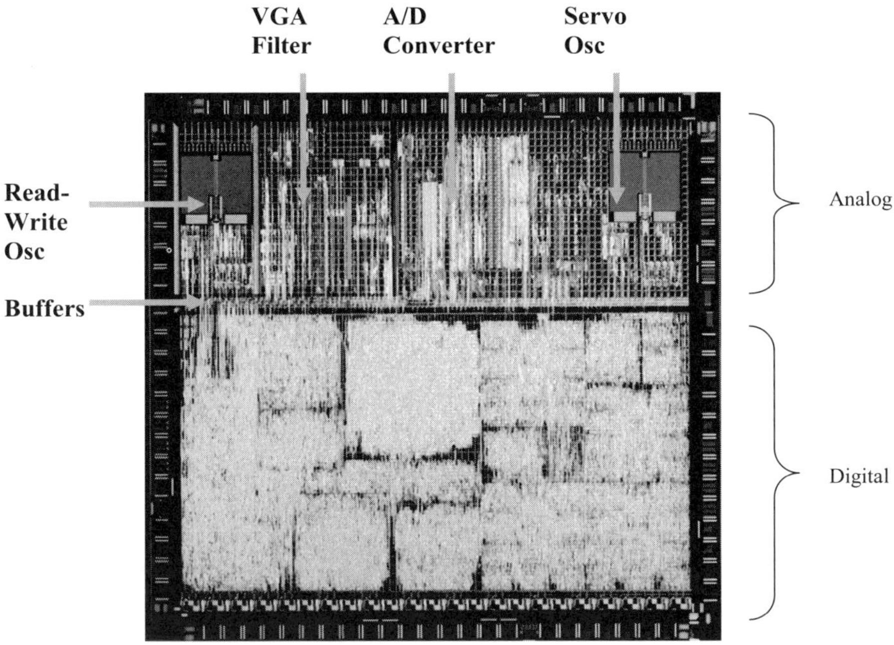

Figure 10 High-speed PRML read channel chip designed in 0.25-μm BiCMOS 6 HP technology offering world-class performance for IBM Storage System Division.

RECENT ACCOMPLISHMENTS IN 2002

Leading-Edge Transistor Performance

In December 2002, at the 2002 International Electron Devices Meeting (IEDM), IBM presented the latest results from its 9HP technology [13]. Figure 12 shows a cross-sectional photograph of the device built. The measurements, as shown in Fig. 13, show the highest reported f_T of 350 GHz for any Si-based transistor, as well as any bipolar transistor. The associated f_{MAX} is 170 GHz, and BV_{CEO} and BV_{CBO} are measured to be 1.4 V and 5.0 V, respectively. Also achieved was the simultaneous optimization of f_T and f_{MAX}, resulting in 270 GHz and 260 GHz, with BV_{CEO} and BV_{CBO} of 1.6 V and 5.5 V, respectively. These results demonstrate the continuing world class leadership of the IBM SiGe process technologies.

A MIXED-SIGNAL SiGe SYSTEM ON A CHIP

IBM has many close relationships with key customers. In 2002, IBM successfully fabricated a true mixed-signal system on a chip (SoC) for Insyte Corporation. The chip photograph is shown in Fig. 14. This design was a first-time success—one of

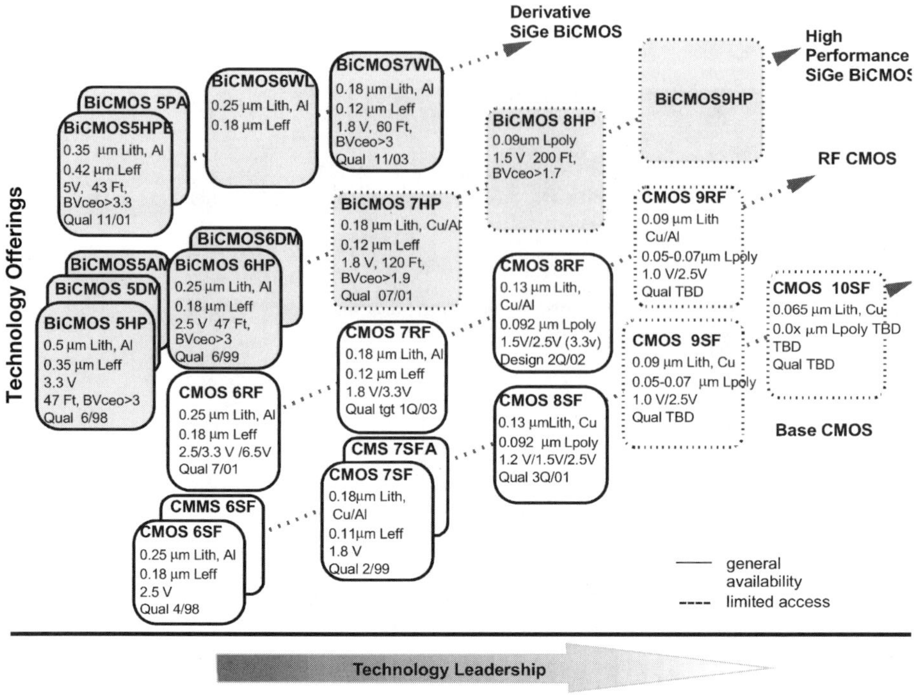

Figure 11 A snapshot view of IBM's foundry offerings.

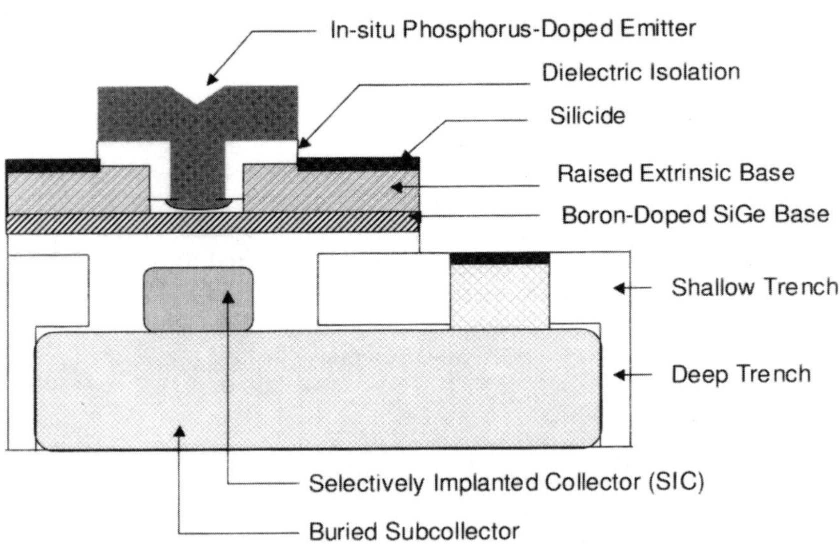

Figure 12 Cross section of the 0.13-µm 9HP HBT.

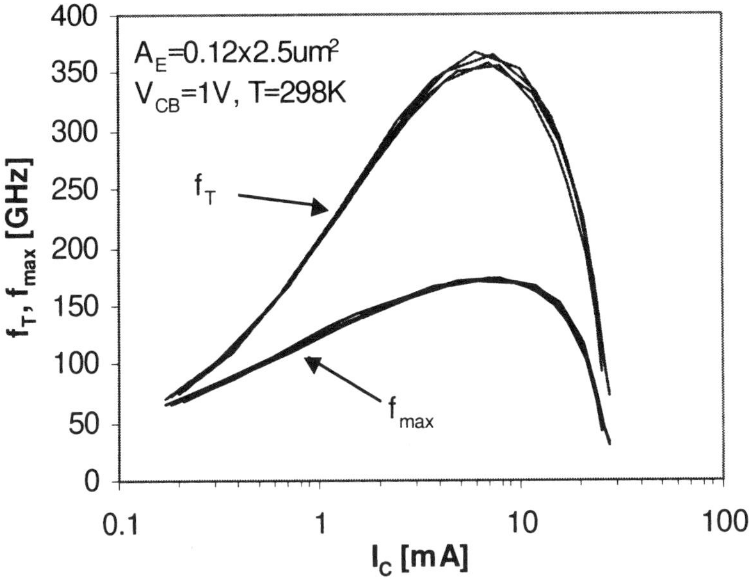

Figure 13 f_T and f_{MAX} of devices optimized primarily for f_T. The curves were taken from four sites across a wafer.

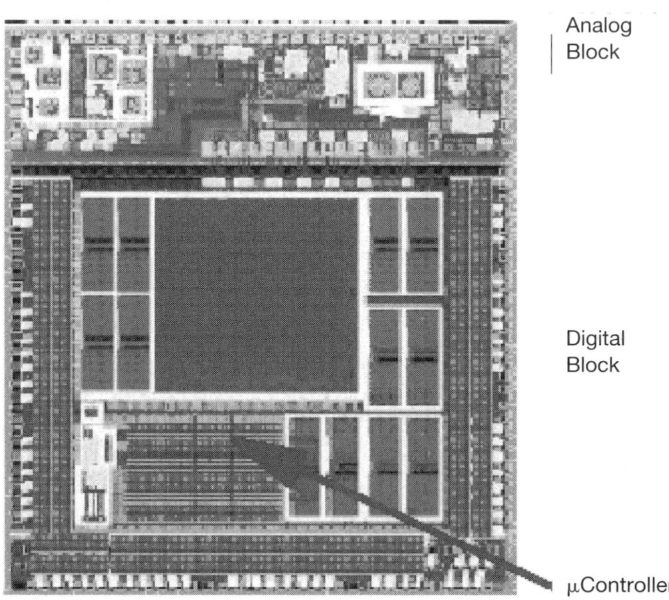

Figure 14 Microphotograph of silicon germanium 7 HP mixed-signal system-on-chip, with 7 M+ transistors. (Printed with kind permission from Insyte Corporation.)

many—demonstrating IBM's consistent ability to meet customer time-to-market and project-cost pressures.

Key characteristics of this design include:

- Proprietary protocol for point-to-point communication at 11-Mbps data rate
- Seven million CMOS transistors, and a BiCMOS RF/analog block
- Six levels of metal (4 CMOS + 2 thick dielectric add-on modules)
- Digital μcontroller clock rate: 100 MHz
- RF/analog block signal rate: 2 GHz

This design not only demonstrates the high levels of integration easily available using SiGe processes, but also shows the ease of compatibility for integrating common intercept points (IP), an increasingly important requirement for successful SoC design.

SUMMARY

There is no doubt about how much of a success story SiGe has been for the communications and IC industries and for IBM. This chapter has brought together many of the key points, hopefully answering the fundamental questions posed in the first paragraph: Where did the idea of SiGe come from? How does it work in simple terms? What are its advantages? What are the latest advances in SiGe and SiGe designs? What has been and currently is IBM's role in this industry? The following sections and chapters in this book will bring out many more details about each of the topics touched upon in this chapter.

REFERENCES

1. Patent numbers 2,502,488 and 2,524,035.
2. H. Kroemer, *Archiv Der Elecktrschen Ubertragung,* vol. 8, pp. 499–504, November 1954.
3. H. Kroemer, "Quasi-Electric and Quasi-Magnetic Fields in Non-Uniform Semiconductors," *RCA Review,* 1957.
4. D. Greenberg, M. Rivier, P. Girard, E. Bergeault, J. Mornz, D. Ahlgran, G. Freeman, S. Subbannu, S. J. Jeng, K. Stein, D. Nguyen-Ngoc, K. Schonenburg, J. Malinowski, D. Colavito, D. L. Harame, B. Meyerson, "Large-Signal Performance of High-BV_{ceo} Graded Epi-Base SiGe HBTs at Wireless Frequencies," *IEDM Tech. Digest,* pp. 799–802, 1997.
5. R Groves, J. Malinowski, R. Volant, D. Jadus, "High Q Inductors in a SiGe BiCMOS Process Utilizing a Thick Metal Add-on Module," in *Proceedings of 1999 BCTM,* 1999.
6. K Stein, J. Kocis, G. Hueckel, E. Eld, T. Barkush, R. Groves, N. Greco, D. Harame, T. Tewksbury, "High Reliability Metal-Insulator-Metal Capacitors for SiGe Analog Applications," in *Proceedings of 1997 BCTM,* p. 191, 1997.

7. R Johnson, "1.8 Million Transistor CMOS ASIC Fabricated in a SiGe BiCMOS Technology," in *Proceedings of 1998 BCTM,* 1998.
8. I. Ranmuthu, P. M. Emersen, K. Maggio, H. Jiang, A. Manjekar, B. E. Bloodworth, and M. Guastaferro, *IEEE J. Solid-State Circuits,* vol. 35, no. 6, pp. 911–913, June 2000.
9. D. P. Swart, and T. J. Schmerbeck, "An 8-Channel, Head Preamplifier for Combination Magnetoresistive Read Elements and Inductive Write Elements," *IEEE Int. Solid-State Circuits Conference, Paper FA 13.4,* pp. 218–219, 1993.
10. R. A. Philpott, R. A. Kertis, R. A. Richetta, T. J. Schmerbeck, and D. J. Schulte, "A 7 Mbyte/s (65 MHz), Mixed-Signal Magnetic Recording Channel DSP Using Partial Response Signaling with Maximum Likelihood Detection," *IEEE J. Solid-State Circuits,* vol. 29, no. 3, pp. 177–184, March 1994.
11. R. J. Reay, K. B. Klassen, and C. S. Nomura, "A Resonant Switching Swrite Driver for Magnetic Recording," *IEEE J. Solid-State Circuits,* vol. 32, no. 2, pp. 267–269, February 1997.
12. S. A. St. Onge, D. L. Harame, J. S. Dunn, S. Subbanna, D. C. Ahlgren, G. Freeman, B. Jagannathan, J. Jeng, K. Schonenberg, K. Stein, R. Groves, D. Coolbaugh, N. Feilchenfeld, P. Geiss, M. Gordon, P. Gray, D. Hershberger, S. Kilpatrick, R. Johnson, A. Joseph, L. Lanzerotti, J. Malinowski, B. Orner, and M. Zierak, "A 0.24 μm SiGe BiCMOS Mixed-Signal RF Production Technology Featuring a 47 GHz f_T HBT and 0.18 μm LEFF CMOS," in *Proceedings of 1999 BCTM,* pp. 117–120, 1999.
13. J.-S. Rieh, B. Jagannathan, H. Chen, K. T. Schonenberg, D. Angell, A. Chinthakindi, J. Florkey, F. Golan, D. Greenberg, S.-J. Jeng, M. Khater, F. Pagette, C. Schnabel, P. Smith, A. Stricker, K. Vaed, R. Volant, D. Ahlgren, G. Freeman, K. Stein, and S. Subbanna, "SiGe HBTs with Cut-off Frequency of 350GHz," *International Electron Devices Meeting,* pp. 771–774, December 2002.

A HISTORICAL PERSPECTIVE AT IBM

INTRODUCTION

Over the last decade, silicon germanium (SiGe) bipolar complementary metal-oxide semiconductor (BiCMOS) technology has become an important technology with many new and exciting product applications, as we discussed in the Introduction and as will be presented in detail in Chapter 4. Once only a research topic, SiGe-based heterojunction bipolar transistor (HBTs) are now found in a wide variety of technology offerings, and are comprehended in the product roadmaps of virtually every major company in telecommunications. The activities of IBM in the invention and commercialization of SiGe HBTs were key to the emergence of SiGe BiCMOS technology. In this chapter, we trace the early development and history of SiGe technology at IBM, and show its evolution into the product offering it is today. SiGe technology is reviewed from the initial motivation and developments to the products that exist today.

MOTIVATION

In the early to mid-1980s, IBM was using ion-implanted base bipolar technology for its mainframe computers. The ion-implanted bipolar device had been, and was being, successfully scaled from generation to generation; but a fundamental limitation of scaling, and thus a major disruption of IBM's technology roadmap, was in the offing. Through the 1970s and 1980s, the basic driver of improved bipolar device performance, and thus enhanced computing power, was the ability to make the bipolar transistor's base region narrower. Conventional silicon bipolar technology formed the base by ion-implanting boron into silicon wafers, and then using high-temperature anneals to drive dopant from a heavily doped polysilicon emitter into the implanted base profile [1]. In theory, the implanted boron profile was a well-

ISBN 0-471-44653-X

behaved Gaussian distribution; but, in fact, boron ions channeled down low-density "alleys" in the silicon crystal, which resulted in a significantly widened base profile with a long boron-channeling tail. This broadening of implanted profiles was further exacerbated by transient enhanced diffusion caused by point defects produced by the implant process itself [2]. An obvious solution to the channeling problem was to reduce the energy of the implant, and therefore the channeling tail. However, reducing the boron implant energy produces a very shallow implanted profile with the peak concentration of boron virtually coincident with the wafer surface. Driving emitter dopant into the now shallow implanted base profile results in a metallurgical emitter base junction with very high dopant concentrations, often in excess of 5×10^{18} atoms/cm^3. Overlapping two heavily doped regions in such a narrow base device would then lead to numerous problems, including excessive junction leakage due to band-to-band tunneling [3], with consequent poor device reliability; greatly increased emitter base capacitance deleterious to circuit performance; and poor base current ideality and the resultant device nonlinearity. There were many attempts to solve these problems with conventional silicon processing techniques [4], such as (a) reducing the total boron implant dose to decrease the field [5] at the emitter base junction, which failed due to high pinched base sheet resistance, high base resistance, and poor control; (b) implanting boron at a higher energy to set it back from the surface, but a wider base and retrograde field reduced carrier mobility, resulting in a slower-than-desired transistor; and (c) using a shallow implanted boron profile and high-pressure oxidation (HiPOX) to reduce the boron at the emitter base junction, which was an interim fix, but not scalable. A new paradigm was required if bipolar performance were to reach the $f_T > 60$-GHz target required for IBM's next-generation bipolar mainframe computers. In the physical sciences area of IBM's T.J. Watson Research Center, significant advances were occurring in silicon epitaxy that would provide a solution for the challenges just described.

THE INVENTION OF UHV/CVD

As practiced in the early 1980s, silicon epitaxy was a high-temperature process involving wafer prebakes in excess of 1100°C for surface preparation/cleaning. To achieve device quality layers, silicon epitaxy was performed at temperatures well in excess of 1000°C. Such thermal cycles are fundamentally incompatible with precision device formation, in that dopant diffusion and strained film relaxation rates are exponential in temperature, and virtually instantaneous at such temperatures. The desire remained, however, to form epitaxial layers of arbitrary dopant and chemical content (e.g., SiGe alloys), which would then enable epitaxial base device technology—the growth of active device regions in situ. In contrast with previous techniques involving ion implantation, a transistor grown at low enough temperatures (in the range $\leqslant$ 800°C) would enable the formation and maintenance of virtually arbitrary dopant and alloy designs. To address the need for low-temperature epitaxy, an effort was launched to understand the origins of the high thermal budget in silicon epitaxy and to develop a method to eliminate it.

Perhaps the most important event that occurred in the course of developing low-temperature epitaxy occurred years earlier, with the observation that bare silicon wafers etched in a high-frequency (HF) solution were hydrophobic (would not wet) for long periods of time subsequent to their being removed from the HF bath. The literature of the day stated that a thin layer of native oxide formed immediately on freshly etched silicon when the silicon was exposed to air, and grew to a terminal thickness within several hours [6]. The observed dewetting of silicon wafers hours after HF etching conflicts with the immediate formation of native oxide (which wets readily), therefore, surface science studies were performed to investigate the dewetting phenomena. We found that HF etching provided a passivation layer consisting of hydrogen-terminated silicon bonds across the silicon surface; this passivation reduced silicon's reactivity with air by more than 13 orders of magnitude. With this knowledge, we were able to eliminate the high-temperature thermal cycle associated with epitaxial growth prebakes, substituting an HF last-etch step.

Similarly, high temperatures were associated with attaining films of extreme crystalline perfection; yet, the values in the literature showed that silicon underwent solid-phase recrystallization at reasonable rates even in the range 500–600°C [7]. The source of imperfections during low-temperature growth was studied further, and a remarkable pattern emerged. We found that defects in silicon epitaxy originated from numerous factors, but low-growth temperature per se was not relevant. Ultimately, bistable conditions for high-quality epitaxy emerged [8]—a surprising and important finding.

If one began with a hydrogen-terminated silicon wafer, readily prepared by wet etching in HF, and exposed the wafer to a silicon-source gas such as silane (SiH_4), there would be no silicon growth until the wafer temperature rose high enough to desorb the initial hydrogen passivation layer. Hydrogen-passivated wafers heated to 500–600°C while exposed to a silicon-source gas produced films of extraordinary perfection well suited to high-performance integrated-circuit applications. In this temperature regime, SiH_4 decomposes on the surface of the wafer, incorporates silicon into the crystal, but *maintains* the hydrogen-passivated layer during growth. The rate at which SiH_4 decomposes and replenishes the hydrogen-passivation layer is higher than the rate of desorption of the hydrogen-passivation layer. The surface is protected from ambient oxygen contamination. Hydrogen-passivated wafers heated to 650°C before beginning film growth produced poor-quality films with very high defect densities. At 650°C the hydrogen-passivating layer is virtually instantaneously removed, allowing even low residual oxygen content in the growth ambient to cause the immediate formation of native oxide on the silicon surface, which causes high defect densities. Therefore, when silicon wafers are heated to temperatures greater than 650°C, initially hydrogen-passivated or not, a high-temperature pretreatment is required to remove the now present native oxide just prior to silicon epitaxial growth to achieve high-quality films. This finding of an unanticipated bistability in conditions for epitaxy is summarized in Figure 1. In the limit of an absolutely perfect vacuum, this bistability would disappear, but in the "real world," the loss of hydrogen passivation at the onset of epitaxy led to sufficient oxidation as to degrade resultant film quality.

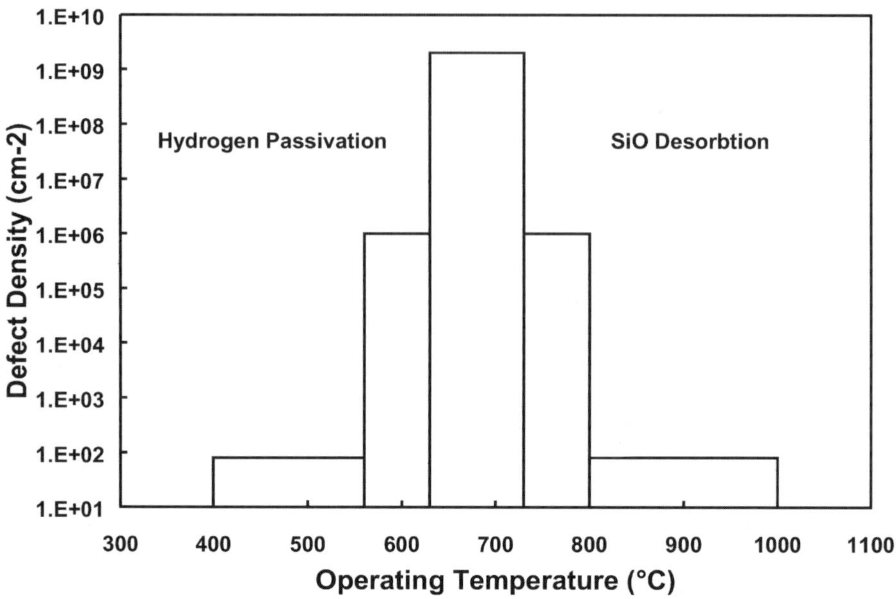

Figure 1 Bistability of epilayer perfection. Note that low density is achieved at low temperatures where the hydrogen passivation layer is maintained or at high temperatures where SiO is desorbed.

Hydrogen passivation meant that wafers could be wet-cleaned and handled in air, and yet have high-quality layers of epitaxial silicon formed upon them by growing at temperatures around 500°C. Utilizing this information and the knowledge of silane-surface and gas-phase chemistry fundamentals, resulted in the development of the ultrahigh vacuum/chemical vapor deposition (UHV/CVD) low-temperature epitaxy (LTE) technique [9]. UHV/CVD was developed based upon the systematic quantification of requirements for epitaxy in terms of silicon surface preparation, gas-phase contaminant limits, and deposition chemistry; it ultimately contributed greatly in providing the basis for the systematic preparation of the layers required to implement the SiGe epitaxial base transistor technology. By the late 1980s UHV/CVD film-growth tooling was mature and software was capable of taking any desired compositional profile of silicon, germanium, and boron, and translating it into the time, temperature, and flow inputs for the equipment.

UVH/CVD LTE and bipolar technology first came together in 1984. The resultant UHV/CVD materials already had been characterized using physical analyses such as secondary ion mass spectroscopy (SIMS) and transmission electron micrograph (TEM) (planar and cross section), and were determined to be of high quality by silicon industry standards. This turned out to be a critical decision; e.g., it set the bar on materials quality at or above the level found in commercial silicon materials of the day. Both prior and much subsequent work on low-temperature epitaxy emphasized analytical techniques such as cross-sectional TEM which, while very important for understanding the nature of defects encountered, were many orders of magnitude below the real coverage needed to determine the ultimate utility of materials so analyzed. To gain a macroscopic understanding of the material properties and quality, low-temperature prepared ($500 < T < 550°C$) epitaxial layers underwent electrical characterization using MOS structures, to measure lifetime and other basic characteristics of these silicon films. The initial data were promising, so the next logical step was transistor fabrication

THE FIRST EPI-BASE TRANSISTOR WITH UHV/CVD

It seemed obvious that the best place to use LTE was in fabricating the base of a bipolar transistor, replacing the conventional ion-implantation step by epitaxially growing the base. This had the advantage of bypassing the limits on scaling the implanted base profile imposed by the ion-implantation process, and boron dopant incorporated by epitaxy could be precisely controlled over an essentially arbitrary range of dopant concentration and dimensions. Given the absence of base-broadening due to channeling and ion-implant damage, if the silicon epitaxial layer could be grown in the form of a narrow dopant "spike," and if it were set back from the wafer surface by an undoped silicon spacer layer, the base dopant would then overlap the emitter dopant at low concentration, greatly reducing the fields at the emitter base junction. Narrow bases could thus be achieved without sacrificing device reliability. Two fundamental points to be proven were whether an in situ doped LTE film could be produced with sufficient dopant control, and if good bipolar device yield could be obtained. After first making various emitter-base and base-collector diodes [10] and designing some fairly simple in situ doped epitaxial base profiles, a transistor run was launched.

A process was put together to make the first epitaxial-base (epi-base) transistor by modifying an existing bipolar structure called nitride self-aligned transistor structure (NTX) [11], which was used in the IBM Research silicon line. After isolation was formed in the initial substrates, the wafers were ready for UHV/CVD epitaxy. As described earlier, UHV/CVD depends on the formation of a hydrogen-passivated surface that is formed by dipping the wafer in a dilute HF solution as a last step (HF last) prior to loading it into the growth chamber. The integrated lot wafers had deep-trench and recessed-field oxide (ROX), and were therefore primarily surface-terminated by hydrophilic regions of silicon dioxide. After the HF dip, residual HF solution had to be blown off the wafers with a nitrogen gun, a less-than-ideal means of removing the stray droplets of HF. The wafers were "blown" dry and loaded into the UHV/CVD system for epi-base growth. After the base was grown, the process was essentially identical to the existing NTX process up through metallization and final anneal, except for a problem with a processing step. One of the polysilicon depositions resulted in anomalously large grains that lead to rough silicon and some pitting in the single-crystal extrinsic base region. Therefore, when the wafers finished processing, it was with a great deal of anticipation (read that *fear*) that the participants (D. L. Harame, B. S. Meyerson, T. Nguyen, J. M. C. Stork) gathered in the lab to measure the current voltage characteristics of the first transistor. A Gummel characteristic measurement was taken in which the collector and base current were plotted on a logarithmically scaled vertical axis versus base emitter voltage on a linearly scaled horizontal axis. The Gummel characteristic plots data over many decades of current, and therefore quickly reveals any nonideal behavior in the transistor. The first transistor probed presented ideal bipolar Gummel characteristics! The SIMS impurity profile data taken on these wafers verified a narrow basewidth of 95 nm at a pinched intrinsic base sheet resistance of 8 kΩ/sq, which was quite impressive compared to the 150-nm basewidths at a much higher pinched-base sheet resistance achieved by advanced ion-implant technology at that time. The SIMS profile is shown in Figure 2. There was much excitement, and we decided to explore the implications of this new class of devices before publication. A paper was published on the fabrication some years later [12]. The different process splits resulted in some transistors with very heavily doped bases set back from the emitter base junction, an ideal aspect for cryogenic operation that had not been previously achievable in ion-implanted devices. These epi-base bipolar tran-

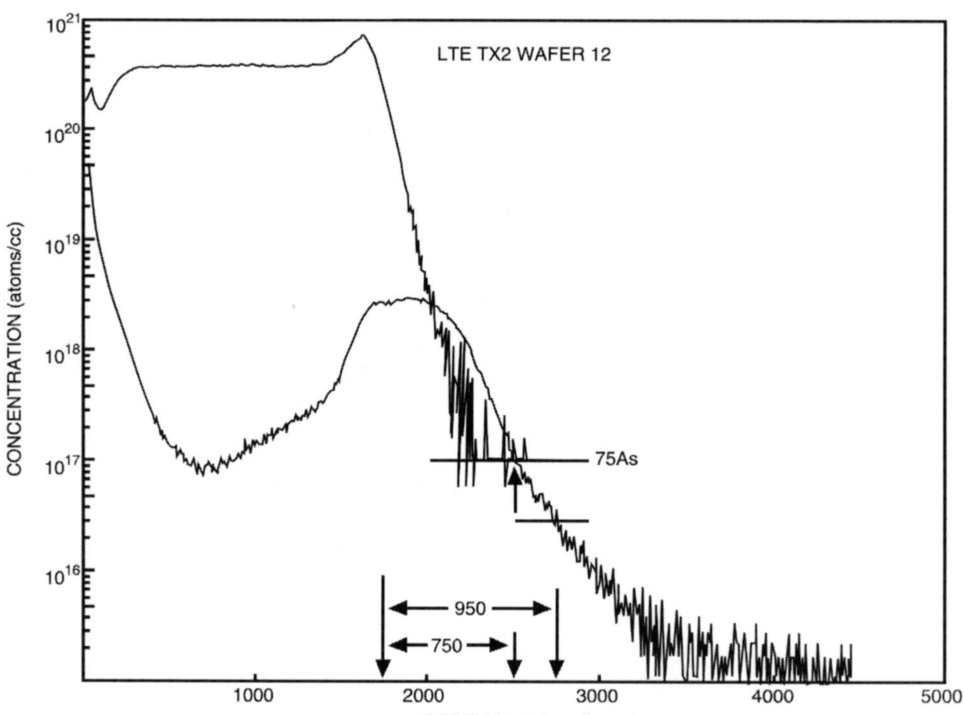

Figure 2 SIMS emitter base profile of a narrow LTE base transistor: Note the basewidth as measured from the notch in the boron profile (emitter junction) to a concentration of 1×10^{16} at/cm^2 is 95 nm.

sistors were characterized at temperatures down to liquid nitrogen, 77 K, and several papers were published without describing the fabrication details [13,14].

The bipolar transistor run initiated activity that used UHV/CVD epitaxy for applications in field-effect transistor (FET) devices. The leverage in an FET device was to replace the doped channel with an undoped Si or SiGe channel [15,16], which significantly improved the performance of the FET transistor but complicated the fabrication process. With a SiGe channel, the pFET was the improved device; the nFET did not increase in performance. This resulted in a lesser impact on CMOS circuit performance. Becasue scaling CMOS becomes more difficult, an epitaxial approach, with strained silicon, for instance, Refs. 17–20, may well become important. To date, the SiGe channel FET process has not had the impact on commercial MOS applications that it has had on bipolar; therefore, it will not be discussed here.

The success of the first epi-base bipolar transistor work was greeted with a mixed response by the IBM bipolar community. There were some who felt that epi-base would never displace conventional ion-implant technology and would forever be relegated to small-volume exotic applications. Others recognized the significance of the work in resolving the scaling limitations of conventional implant technology [4]. Many raised the legitimate question of whether any epi-base technology could achieve high yields when transferred to a production environment. This was the beginning of a friendly competition between epi-base and ion-implant technology that would continue inside IBM for several years. This internal competition helped develop epi-base technology by requiring that all the work be virtually common to the installed conventional silicon tool set and judged by the same standards.

THE FIRST SiGe BASE TRANSISTORS AT IBM

Having fabricated the first silicon epi-base bipolar transistor, the emphasis shifted to developing SiGe epi-base transistors. A secondary but related activity was optimizing transistor structures and process integration for epi-base transistors. In 1988, the materials growth techniques for SiGe were quite poor; there was a general preoccupation with the stability of SiGe layers [21], the processing temperatures they could withstand [22], and the growth techniques used to grow the layers. In IBM, three SiGe deposition techniques were being investigated: molecular beam epitoxy (MBE), CVD epitaxy using conventional CVD tools, and ultrahigh vacuum epitaxy (UHV/CVD or LTE).

The first SiGe-base mesa transistor inside of IBM was fabricated using MBE and low-temperature processing. The collector, base, and emitter layers were epitaxially grown by MBE without breaking vacuum. A mesa-defined transistor structure was fabricated using dry-etching techniques. A 6× increase in the collector current was measured for a device with 12% uniform germanium across a 100-nm basewidth, which corresponded to a total bandgap shrinkage of 45 meV [23]. This work was significant in that an enhancement in collector current was observed with the SiGe base, confirming the general predictions from physics. Later values of bandgap shrinkage for the SiGe layers used in this experiment would have predicted about 90 meV (75 meV for 10% Ge [24]), but poor material quality and possible relaxation may have altered these results. Although the thermal cycle was kept relatively low (<780°C), the current–voltage (*I-V*) characteristics, usually viewed as a Gummel plot when the collector and base current is plotted on a logarithmically scaled vertical axis versus base emitter voltage on a linearly scaled horizontal axis, were very nonideal and essentially unusable. To some observers, this confirmed that SiGe materials continued to be intractable and, ultimately, of little use. It quickly became a bit of a competition as to who would fabricate the first SiGe HBT with ideal Gummel characteristics. The literature was closely watched for outside competitors while internal work on all of the deposition techniques continued.

The first SiGe HBTs with a UHV/CVD-deposited SiGe base were fabricated about a year after the first MBE SiGe-base HBTs. A key decision in the device design was to build a conventional bipolar structure on patterned substrates and to use an in situ–doped graded SiGe-base profile, readily accomplished with the CVD technique. The processing and structures employed the same isolation and polysilicon emitter processes found in conventional bipolar technology, setting a new and important direction for all SiGe epitaxial base work—full compatibility with the silicon technology base of the day. A non-self-aligned transistor structure, in which the emitter opening was spaced away from extrinsic base or linkup regions, was used to simplify extracting vertical profile information from the resultant devices. The structure, shown in Figure 3, borrowed many process steps from the conventional double-polysilicon self-aligned transistor technology and addressed the HF last LTE preclean problem.

This modified process was identical to that of IBM's standard double-polysilicon self-aligned (DPSA) bipolar transistor process up through device isolation, reachthrough, and FET well implants. However, after the FET well implant steps, a thin layer of polysilicon was deposited over the entire wafer, making contact with the exposed silicon much the same as in a DPSA extrinsic base polysilicon process

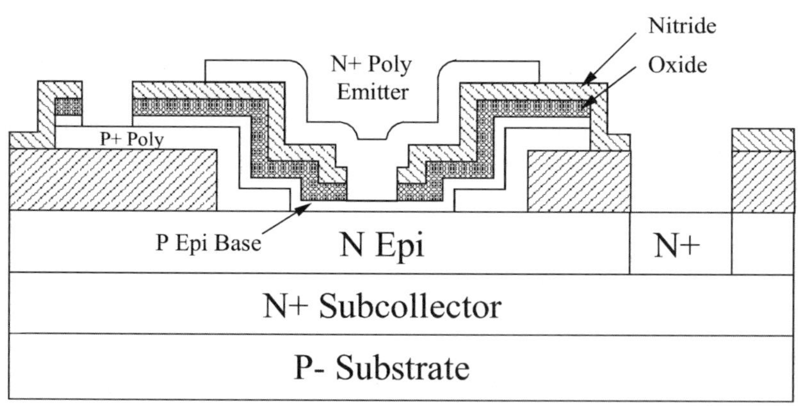

Figure 3 Schematic cross section of the non-self-aligned double-polysilicon structure.

[25]. With the modified process, the polysilicon is implanted with boron, patterned, and etched away from active areas. At this point only silicon is exposed on the wafer, rendering the entire wafer surface hydrophobic. With an all-silicon surface exposed, the LTE preclean HF dip is rendered a routine process, because the now-hydrophobic wafer surface dewets readily, eliminating residual liquid HF and allowing automation of what had been a manual process. The 500–550°C nonselective epitaxial growth forms single-crystal regions above the open base windows, and polysilicon over all other areas. The SiGe-base growth is followed by depositing dielectrics and forming a non-self-aligned emitter opening inside the extrinsic base polysilicon features. The emitter polysilicon, contacts, and interconnect modules completed the process. The term "polyprotect" layer was coined for the general use of polysilicon over the wafer surface to protect underlying structures during the HF last LTE preclean. A full BiCMOS version of this process was designed based on the strategy of growing the gate oxide, covering it with a thin layer of polysilicon, removing the polysilicon over the bipolar area, and subsequently growing a nonselective silicon layer, which would merge with the polyprotect layer to complete the polysilicon stack over the FET devices.

Even though these graded SiGe-base NPN HBTs were subjected to the same thermal cycle as found in conventional polysilicon emitter processes (including a furnace anneal at 850°C, and a rapid thermal anneal (RTA) at around 1000°C), ideal *I-V* characteristics or Gummel plots were achieved. The survival of high-quality devices largely debunked the previously widespread notion that the SiGe base was "fragile" and would tolerate only low-temperature processing subsequent to the base layer's formation. In fact, a SiGe layer designed to be unconditionally stable could be treated in the same manner as a conventional silicon layer [26,27].

The direct current (DC) characteristics of this run were nicely summarized by a plot of collector current density versus pinched base sheet resistance for the homojunction control wafers and the SiGe-base transistors shown in Figure 4 [28]. The collector current is linearly proportional to the pinched-base sheet resistance over the range plotted here, and the SiGe HBT collector current is enhanced over the silicon homojunction device by a factor of 10. Unfortunately, alternating current (AC) characteristics could not be obtained on these wafers due to inadequate calibration structures. In the same time frame, a joint Stanford University and Hewlett-Packard team were working on a new growth technique that controlled the growth by rapidly controlling the wafer temperature [29–32]. The technique, referred to as limited reaction processing (LRP), was said to be capable of growing arbitrary profiles with limited thermal substrate exposure [33]. With LRP, the Stanford HP group was the first to report a SiGe HBT with *ideal* Gummel characteristics at the device research conference [34]. This was a bit of a surprise, because the growth system was not load-locked and was known to have high oxygen concentrations in epitaxial grown films [35–37], which did not seem to affect the ideality of the device. Clean AC measurements on a SiGe HBT had still not been achieved and this served as the next target milestone.

A small boron interfacial peak was found at the initial growth interface of the LTE base, as reported in the SIMS profile [28], which was later traced to a low-lev-

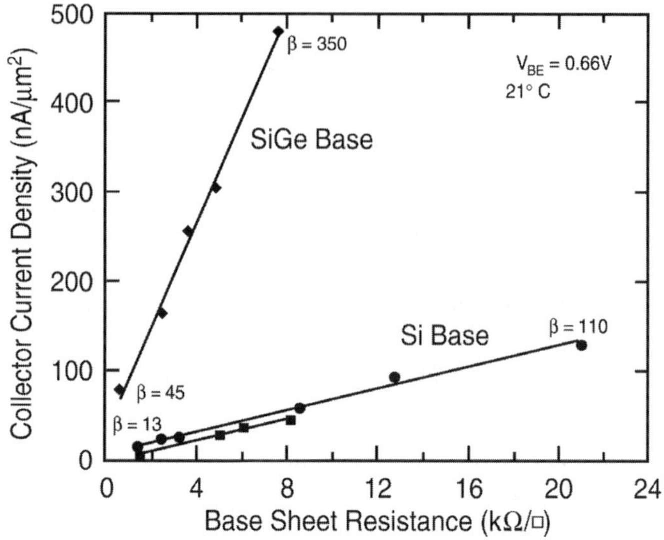

Figure 4 Base resistance leverage of SiGe-base devices. A 0–15% graded Ge profile results in a collector current that is 10 times higher than that in Si devices having the same process and pinched-base sheet resistance.

el leak in a boron flowmeter. In high injection, the presence of the boron created barrier effects that were studied in detail. Both phenomena were suppressed, but not without some invention.

THE BANANA

The presence of a boron interfacial peak was a great concern and considered to possibly be a significant problem for the SiGe technology. Although interfacial dopant contamination is well known in many growth techniques, it had not been observed in prior work with UHV/CVD. The challenge lay in proving the origin to be a flowmeter "through" a leak in the boron source diluted to such an extent that it was undetectable in the system, even with the mass spectrometer integrated within the growth apparatus. Although dopant sources at that time were typically diluted in hydrogen, helium gas was employed as the diluent for dopants employed in UHV/CVD. Helium is readily detected and has no natural abundance. The helium diluent ultimately enabled the fault tracing of what would otherwise have been accepted as a well-known interfacial contamination effect seen in many MBE and CVD processes. Although this deleterius effect was comical once resolved, it occupied the time of a large technical team for several months and ultimately became known as the "BANANA," the boron artifact nonartifact nefarious anomaly. We are not making light of this episode; rather, we are pointing out that in the course of taking this technology from the first successful demonstrations to general commercial acceptance, there were many instances where the combined technical efforts of a large team of experts, rather than the talent of an individual or two, was required to flatten what was initially cited as a fundamental technical showstopper.

PNP SiGe TRANSISTORS

PNP SiGe base transistors were also fabricated in the same time frame as the first NPN SiGe-base transistors. Epi-base PNP transistors were particularly difficult to fabricate, because it required achieving abrupt N-type dopant profiles reaching peak concentrations around 1×10^{18}–1×10^{19} atoms/cm^3. The first SiGe HBT PNP transistors were grown with an N-type MBE-deposited epi-base and a single crystalline boron-doped emitter deposited by UHV/CVD. The MBE N-type base films, as with the MBE boron doped layer before, had very high defect densities, which caused considerable leakage and poor Gummel characteristics. However, the f_T was measured on the SiGe-base PNP transistors, and it was higher than the control homojunction Si-base PNP transistors, or any other existing PNP transistors. This was the first known instance of a SiGe-base enhancing the AC performance of a bipolar transistor. It was also the highest f_T published for any silicon PNP [38]. Later work on PNPs fabricated entirely with UHV/CVD led to a series of 55-GHz PNPs [39], a record that still stands.

75 GHz f_T SiGe-BASE TRANSISTORS

Work continued on the SiGe-base NPN with an improved set of calibration structures designed to capture the AC performance. An extensive and detailed modeling effort using a custom one-dimensional simulator [40] and existing 2-dimensional modeling programs (2DP [41] and Fielday [42,43]) redesigned the SiGe profiles. Again, the structure chosen for the SiGe base NPN run was a simple non-self-aligned structure. We decided to use the UHV/CVD technique to grow the SiGe-base for this experiment and all future bipolar experiments. The device run finished and good DC characteristics were achieved. The AC performance was measured for a week with the new AC calibration structures to guarantee accurate results. Recall that this was 1988, and most silicon f_T measurements were only taken to 3 GHz and then extrapolated to the frequency where device gain degraded to unity, the definition of f_T. The higher-frequency measurement equipment and structures employed here were more typical of III–V efforts than silicon work of that time. The calibration structures worked well, and f_T values of 75 and 52 GHz were achieved, respectively, for SiGe and Si-base bipolar transistors with intrinsic-base sheet resistances in the 10–17 kΩ/sq range. These results represented the highest reported values ever achieved in silicon technology by almost a factor of 2, and extended the speed of silicon bipolar devices into a regime previously reserved for gallium arsenide (GaAs) and other compound semiconductor technologies. The magnitude of this

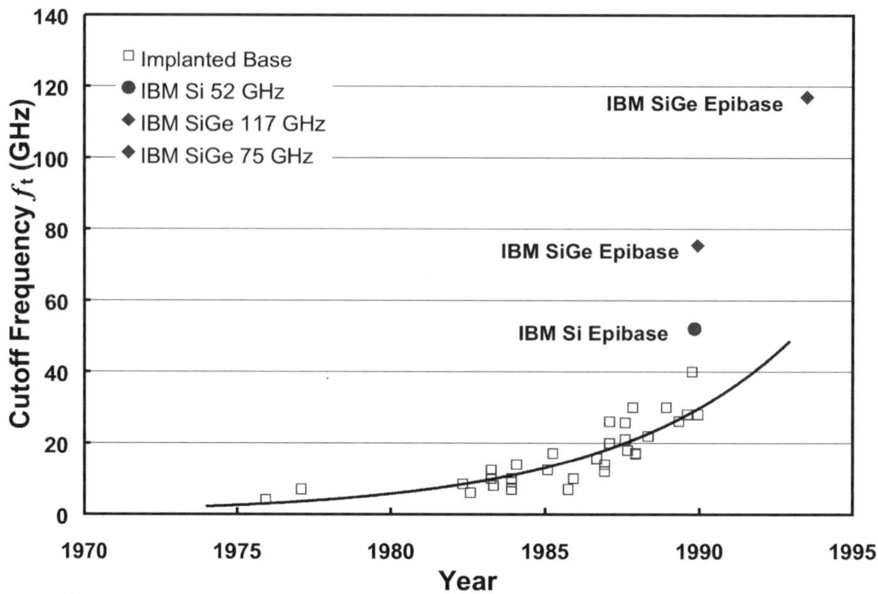

Figure 5 Unity-gain cutoff frequency for conventional implanted and epi-base bipolar transistors. Note that around 1989 there was a radical departure from the trend line of conventional implanted transistor technology.

achievement is nicely shown in Figure 5, which is a plot of bipolar f_T values versus year for conventional ion-implanted and epi-base technologies. Notice how the epi-base f_T results clearly depart from the conventional ion-implant bipolar curve. The work was extremely important, providing the first evidence of viable SiGe device technology, and it was the report of these results that launched the modern version of what is now accepted as commercial SiGe technology [44].

During a celebration of the 75-GHz f_T results, Dr. John Kelly of the IBM Microelectronics Division issued a new challenge: he wagered a steak-and-lobster luncheon against the technology bringing in a silicon-based bipolar transistor with an f_T over 100 GHz. This became the new holy grail for the evolution of this technology and its further march into the performance domain of compound semiconductors.

THE EPITAXIAL-BASE TRANSISTOR (ETX)

The 75 GHz result changed the mindset of IBM, and attention turned toward using the technology in practical circuit applications. Because IBM was still in the digital bipolar mainframe computer business, research focused on using SiGe for emitter-coupled logic bipolar circuits. The objective was to first establish an optimized device structure for use in bipolar logic, and then to work on BiCMOS integration at a later time. Outside of IBM there was an increasing number of publications on SiGe

HBTs [45–49], but all of these publications used MESA structures and did not emphasize developing structures compatible with conventional planar processes that could be easily incorporated into a BiCMOS process. Initially, IBM was pretty much on its own. Some of the different kinds of structures developed at IBM during this time included a simple non-self-aligned structure very similar in approach to that being used today by several manufacturers [50,51]: a DPSA structure [52] that incorporated a SiGe epi-base in the structure to replace the implanted base; a fully-self aligned MSST [53] which required a tall stack of sacrificial layers to self align the isolation and the extrinsic base; the selective epitaxial-emitter window (SEEW) structure utilizing a UHV/CVD intrinsic epi-base growth followed by a dielectric landing-pad definition, selective epitaxial overgrowth, and an emitter opening formed by oxidation of the selective epitaxy overgrowth [54]; a structure in which the extrinsic and intrinsic base was deposited in one deposition and an emitter formed in a very similar fashion to a DPSA structure, but with a "blind etch" and selective oxidation step (the structure and process were patented [55], but never published); and the epitaxial transistor structure (ETX [56]), which was eventually selected as the IBM product structure and will be described in detail next.

The ETX structure was discovered by modifying an existing single-polysilicon self-aligned bipolar process (NTX) [11] to incorporate an epitaxial SiGe base. The key to the emitter definition in the original NTX process was the formation of a nitride pedestal, followed by an oxidation that defined a thick extrinsic base dielectric outside the emitter opening, as determined by the nitride pedestal, that is, it used a local oxidation silicon (LOCOS) process for self-alignment. The process flow is shown on the left-hand side of Figure 6. In a SiGe epi-base process, the oxidation could not be allowed to consume the thin silicon cap layer over the SiGe base, which limited how much oxide could be grown during oxidation. This limitation led to a general "oxide budget problem" for an epi-base implementation of the process. A simple and elegant fix was proposed. The ETX process is as follows (picking up from the growth of the epitaxial base). First, a series of films is formed: (1) a *passivation oxide* is grown; (2) a layer of *barrier nitride* is deposited to form a protective layer against subsequent oxidations; (3) a layer of *conversion polysilicon* is deposited on top of the nitride to provide silicon for the LOCOS self-alignment oxidation; (4) a layer of *pedestal nitride* is deposited, which blocks the LOCOS oxidation that defines the emitter opening; and, finally, (5) a thick *pedestal oxide* layer is patterned at the same time as the nitride, which gives the height for an oxide sidewall to self-align the extrinsic base implant to the emitter opening. After the stack layers are all formed, the emitter pedestal oxide and nitride are patterned and etched, stopping on the conversion polysilicon. It was observed that the polysilicon between the patterned pedestal nitride and that the blanket barrier nitride did not oxidize during subsequent oxidations, causing a sharply defined LOCOS oxide around the emitter pedestal nitride. The pedestal oxide was added to increase the height of the emitter pedestal and an oxide sidewall added to self-align an extrinsic base implant [57]. After some optimization of the layer thicknesses (a process that actually took years) and the addition of a polysilicon emitter, the single poly self-alignment scheme was completed. The process flow is shown in Figure 6. This structure and the first emit-

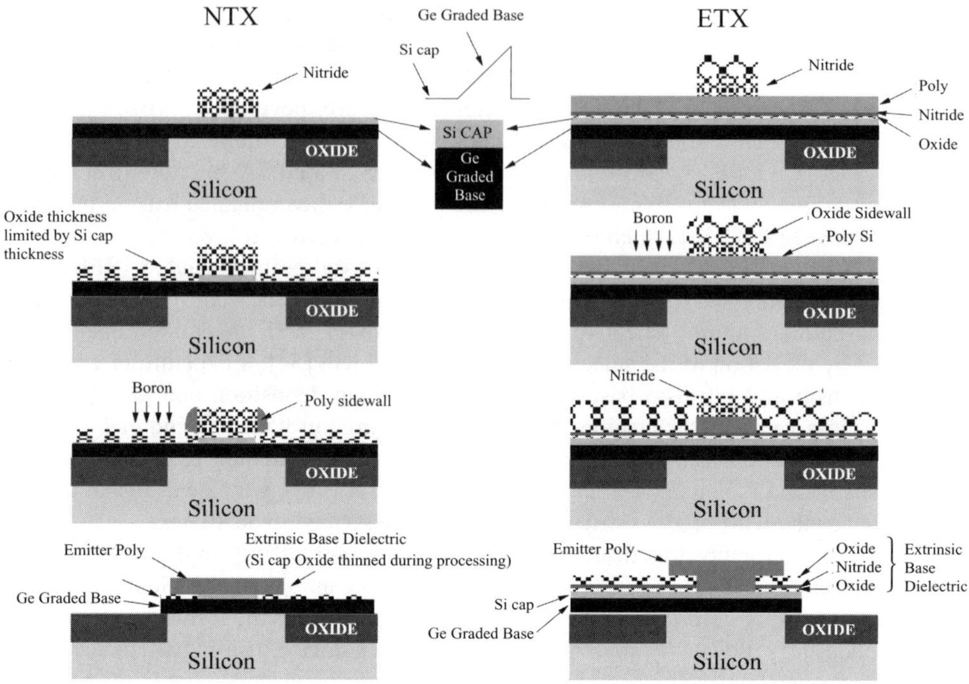

Figure 6 Schematic cross section of process flows for the NTX and ETX transistor structures.

ter-coupled logic (ECL) ring oscillator results was published in the 1990 *IEDM Technical Digest* [58]. In fact, at the 1990 IEDM there were six SiGe epi-base technology papers [56,59–63], all detailing different aspects of the work at IBM. The research activity in epi-base work was peaking at IBM and gaining notoriety outside of the company. The activity now involved a group of about 15 researchers at the Thomas J. Watson Research Laboratory. It was growing, but still largely confined to the research laboratory.

SiGe BICMOS FOR DIGITAL APPLICATIONS: HPT6

The first targeted application of SiGe technology at IBM was that of a silicon BiCMOS technology for mainframe computers. Using SiGe rather than conventional implanted technology was hotly contested, because there was virtually no yield information on SiGe. However, the $f_T > 60$-GHz performance target was very aggressive (as noted earlier), so the conventional ion-implanted technology was eliminated from consideration.

The mainframe program was named H10C, and the technology was dubbed high performance transistor generation 6 (HPT6). During the program's initial phase, there was no general agreement on which structure or process to use with SiGe. Several proposals were eventually narrowed down to either using ETX (discussed earlier) or an alternative proposal that we at IBM referred to as NPT. The NPT proposal was a very clever proposal that depended on the selective oxidation of very heavily doped silicon. The structure was possible because UHV/CVD was used, which has the capability of depositing fully activated boron-doped films in the 10^{20} atoms/cm^3 range. However, the structure required a timed etch in silicon, which, at least initially, was difficult to control. Working ETX hardware was demonstrated first, and it was selected as the process for HPT6.

Both processes emphasized planarity at the base deposition step, to reduce the impact of problems during the pre-LTE "HF dip." The process sequence was as follows: deep and shallow trench isolation, FET wells, reachthrough diffusion, gate oxidation, deposition of a thin undoped layer of polysilicon referred to as the "polyprotect" layer, and finally patterning and etching the polyprotect layer from the NPN area. After this sequence, the surface of the wafer was planar and hydrophobic from the polyprotect polysilicon. The LTE preclean step was now easily done and good single-crystal epitaxy grew over the active base region where single-crystal silicon was exposed. This was the first actual application of the polyprotect layer in a BiCMOS process.

In its heyday, the HPT6 SiGe program involved more than a hundred people (technology, circuit design, and support groups) from both the IBM Yorktown and East Fishkill facilities. The program's first checkpoint for early performance and yield demonstrations was scheduled for September 1991. Just prior to that time there were problems achieving good demonstration hardware. Finally, just a few days before checkpoint, a last-effort run was completed, which resulted in superb DC characteristics for the bipolar and CMOS devices, but very poor AC characteristics. Boron from the back side of the wafers had contaminated the front side of the wafers during an anneal, which resulted in very wide bases with SiGe barrier effects. In spite of the barrier effects, the yield on this run was spectacular. In fact, the SiGe HPT6 program achieved its technology feasibility checkpoint, mostly based on the superb yield results from a single very high yielding run! As it turned out, passing this major checkpoint was moot. In 1992, IBM made a tactical decision to cancel all bipolar development and use CMOS and parallel architecture for all future mainframe programs. The HPT6 program was canceled along with all bipolar and SiGe activity. The H10C machine was never built. But, the final results of the ECL BiCMOS were published in 1992 at the IEDM conference [64]. After the cancellation of the bipolar programs, the participants were reassigned to other areas in IBM.

As the program wound down, Dr. Kelly's challenge of a "steak and lobster" wager to the team that produced the first bipolar device to break the 100 GHz barrier had not yet been met. The ultimate winning approach used 0-25% Ge-graded base designs, sub 50nm final base widths, greatly enhanced SIMS techniques to characterize the thin epitaxial base depositions, in-situ doped phosphorus emitters annealed at 800C, and very low temperature processing. The run was completed; but, given other activities at the time, it was only partially measured. Preliminary measurements indicated that the early voltage was very high with the high Ge content fully graded base, despite the presence of very heavily doped collectors

($>10^{17}$atoms/cm^3); this knowledge later be put to good use. Incredibly, when the initial AC measurements yielded devices under 100 GHz the wafers were set aside without completing their characterization and work ceased.

NEW LIFE: ANALOG APPLICATIONS

In the 1990 time frame, it was observed that the fundamental characteristics of these HBTs, notably their remarkably low base resistance for such high-performance transistors, made them ideally suited (even though not yet optimized) for analog and mixed-signal applications. With the high-speed, low-noise, and low-power capabilities (trading off excess speed for reduced I_C) of the SiGe HBTs, IBM possessed a world-class technology orphan. This presented the opportunity for IBM to take new technological and business directions, becoming an original equipment manufacturer (OEM) player in communications. However, IBM did not have the skills internally to leverage this new analog and mixed-signal technology in the marketplace, or even demonstrate the capability at the circuit level. To continue development, and leverage all the earlier work, a strategic decision was made to seek external alliances with other companies possessing appropriate expertise in analog and mixed-signal design, application selection, and marketing. The first of many such alliances was formed with Analog Devices of Waltham, Massachusetts. It was an alliance to investigate the leverage of SiGe for analog applications, initially for data conversion, sharing the risks and rewards always present in such a new gambit.

To best leverage SiGe HBT technology in its new marketplace, the SiGe profile had to be reoptimized for analog applications. The profile developed for HPT6, the digital profile, put the grade across the highest doped region of the base to achieve the highest f_T, but had relatively poor early voltage enhancement [65]. The new analog profile graded the base fully across the neutral base to achieve a high f_T and high early voltage. A test chip was designed with circuits supplied by Analog Devices engineers, and technology characterization structures supplied by the IBM team. The first analog SiGe IC was attempted on this run, that being a 12-bit digital-to-analog converter (DAC) assembled from roughly 3000 HBTs, 2000 resistors, and other elements.

Fabricating this SiGe IC proved challenging, but not due to technology issues; rather, it was because at that time, the sole SiGe-capable fabrication line (the Yorktown silicon facility) was being shut down as part of a consolidation effort with other development facilities around the company. Tools were literally torn out of the line and disposed of as SiGe product was being run through the facility. We can recall inspecting a wafer and looking up to see a just-used semiconductor process tool being wheeled out of the facility for disposal! Clearly there would be no backup runs, and no margin for error. This run and the fabrication-facility disassembly were completed virtually simultaneously, with the wafers one process step ahead of the wreckers.

As for the SiGe IC run, the 12-bit DAC not only worked, but was clocked faster than 1 giga sample per second (GSPS), which was 10 times faster in the SiGe process than in the Analog Devices process in which it was initially designed. Later work on noise established the 12-bit DAC as having greatly reduced phase noise at significantly lower power than the competing III–V-based devices of the day. The 1-GSPS 12-bit DAC was a spectacular result that established the ability to make medium-scale ICs in SiGe; the work was presented at the 1993 IEDM conference [65], and conference attendees from Hughes saw the paper and concluded they too could leverage this high-speed technology.

It is interesting to note that, at this point, the SiGe group was composed of only a couple of people, whereas at its peak over a hundred had been involved. Many fine engineers had either joined CMOS logic or dynamic random-access memory (DRAM) programs, or left the company after the dissolution of the HPT6 bipolar effort. The IBM participants now included three people who were in the program from its inception and stayed with it throughout: one technologist who defined the technology, ran the experiments in the line, wrote ground rules, interfaced with the circuit designers, and assembled and designed the test sites; one applications engineer who ran the UHV/CVD systems and supported the fledgling business applications, and one IBM manager, the inventor of UHV/CVD, who took on the task of selling the concept of SiGe to IBM and to external executive management, driving the program to an external customer focused OEM business. Ultimately, IBM was persuaded to invest in this new direction, leveraging SiGe in concert with expert external companies, kicking off an "entrepreneurial" program that spanned three years, from launch to broad internal acceptance of the new business concept. While it is true that IBM had officially "canceled" the program, the culture was such that a core of management, from first-level fabrication managers to senior executives, always found a way to aid and abet what at times was a remarkable guerilla operation. Few, if any, organizations would have tolerated the radical and unsolicited bottom-up-driven reorganization of an entire class of existing technology. Although most members of the original HPT6 program had moved on, they continued to provide technical and management support, and their willingness to spend personal time and their program resources to ensure the success of this fledgling business figured greatly in its ultimate success.

At about the same time as the new business concept was being accepted, the remaining wafers from the "steak and lobster run" were finally characterized. "Fast" transistors, easily winning the bet, were obtained from a previously unmeasured split of 0–25% Ge-graded base devices with in situ doped phosphorus emitters. An early voltage of 110 V with a peak f_T of 113 GHz, together with a Beta-VA product of 48,400, was reported. Steak and lobster were finally collected by the entire original HPT6 team, and the results presented at the 1993 IEDM conference [66].

TRANSFER TO THE ASTC AND THE BIPOLAR QUALIFICATION

After the closure of the Yorktown silicon facility, the SiGe HBT process had to be recreated in a new facility, and the Advanced Semiconductor Technology Center (ASTC) in East Fishkill, New York, was selected. Fortunately, UHV/CVD tooling had already been installed, so the challenge of installing a SiGe process in the ASTC came down to recreating the process quickly and with still-limited resources. An engineer was added to the group, bringing the total number of workers up to three. A new BiCMOS test site was designed in addition to a new Analog Devices test chip with a redesigned version of the original 12-bit DAC. Working DACs were obtained on the first silicon hardware processed through the ASTC line, in spite of many process issues (which were flattened on the fly). This was quite a statement about the robustness and manufacturability of the process. As results continued to propagate from the early manufacturing environment of the ASTC, program visibility and viability greatly increased. New partnerships were quickly formed with Northern Telecom and Hughes Electronics. Nortel completed a test site in early 1994 that focused on circuits in the 1–2-GHz range. Hughes was interested in very-high-frequency applications up to 5–30-GHz, and, for many of the circuits, added an additional layer of thick polyimide and metal to get around lossy transmission lines on silicon. Because the two companies covered such a wide range of frequencies and applications, they were able to extensively evaluate the applications of the SiGe technology. The test chips for both companies were completed around December 1994, and the excellent results from the years of work can be seen in summary papers from the 1996 ISSCC conference [67].

At this point, the program was largely oriented around fabrication for alliance customers and the development of a suite of devices suitable for their needs. Our technology group had now grown to around 10 people, and it was enjoying the benefit of support from numerous device and fabrication support groups in the ASTC. The customers received a design manual, a tape of device layouts, and a set of simulation program with integrated-circuit emphasis (SPICE) models for the device layouts. However, a complete design system was still unavailable for circuit design and simulation; P-cell layouts with models attached to the P cells, parasitic extraction, and automated design rule checking was still missing. This was a serious impediment to designing higher-level integration circuits. To get over this stumbling block, the early alliance partners each developed their own design systems and models with help and support from IBM.

A motto of this SiGe program from its earliest stages has been that "data wins," a simple acknowledgment that all obstacles will fall before an onslaught of compelling data. As the OEM program grew, partners and internal efforts generated ever more compelling data, so that, in August 1995, IBM internally funded the first SiGe technology qualification. The decision to undertake a formal production qualification was the final step in bringing the orphan program once again into the mainstream. The "bipolar" process was really more of a bipolar FET (BiFET) qualification with pFETs, LPNPs, resistors, capacitors, inductors, diodes, and HBTs. A full BiCMOS qualification would come later. The bipolar technology qualification was completed in September 1996 in the ASTC. The importance of this manufacturing qualification cannot be understated, as this SiGe technology was held to the same commercial standards met by any and all of IBM's technology offerings. The rigor and scope of this first qualification enabled IBM's launch of commercial SiGe chip shipments years earlier than anyone else in the industry; and, since 1996, this exercise has set a high standard to which all subsequent generations of SiGe technology have been held.

BiCMOS TECHNOLOGY

After the bipolar qualification, attention turned toward qualifying a BiCMOS technology. A 2.5-V BiCMOS process based on the bipolar process was available [68], but IBM received customer requests for a 0.5-mm 3.3-V BiCMOS offering. Therefore, development began on a new 3.3-V BiCMOS compatible with IBM's CMOS 5S, a 0.5-μm BiCMOS process. At this time there was no revenue stream coming from SiGe products, so funding for the new technology was limited. Additional support was obtained from the Defense Advanced Research Projects Administration (DARPA) [69] to help with the design kits and modeling of this first attempt to merge a heterojunction-based silicon technology with highly integrable silicon CMOS. Early success in this effort enabled the project's expansion to encompass the fabrication of a number of multiproject wafer runs, with university and government participation.

It was now time to move the process to a large-volume fabricator; so, in parallel, work began to have the process installed at IBM's large-scale fabricator in Essex Junction, Vermont. Transferring the 3.3-V BiCMOS process before the definition was completed was advantageous because it allowed the process definition to have as much commonality as possible between the two sites. The technology transfer to manufacturing was also aided by having the same person who formed the East Fishkill SiGe group move to Essex Junction, Vermont, to accomplish the technology transfer. In the fall of 1996, D. Harame transferred to Essex Junction, and formed a new group in the manufacturing organization, SiGe Product Development and Manufacturing. This was a very important step, because IBM's manufacturing organization had now assumed ownership of the SiGe Technology and Process group, expanding the SiGe effort outside of IBM's research and development organizations.

The first BiCMOS process qualification was for a 0.5-μm, 3.3-V BiCMOS in January 1998 at the ASTC, and, in June 1998 the process was qualified in Essex Junction. Production-level products in this technology are now being shipped to both internal IBM customers and external customers.

Until 1998 there were actually two BiCMOS development organizations in IBM: the SiGe BiCMOS technology group, both at the ASTC in East Fishkill and Essex Junction, and the Analog and Mixed-Signal group developing ion-implanted homojunction BiCMOS technology at Essex Junction. On a grassroots level, the technology engineers merged forces to develop the next-generation BiCMOS, which would be used for storage, wired, and wireless applications across the company. That merging of forces was very successful; and, toward the end of 1998, the groups formally merged under one BiCMOS development organization. In June 1999, the second generation of SiGe BiCMOS inside IBM, the 0.25-μm 2.5-V SiGe BiCMOS technology, was qualified in the Essex Junction plant. Starting from that point, all bipolar/BiCMOS development at IBM will be SiGe bipolar/BiCMOS. SiGe technology continues to mature inside IBM, with the addition of numerous support groups for models, design kits, marketing, product applications, device design aids, test-site layout, product engineering, and product development engineering.

SUMMARY

The history of SiGe at IBM is a story of persistence. The program began with an idea to replace a conventional implantation step used in every silicon semiconductor bipolar process by growing an in situ doped alloy (SiGe base). Many people thought the idea was of value for only very small exotic niche research applications. But the SiGe story is about a small group of people who persuaded a large digital computer manufacturer to invest in a new unproven technology for telecommunication applications in a field that the company knew little about. It is a story of success, as the technology is now the only BiCMOS in development in IBM, and is in the roadmap of every major telecommunication company.

As SiGe technology rapidly becomes pervasive, many players will undoubtedly emerge, and the pace of advancement may accelerate even further. Nonetheless, this once-orphaned technology has become a leading contender in the high-volume communications marketplace. Suffice it to say that the small core team that took this project forward could not have succeeded without the support of others too numerous to mention, but many of whom came and went from the program, leaving its success as their legacy.

REFERENCES

1. K. K. Ashok and D. J. Roulston, *Polysilicon Emitter Bipolar Transistors,* IEEE Press, New York, 1989.
2. M. J. van Dort, W. van der Wel, J. W. Slotboom, N. E. B. Cowern, M. P. G. Knuvers, H. Lifka, and P. C. Zalm, "Two-Dimensional Transient Enhanced Diffusion and Its Impact on Bipolar Transistors," *IEDM Tech. Digest,* pp. 865–868, December 1994.
3. J. M. C. Stork, and R. D. Isaac, "Tunneling in Base-Emitter Junction," *IEEE Trans. Electron Devices,* vol. 30, no. 11, pp. 1527–1534, November 1983.
4. J. D. Warnock, "Silicon Bipolar Device Structures for Digital Applications: Technology Trends and Future Directions," *IEEE Trans. Electron Devices,* vol. 42, no. 3, pp. 382–383, March 1995.
5. K. Suzuki, "Optimum Base Doping Profile for Minimum Base Transit Time," *IEEE Trans. Electron Devices,* vol. 38, no. 9, pp. 2128–2133, September 1991.
6. S. K. Ghandhi, "Native Oxide Films," in *VLSI Fabrication Principles,* Wiley, pp. 373, 1983.
7. R. B. Fair, "Low-Thermal-Budget Process Modeling with PREDICT Comptuer Program, *IEEE Trans. Electron Devices,* vol: 35, no. 3, pp. 285–293, March 1988.

8. B. S. Meyerson, F. Himpsel, and K. J. Uram, " Bistable Conditions for Low Temperature Silicon Epitaxy," *Appl. Phys. Lett.* vol. 57, pp. 1034–1036, 1990.
9. B. S. Meyerson, " Low Temperature Silicon Epitaxy by Ultra-high Vacuum/Chemical Vapor Deposition," *Appl. Phys. Lett.* vol. 48, pp. 797–799, 1986.
10. T. Nguyen, D. L. Harame, J. M. C. Stork, F. K. LeGoues, and B. S. Meyerson, "Diodes Fabricated from UHV/CVD," *IEDM Tech. Digest,* pp. 304–307, December 1986.
11. G. P. Li, T.-C. Chen, C.-T, Chuang, J. M. C. Stork, D. D. Tang, M. B. Ketchen, and L.-K. Wang, "Bipolar Transistor with Self-Aligned Lateral Profile," *IEEE Electron Devices Lett.,* vol. 8, no. 4, pp. 338–340, August 1987.
12. D. L. Harame, J. M. C. Stork, B. S. Meyerson, T. N. Nguyen, and G. J. Scilla, "Epitaxial Base Transistors with Ultrahigh Vacuum Chemical Vapor Deposition (UHV/CVD) Epitaxy: Enhanced Profile Control for Greater Flexibilty in Device Design," *IEEE Electron Devices Lett.,* vol. 10, no. 4, pp. 156–158, April 1989.
13. J. M. C. Stork, "D. L. Harame, B. S. Meyerson, and T. N. Nguyen, "High Performance Operation of Silicon Bipolar Transistors at Liquid Nitrogen Temperature," *IEDM Tech. Digest,* pp. 405–407, December 1987.
14. J. M. C. Stork, D. L. Harame, B. S. Meyerson, and T. N. Nguyen, "Base Profile Design for High-Performance Operation of Bipolar Transistors at Liquid-Nitrogen Temperature," *IEEE Trans. Electron Devices,* vol. 36, no. 8, pp. 1503–1509, August 1989.
15. S. Subbanna, V. P. Kesan, M. J. Tejwani, P. J. Restle, D. J. Mis, and S. S. Iyer, "Si/SiGe P-Channel MOSFETs," in *Digest Tech. Papers 1991 Symposium on VLSI Technology,* pp. 103–104, 1991.
16. S. Verdonckt-Vandebroek, E. F. Crabbe, B. S. Meyerson, D. L. Harame, P. J. Restle, J. M. C. Stork, A. C. Megdanis, C. L. Stanis, A. A. Bright, G. M. W. Kroesen, and A. C. Warren, "Graded SiGe-Channel Modulation-Doped p-MOSFETS," in *Digest Tech. Papers 1991 Symposium on VLSI Technology,* pp. 105–106, May 1991.
17. K. Ismail, B. S. Meyerson, S. Rishton, J. Chu, S. Nelson, and J. Nocera, "High-Transconductance n-Type Si/SiGe Modulation-Doped Field-Effect Transistors," *IEEE Electron Devices Lett.,* vol. 13, p. 229, 1992.
18. S. J. Koester, J. O. Chu, and R. A. Groves, "High-fT n-MODFETs Fabricated on Si/SiGe Heterostructures Grown by UHV-CVD," *Electronics Lett.,* vol. 35, p. 86, 1999.
19. R. Hammond, S. J. Koester, and J. O. Chu, "Extremely-High Transconductance Ge/Si0. 4Ge0. 6 pMODFETs Grown by UHV-CVD," *IEEE Electron Devices Lett.,* vol. 21, no. 3, March 2000.
20. K. Ismail, "Si/SiGe High-Speed Field-Effect Transistors," *IEDM Tech. Digest,* p. 509, December 1995.
21. R. People, "Indirect Band Gap of Coherently Strained Si1-xGex Bulk Alloys on <001> Silicon Subsrtrates," *Phys. Rev. B.,* vol. 32, no. 2, pp. 1405–1408, December 1985.
22. E. Kasper, H. Dambkes, and J.-F. Luy, "Very Low Temperature MBE Process for SiGe and Si Device Structures," *IEDM Tech. Digest,* pp. 558–561, December 1988.
23. S. S. Iyer, G. L. Patton, S. S. Delage, S. Tiwari, and J. M. C. Stork, "Silicon-Germanium Base Heterojunction Bipolar Transistors by Molecular Beam Epitaxy," *IEDM Tech. Digest,* pp. 874–876, December 1987.
24. G. L. Patton, D. L. Harame, J. M. C. Stork, B. S. Meyerson, G. J. Scilla, and E. Ganin, "Graded-SiGe-Base, Poly-Emitter Heterojunction, Bipolar Transistors," *IEEE Electron Devices Lett.,* vol. 10, no. 12, pp. 534–536, 1989.

25. D. D. Tang, P. M. Soloman, T. H. Ning, R. D. Isaac, and R. E. Burger, "1. 25 μm Deep-Groove-Isolated Self-Aligned Bipolar Circuits," *IEEE J. Solid-State Circuits,* vol. SC–17, pp. 925–931, 1982.

26. J. W. Matthews and A. E. Blakeslee, "Defects in Epitaxial Multilayers I. Misfit Dislocations in Layers," *J. Crystal Growth,* vol. 27, pp. 118–125, 1974.

27. S. R. Stiffler, J. H. Comfort, C. L. Stanis, D. L. Harame, E. deFresart, and B. S. Meyerson, "The Thermal Stability of SiGe Films Deposited by Ultra-High Vacuum Chemical Vapor Deposition," *J. Appl. Phys.,* no. 70, p. 1416, 1991.

28. G. L. Patton, D. L. Harame, J. M. C. Stork, B. S. Meyerson, G. J. Scilla, and E. Ganin, "Graded-SiGe-Base, Poly-Emitter Hetrojunction, Bipolar Transistors, *IEEE Electron Devices Lett.,* vol. 10, no. 12, pp. 534–536, 1989.

29. J. F. Gibbons, C. M. Gronet, and K. E. Williams, "Limited Reaction Processing: Silicon Epitaxy," *Appl. Phys. Lett.,* vol. 47, no. 7, pp. 721–723, 1985

30. C. M. Gronet, J. C. Sturm, K. E. Williams, J. F. Gibbons, and S. D. Wilson, "Thin, Highly Doped Layers of Epitaxial Silicon Deposited by Limited Reaction Processing," *Appl. Phys. Lett.,* vol. 48, no. 15, pp. 1012–1014, April 1986.

31. J. C. Sturm, C. M. Gronet, and J. F. Gibbons, "Minority-Carrier Properties of Thin Epitaxial Silicon Films Fabricated by Limited Reaction Processing," pp. 4180–4182, *J. Appl. Phys.,* vol. 59, no. 12, 1986.

32. C. A. King, J. L. Hoyt, C. M. Gronet, J. F. Gibbons, M. P. Scott, S. J. Rosner, G. Reid, S. Laderman, K. Nauka, and T. I. Kamins, "Characterization of p-N Si/sub 1-x/Ge/sub x//Si Heterojunctions Grown by Limited Reaction Processing," *IEEE Trans. Electron Devices,* vol. 35, no. 12, p. 2454, December 1988.

33. C. M. Gronet, C. A. King, A. W. Opyd, J. F. Gibbons, S. D. Wilson, and R. Hull, "Growth of GeSi/Si Strained-Layer Superlattices using Limited Reaction Processing," *J. Appl. Phys.,* vol. 61, no. 6, pp. 2407–2409, March 1987.

34. J. F. Gibbons, C. A. King, J. L. Hoyt, D. B. Noble, C. M. Gronet, M. P. Scott, S. J. Rosner, G. Freid, S. Laderman, K. Nauka, J. Turner, and T. I. Kamins, "Si/$Si_{1-x}Ge_x$ Heterojunction Bipolar Transistors Fabricated by Limited Reaction Procesing," *IEDM Tech. Digest,* pp. 566–569, December 1988.

35. C. A. King, C. M. Gronet, J. F. Gibbons, and S. D. Wilson, "Electrical Characterization of In-Situ Epitaxially Gown p-n Junctions Fabricated Using Limited Reaction Process," *IEEE Electron Devices Lett.,* vol. 9, no. 5, pp. 229–231, May 1988.

36. C. A. King, J. L. Hoyt, C. M. Gronet, J. F. Gibbons, M. P. Scott, and J. Turner, "Si/Si1-xGex Heterojunction Bipolar Transistors Produced by Limited Reaction Processing," *IEEE Electron Devices Lett.,* vol. 10, no. 2, pp. 52–54, February 1989.

37. C. A. King, J. L. Hoyt, and J. F. Gibbons, "Bandgap and Transport Properties of Si1-xGex by Analysis of Nearly Ideal Si/Si1-xGex/Si Heterojunction Bipolar Transistors," *IEEE Trans. Electron Devices,* vol. 36, no. 10, pp. 2093–2104, October 1989.

38. D. L. Harame, J. M. C. Stork, S. S. Iyer, B. S. Meyerson, G. J. Scilla, E. F. Crabbe, and E. Ganin, "High Performance Si and SiGe Base PNP Transistors," *IEDM Tech. Digest,* pp. 889–891, December 1988.

39. D. L. Harame, B. S. Meyerson, E. F. Crabbe, C. L. Stanis, J. M. Cotte, J. M. C. Stork, A. C. Megdanis, G. L. Patton, R. Stiffler, J. B. Johnson, J. D. Warnock, J. H. Comfort, and J. Y-C. Sun, "55 GHz Polysilicon-Emitter Graded SiGe-Base PNP Transistors," in *Tech. Digest 1991 Symposium on VLSI Technology,* pp. 71–72, May 1991.

40. G. L. Patton, “Explorer” a 1D device simulator implemented in both Lotus123 and Fortran.

41. S. P. Gaur, P. A. Habitz, Y. J. Park, R. K. Cook, Y. S. Huang, L. F. Wagner, “Two-Dimensional Device Simulation Program: 2DP,” *IBM J. Res. Develop.,* vol. 29, no. 3, pp. 242–253, May 1985.

42. E. M. Buturla, P. E. Cottrell, B. M. Grossman, and K. A. Salzberg, “Finite Element Analysis of Semiconductor Devices: The FIELDAY Program,” *IBM J. Res. Develop.,* vol. 25, no. 4, pp. 218–231, 1981.

43. E. M. Buturla, J. Johnson, S. Furkay, and P. Cottrell, “A New 3D Device Simulation Formulation,” in *Nascode VI Proceedings,* J. J. H. Miller, ed., Boole Press, Dublin, pp. 291–302, 1989.

44. G. L. Patton, J. H. Comfort, B. S. Meyerson, E. F. Crabbe, G. J. Scilla, E. de Fresart, J. M. C. Stork,J. Y. C.-Sun, D. L. Harame, and J. N. Burghartz, “63–75 Ghz fT SiGe-Base Heterojunciton Bipolar Tecnology,” in *Digest Tech. Papers 1990 Symposium on VLSI Technology,* pp. 49–40, 1990.

45. E. J. Prinz, P. M. Garone, P. V. Schwartz, X. Xiao, and J. C. Sturm, “The Effect of Base-Emitter Spacers and Strin-Dependent Densities of States in $Si/Si_{1-x}Ge_x$ Heterojunction Bipolar Transistors,” in IEDM Tech. Digest, pp. 639–645, December 1989.

46. H. U. Schreiber and B. G. Bosch, “Si/SiGe Heterojunction Bipolar Transistors with Current Gains of Up to 5000,” *IEDM Tech. Digest,* pp. 643–646, December 1989.

47. T. I. Kamins, K. Nauka, L. H. Camnitz, J. B. Kruger, J. E. Turner, S. J. Rosner, and M. P. Scott, and J. L. Hoyt, C. A. King, and D. B. Noble, and J. F. Gibbons, “High Frequency $Si/_{1-x}Ge_x$ Heterojunction Bipolar Transistors,” *IEDM Tech. Digest,* pp. 647–650, 1989.

48. R. C. Taft, and J. D. Plummer, “Advanced Heterojunction Ge_xSi_{1-x} /Si Bipolar Transistors,” *IEDM Tech. Digest,* pp. 655–658, December 1989.

49. P. Narozny, M. Hamacher, H. Dambkes, H. Kibbel, and E. Kasper, “Si/SiGe Heterojunction Bipolar Transistor Made by Molecular Beam Epitaxy,” *IEDM Tech. Digest,* pp. 562–565, December 1988.

50. A. Chantre, M. Marty, J. L. Regolni, M. Mouis, J. de Pontcharra, D. Dutarte, C. Morin, D. Gloria, S. Jouan, R. Pantel, M. Laurens, “A High Performance Low Complexity SiGe HBT for BiCMOS Integration,” in *Proceedings of 1998 BCTM,* pp. 93–96, 1998.

51. D. Knoll, B. Heinemann, J. J. Osten, K. E. Ehwald, B. Tillack, P. Schley, R. Barth, M. Matthes, K. S. Park, Y. Kim, and W. Winkler, “Si/SiGe:C Heterojunction Bipolar Transistors in an Epi-Free Well, Single-Polysilicon Technology,” *IEDM Tech. Digest,* pp. 703–706, 1998.

52. E. Ganin, T. C. Chen, J. M. C. Stork,B. S. Meyerson, J. D. Cressler, G. Scilla, J. Warnock, D. L. Harame, G. L. Patton, and T. H. Ning, “Epitaxial-Base Double-Poly Self-Aligned Technology,” *IEDM Tech. Digest,* pp. 603–605, 1990.

53. R. Schulz, M. Jost, D. L. Harame, G. J. Scilla, B. S. Meyerson, and G. B. Bronner, “A Fully Self-Aligned Epitaxial-Base Transistor,” in *Tech. Digest of 1989 Symp. on VLSI Technology,* pp. 89–90, 1989.

54. J. N. Burghartz, J. H. Comfort, G. L. Patton, B. S. Meyerson, J. Y.-C. Sun, J. M. C. Stork, S. R. Mader, C. L. Stanis, G. J. Scilla, and B. J. Ginsberg,”Self-Aligned SiGe-Base Hetrojunction Bipolar Transistors by Selective Epitaxy Emitter Window (SEEW) Technology,” *IEEE Electron Devices Lett.,* vol. 11, no. 7, p. 288–290, July 1990.

55. J. L. Blouse, I. G. Fulton, R. C. Lange, B. S. Meyerson, K. A. Nummy, M. Revitz, and R. Rosenberg, "US5132765: Narrow Base Transistor and Method of Fabricating Same," Issued July 21, 1992.

56. J. H. Comfort, T. C. Chen, P. F. Lu, B. S. Meyerson, Y. C. Sun, and D. D. Tang, "US5117271: Low Capacitance Bipolar Junction Transistor and Fabrication Process Therefor," Issued May 26, 1992.

57. S. Jeng, D. Greenberg, M. Longstreet, G. Hueckel, D. L. Harame, and D. Jadus, "Lateral Scaling of the Self-Aligned Extrinsic Base in SiGe HBTs," in *Proceedings of the 1996 BCTM,* pp. 15–18.

58. J. H. Comfort, G. L. Patton, J. D. Cressler, W. Lee, E. F. Crabbe, B. S. Meyerson, J. Y.-C. Sun, J. M. C. Stork, P.-F. Lu, J. N. Burghartz, J. Warnock, G. Scilla, K.-Y. Toh, M. D'Agostino, C. Stanis, and K. Jenkins, "Profile Leverage in a Self-Aligned Epitaxial Si or SiGe Base Bipolar Technology," *IEDM Tech. Digest,* pp. 21–24, 1990.

59. G. Patton, J. Stork, J. Comfort, E. Crabbbe, B. Meyerson, D. Harame, and J. Sun, "SiGe-Base Heterojunction Bipolar Transistors: Physics and Design Issues," *IEDM Tech. Digest,* pp. 13–16, December 1990.

60. E. F. Crabbe, G. Patton, J. Stork, J. Comfort, B. Meyerson, and J. Sun, "Low Temperature Operation of Si and SiGe Bipolar Transistors," *IEDM Tech. Digest,* pp. 17–20, December 1990.

61. D. Harame, J. Stork, B. Meyerson, E. Crabbe, G. Scilla, C. Stanis, A. Megdanis, G. Patton, J. Comfort, A. Bright, E. de Fresart, J. Johnson, and S. Furkay, "30 GHz Polysilicon-Emitter and Single-Crystal Emitter Graded SiGe-Base PNP Transistors," *IEDM Tech. Digest,* pp. 33–36, December 1990.

62. J. Burghartz, J. Comfort, G. Patton, J. Cressler, B. Meyerson, J. Stork, J. Sun, G. Scilla, J. Warnock, B. Ginsberg, K. Jenkins, K. Toh, D. Harame, and S. Mader, "Sub–30 ps ECL Circuits Using High-f_T Si and SiGe Epitaxial Base SEEW Transistors," *IEDM Tech. Digest,* pp. 297–300, December 1990.

63. E. Ganin, T. C. Chen, J. M. C. Stork, B. S. Meyerson, J. D. Cressler, G. Scilla, J. Warnock, D. L. Harame, G. L. Patton, and T. H. Ning, "Epitaxial-Base Double-Poly Self-Aligned Bipolar Transistors," *IEDM Tech. Digest,* pp. 603–606, December 1990.

64. D. L. Harame, E. F. Crabbe, J. D. Cressler, J. H. Comfort, J. Y.-C. Sun, S. R. Stiffler, E. Kobeda, J. N. Burghartz, M. M. Gilbert, J. C. Malinowski, and A. J. Dally, "A High Performance Epitaxial Base SiGe ECL BiCMOS technology," *IEDM Tech. Digest,* pp. 19–22, December 1992.

65. D. L. Harame, J. M. C. Stork, B. S. Meyerson, K. Y.-J. Hu, J. Cotte, K. A. Jenkins, J. D. Cressler, P. Restle, E. F. Crabbe, S. Subbanna, T. E. Tice, B. W. Scharf, and Y. A. Ysaitis, "Optimization of SiGe HBT Technology for High Speed Analog and Mixed-Signal Applications," *IEDM Tech. Digest,* pp. 71–74, December 1990.

66. E. F. Crabbe, B. S. Meyerson, J. M. C. Stork, and D. L. Harame, "Vertical Profile Optimization of Very High Frequency Epitaxial Si- and SiGe-base Bipolar Transistors," *IEDM Tech. Digest,* pp. 83–86, December 1993.

67. J. H. Long, M. A. Copeland, S. J. Kovacic, D. S. Mahli, and D. L. Harame, "RF Analog and Digital Circuits in SiGe Technology," *ISSCC Digest of Tech. Papers,* pp. 82–83, 1996.

68. D. Nguyen-Ngoc, D. L. Harame, J. C. Malinowski, S. J. Jeng, K. T. Schonenberg, M. M.

Gilbert, G. Berg, S. Wu, M. Soyuer, K. A. Talman, K. J. Stein, R. A. Groves, S. Subbanna, D. Colavito, D. A. Sunderland, and B. S. Meyerson, "A 200 mm SiGe-HBT BiCMOS Technology for Mixed Signal Applications," in *Proceedings of the 1995 BCTM,* pp. 89–92, 1995.

69. DARPA contract No. N66001-96-C-8606, administered by SPAWARYSCEN, San Diego, February 1996–March 2000.

1

TECHNOLOGY DEVELOPMENT

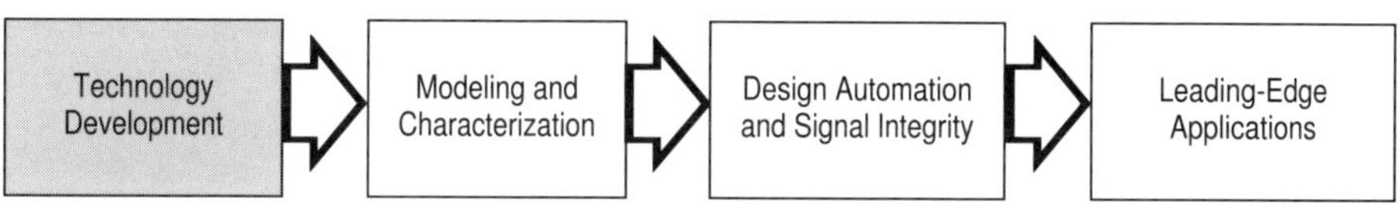

Active devices
- HBT, FET

Advanced passives and ESD
Process development
Technology development implications

Predictive modeling
Model characterization
Compact modeling
- Active devices
- Advanced passives

Design automation overview
- RF Simulation
- ESD computer-aided design (CAD) solutions

Signal integrity effects
- Interconnect extraction & modeling
- Substrate coupling & modeling

Wireless communications
- WCDMA transceiver
- Power amp

Wired communications
- OC768 SERDES

Memory design

OVERVIEW

This chapter provides a detailed description of IBM's silicon germanium (SiGe) bipolar complementary metal-oxide transistor (BiCMOS) technology development program. This family of technologies provides high-performance SiGe heterojunction bipolar transistors (HBTs) combined with advanced CMOS enablement, and a variety of advanced passive devices critical for realizing an integrated analog and mixed-signal (AMS) system on a chip (SoCs). The technologies have been utilized by internal and external customers through IBM's foundry offerings to produce integrated circuits (ICs) in a wide-ranging variety of applications, as discussed throughout the book.

This chapter also reviews the IBM process development and integration methodologies, as well as the device characteristics. The discussions describe how the development and device selection is geared toward usage in mixed-signal IC development.

Silicon Germanium: Technology, Modeling, and Design. By Singh, Harame, and Oprysko

ISBN 0-471-44653-X

- Section 1.1 discusses the development of active devices, namely bipolar transistors with an N-type emitter, P-type base, and N-type collector (NPN) and field-effect transistors (FETs).
- Section 1.2 discusses the development of advanced passive devices, such as resistors, capacitors, and inductors, as well as electrostatic-discharge (ESD) protection devices.
- Section 1.3 overviews many of the issues in process integration, including manufacturing—namely, predictability, reliability, and yield.
- Section 1.4 discusses the technology implications of the different implementation choices.

1.1 ACTIVE DEVICES

Radio frequency (RF) designers rely on the quality of the models for accurate simulations. Typically, the first performance measure of a process technology is the active-device performance. In this section, we discuss the development of active-device SiGe process technologies, specifically the HBT and FET devices. For the HBT, we present an overview and details of the design of the device. A summary of details of IBM's SiGe and RF-CMOS offerings is provided in the Appendix.

1.1.1 The SiGe HBT

Both the HBT and the FET are commonly available to designers as active devices for analog and RF applications. Whereas the FET receives significant attention in analog design due to its technology accessibility and cost perspective, the bipolar junction transistor (BJT) is often the favorite among RF designers, and is often the only option when demanding specifications are to be met. Fundamental differences between devices favor one or the other for certain applications. Noise, current drive, voltage gain, and repeatability are the strong suits for the BJT. Because the noise in the BJT is driven by bulk (not surface) electron and hole generation and recombination, compared to defect-dominated surface physics in FETs, the BJT is often favored for its low-frequency noise properties. This is discussed in detail in Section 1.4. Transconductance is also very different between devices. In a BJT, the output current changes exponentially with the input voltage, compared to a linear relationship with a FET. At its peak operating point, the BJT is found to achieve about three times the transconductance, and thus three times the drive capability compared to the FET. This factor translates directly to the operating frequency and the gain of the device in real-world applications. Figure 1.1 illustrates the point that the high current drive capability of the bipolar device is an advantage with a significant capacitive load. Voltage gain, associated with the transistor "early voltage" or flatness of the output current versus output voltage characteristics, also favors the BJT due to fundamental structural differences. With higher voltages on the output terminal of the device, a BJT will deplete less into the heavily doped base region of the de-

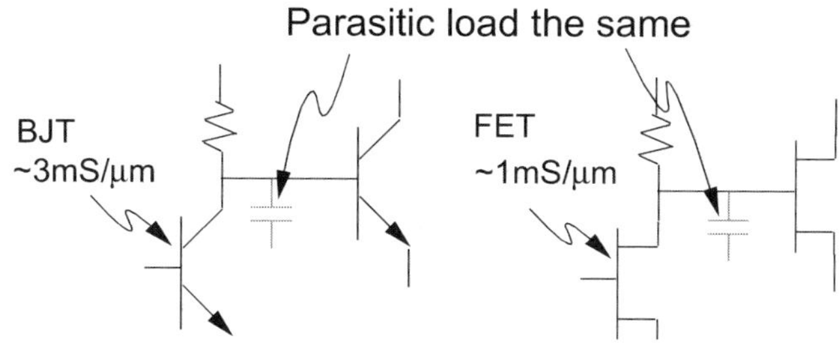

Figure 1.1 Bipolar junction transistor (left), with three times higher transconductance, more effectively drives a parasitic load than a field-effect transistor (right).

vice, compared to the depletion into the medium-doped channel region of the FET, where this is commonly known as drain-induced barrier lowering (DIBL). Scaled CMOS devices exhibit larger DIBL problems. When the output voltage affects the output current, the voltage gain of the device is compromised. Repeatability of the turn-on voltage is also a differentiator. A BJT turn-on voltage is determined principally from dopant properties, such as the total dopant in the base region of the device, and certain dopant properties at the junction, and is logarithmically related to the lithographic device dimensions. The FET turn-on voltage is a function of many more factors, including gate oxide properties, gate oxide thickness, dopants in the polysilicon gate conductor, dopants in the channel region, and above all, is linearly related to the channel dimension, which is determined from small-dimension photolithography, etch, and diffusion. As a result, design with the BJT, incorporating real-world tolerances, is more straightforward compared to the FET.

Higher voltage limits are sometimes beneficial to the designer utilizing the BJT for power applications, such as driving laser modulators or antennas for cell phones. In FETs, hot carriers and gate oxide tunneling limit the voltage that can be applied to the device. Designers often view BJT limits as BV_{CEO}, defined as the collector-emitter voltage, with the base terminal *open,* that causes a dramatic increase in the collector current. This value is typically higher than the comparable FET technology voltage limit when normalized to the same f_T. Yet, the perception of BV_{CEO} as a limit is, in fact, based on conservative concerns for the electrical behavior of the device rather than concern for device degradation. No degradation mechanism has been associated with this BV_{CEO} value, and, in fact, careful studies have found other limits in voltage that approach values closer to BV_{CBO}, which is typically about three times higher than BV_{CEO} [1]. This means that BJTs may have substantially higher voltages applied to the terminals and offer significant flexibility in high-voltage applications. While f_T and f_{MAX} can be scaled in CMOS, designers have significant problems designing RF/analog circuits with the lower supply voltages.

Another less obvious benefit designers have found is the outstanding linearity that may be found in the BJT. This is less a fundamental difference, as it is a function of the device design. It has been theorized that the nonlinear behavior of the device avalanche is canceled by capacitive effects in the device [2]. The bipolar device in IBM's BiCMOS 5HP technology, used extensively in wireless applications, has been recognized for its linearity, making it an outstanding choice in sensitive wireless signal paths.

HBT Device Design Today, IBM is engaged in a number of SiGe HBT device design activities, driven by markets with differing requirements (see summary of HBT characteristics in the Appendix). Even though high-speed performance often gets the attention in SiGe HBT device developments, development is taking place to address applications that do not demand higher speed, but rather higher voltage operation or lower costs. Semiconductor chips used in wireless applications, such as in cell phones, wireless networks, and Global Positioning Systems (GPS) are required to be inexpensive. The number of masks and the complexity of processing affects wafer cost and yield, and therefore the final packaged part cost. To achieve cost reductions wafer processing is simplified by reducing the number of process steps. These reductions can take the form of eliminating a portion of the structure (deep-trench isolation) to consolidation of masking steps to changes to the HBT structure (non-self-aligned (NSA) extrinsic base). In all situations, the device performance is altered as part of the device customization required for a particular end use.

Various improvements in device design were incorporated in the SiGe 5MR technology, which was tailored for the $\pm$ 5-V supply voltage used by hard disk drive preamplifiers. These enhancements were also incorporated into SiGe 5HPE. Again the cost for die is required to be low, and the challenge was to meet both the cost and use voltage criteria simultaneously. While the higher BV_{CEO} target (9.6 V) was met by increasing the lightly doped collector epitaxial layer thickness, the output characteristics of the high-breakdown HBT suffered from barrier effects [3] caused by base broadening, as shown in Fig. 1.2A. The usual high values of early voltage are compromised. Two approaches were taken to improve transistor performance: (1) improve the base germanium profile by introducing the boron within the germanium base layer, and (2) increase the lateral spacing between the extrinsic base implant with respect to the emitter opening, thus decreasing enhanced diffusion of the intrinsic base caused by the extrinsic-base implant [4]. These two improvements resulted in a substantially improved V_A, as shown in Fig. 1.2B. In addition, the peak frequency performance was improved from 14 GHz to 19 GHz.

With an increased distance from the extrinsic base implant to the emitter opening, it is more cost effective to simplify the usual self-aligned extrinsic-base structure to an NSA version, whereby the emitter polysilicon itself is used as the mask for the extrinsic-base implant. A 7% reduction in processing time and equally substantial reduction in wafer cost was achieved. These device improvements were feasible because the circuit designers were willing to trade off higher base resistance for increased frequency performance and early voltage. Due to the less complex emitter definition process, the V_{BE} matching is also markedly improved. These device trade-offs meet both the circuit design and wafer-cost requirements.

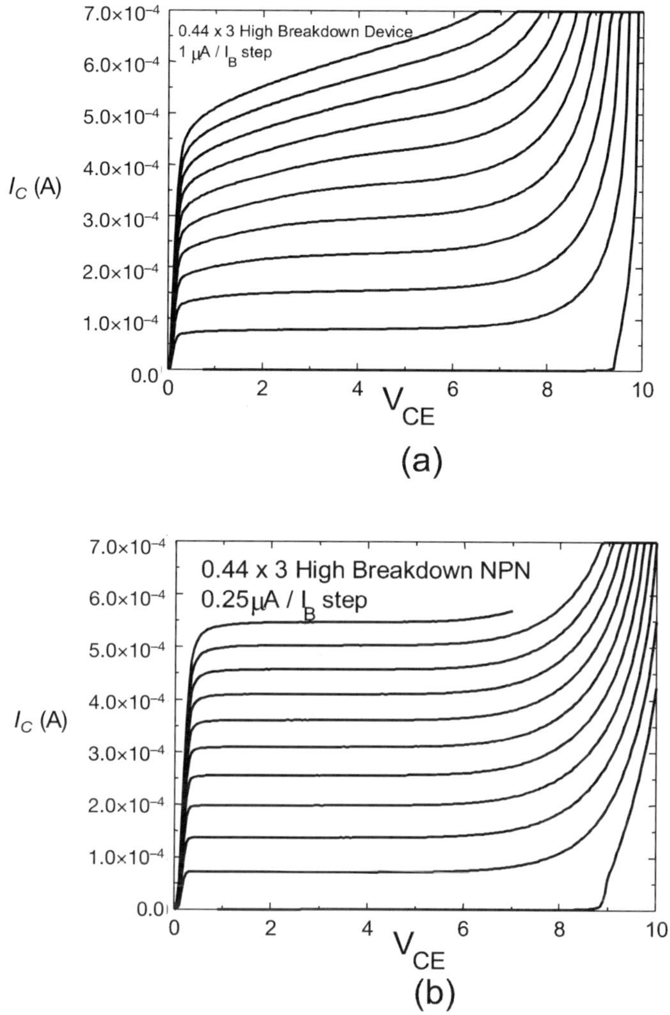

Figure 1.2 Output characteristics (a) prior to and (b) following germanium profile and extrinsic-base modifications in the 5HPE technology.

In the high-speed arena, SiGe HBTs today are surpassing even the fastest III–V production devices. The key to this achievement is the superior within-device parasitic-shaping technology available to the device designer, compared to what is available to the III–V device designer. With lithographically defined implants, trench isolation, self-aligned low-resistive regions, and such options as spacer technology, the silicon device designer has myriad tools at their disposal. The most common measure of performance is f_T, which is the maximum frequency that the transistor demonstrates useful (i.e., above unity) *current* gain. The components of f_T

are the diffusion capacitance charging relation $kT/qI_E(C_{EB} + C_{CB})$, transit times across the device (principally consisting of base transit time τ_B and collector space-charge transit time, τ_C), and the collector resistance–collector base capacitance $R_{C\text{-}}C_{CB}$ charging time, as shown in the expression in Equation (1.1):

$$\frac{1}{2\pi f_T} = \tau_{EC} \cong kT/qI_E(C_{EB} + C_{XB}) + R_C C_{CB} + \tau_B + \tau_C \qquad (1.1)$$

Reducing the layer thickness in each of the layers through which the electrons must travel improves f_T: the neutral base (i.e., affecting τ_B) and the collector space-charge region (i.e., affecting τ_C), as well as reducing the resistance–capacitance (*RC*) charging terms for the parasitic capacitances in the device. This concept is shown in Fig. 1.3A. The boron dopant (which makes up the base) is made narrower; the Ge is also made narrower and the grade is increased; and the collector concentration is increased, which reduces the space-charge region thickness, and at the same time reduces the collector access resistance. We refer to this as vertical scaling of the transistor, since these aspects are not related to the lateral dimensions of the device. The result of vertical scaling is to reduce the transit time, and increase the maximum operating current density in the device (i.e., for the same-size device, the current to reach maximum f_T performance is increased). Figure 1.3B shows this effect of vertical scaling on a plot of f_T versus current density.

The second common measure of performance is f_{MAX}, which is the maximum frequency where the transistor has useful (i.e., above unity) *power* gain. A shown in Equation (1.2), f_{MAX} follows closely the well-known relationship between f_T and parasitics:

$$f_{MAX} \cong \sqrt{f_T/8\pi R_{BB} C_{CB}} \qquad (1.2)$$

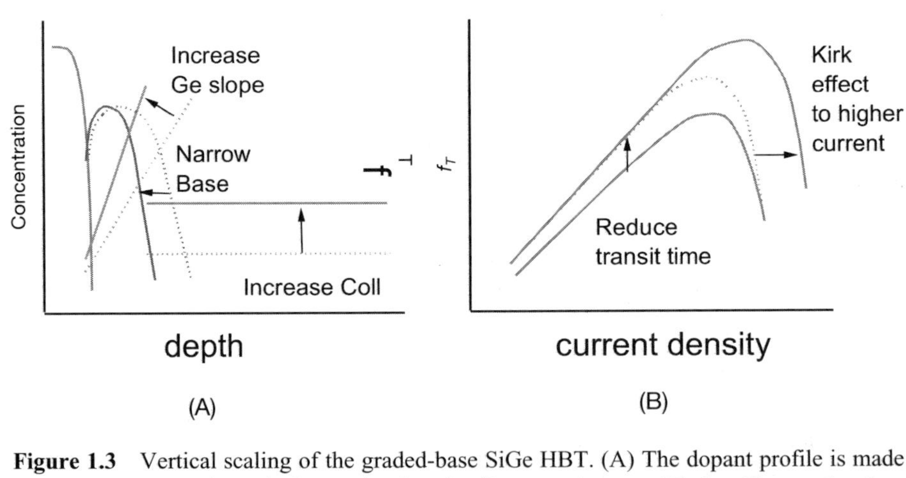

Figure 1.3 Vertical scaling of the graded-base SiGe HBT. (A) The dopant profile is made narrower for reduced transit time and reduced collector resistance; (B) the effect on the electrical properties f_T vs. current density.

where R_{BB} and C_{CB} are the parasitic base resistance and collector-base capacitance, respectively. For most applications, it is required that the f_{MAX} value be at least comparable to the f_T value for optimal circuit performance. Achieving high f_{MAX} is a challenge from both a device design and process point of view, since it is a strong function of the device structure, which largely determines the values of the base resistance R_{BB} and C_{CB}. This is because the majority of R_{BB} and C_{CB} are present in the extrinsic part of the device, or that region of the device that is not an essential part of the carrier transport. This fact provides the expectation that f_{MAX} will continue to be substantially improved with new device structures.

Comparison of IBM's SiGe HBT device structures illustrates how improvements in device structure can provide increases in the f_{MAX} figure of merit. Through several generations of technology, IBM has utilized the same device structure, often referred to as epitaxial transistor structures (ETX). Its identifiable structural characteristic is an extrinsic base implanted into the SiGe epitaxial film. Through careful analysis, including two-dimensional (2D) simulations [5], it was determined that this structure has some significant limitations. Implants into the silicon create lattice defects, which affect the diffusion of the intrinsic-base boron, increasing τ_B, and thus reducing device performance. This limits the proximity to which the implant may be placed to the intrinsic device, and therefore creates a lower limit on the achievable base resistance, R_{BB}. The implanted extrinsic base also extends deep into the silicon, and intersects the collector implants at a high concentration. This results in high C_{CB}. Shown in Figure 1.4 shows the ETX structure and the structure IBM is

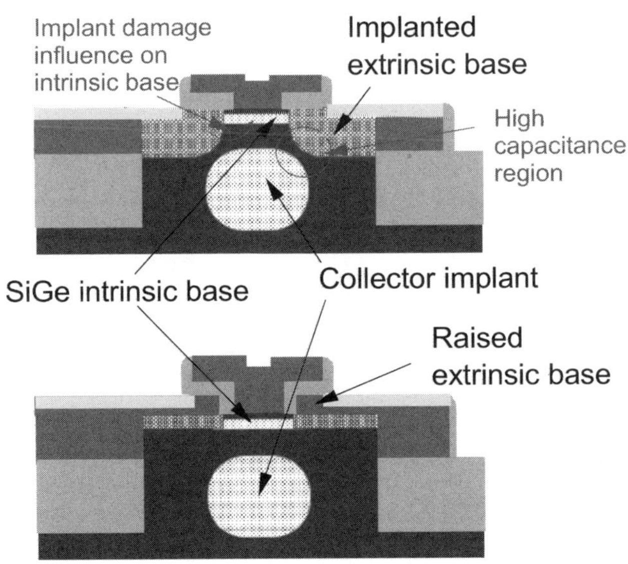

Figure 1.4 ETX structure (top) and RXB structure (bottom). The RXB structure eliminates the unwanted effects from the deep implant, including the excess capacitance and diffusion effects on the intrinsic-base region.

pursuing with a raised extrinsic base (*RXB*) to substantially reduce R_{BB} and C_{CB}. The raised extrinsic base has less influence on the intrinsic dopant diffusion, and may be placed in close proximity to the intrinsic device and therefore reduce R_{BB} without impact on f_T. It also has a minimal junction depth, and as such, has a relatively small C_{CB}.

Initial results on the structure demonstrate its benefits over the ETX structure. While retaining the f_T performance of a structure without a self-aligned extrinsic base (indicating that the extrinsic base has no influence on the intrinsic base), the base resistance has been reduced by approximately a factor of 2 and C_{CB} has been maintained constant compared to a similar area previous-generation device, with lower f_T. This results in simultaneous f_T and f_{MAX} improvements between generations of greater than 80%, shown in Figure 1.5.

1.1.2 FETs and Their Utility

The CMOS device takes on different roles when offered as part of a BiCMOS technology, where the bipolar device is available for analog functions and when part of a CMOS-only technology, where the FET devices must take a primary role in analog functionality. This differentiation influences technology development from de-

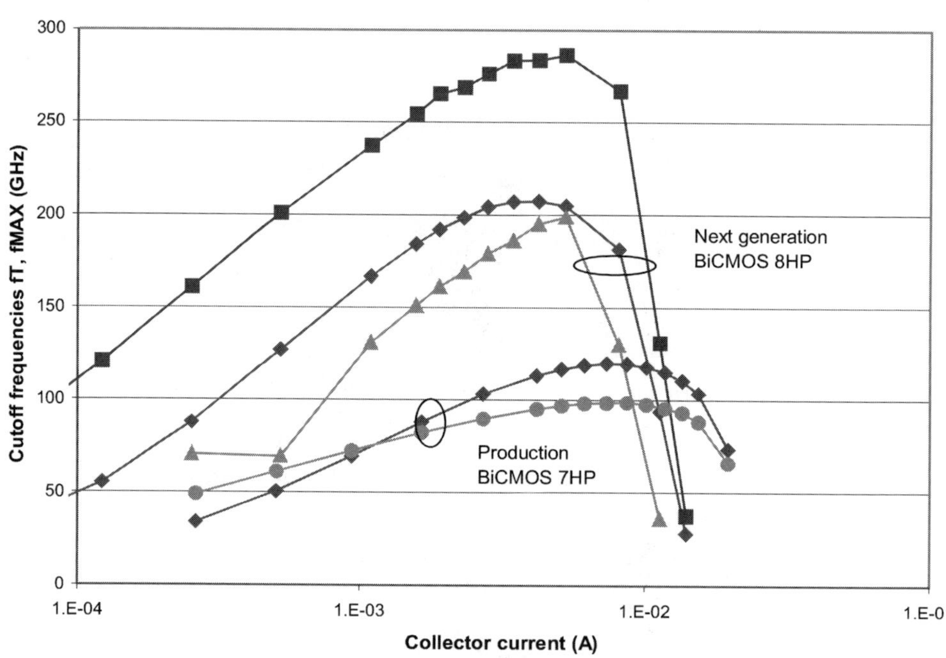

Figure 1.5 f_T and f_{MAX} comparison between prior generation f_T = 120-GHz device with ETX structure and next generation f_T = 210-GHz device, with vertically scaled profile and new RXB structure.

vice design, device layout, process development, characterization, and modeling. In CMOS technologies, one must not only consider the digital design aspects of FET devices, but also aspects that are driven by analog requirements, such as the ability of these devices to withstand higher voltages, or have lower body effect or higher self-gain (g_m/g_0). These aspects are less important in BiCMOS processes because of the presence of the bipolar device. Thus, one finds that additional masks and complexity are required in CMOS processes to allow a greater variety of FET devices to be offered to designers.

When offered as part of IBM's SiGe BiCMOS technologies, the CMOS devices are primarily used for integrating digital logic functions with high-speed bipolar analog circuits. This allows fully integrated "system on a chip" (SoC) products with the CMOS performing the lower-frequency baseband signal processing, as discussed in the Introduction. The CMOS devices can also be used for low-frequency analog functions such as analog-to-digital (A/D) converters, multiplexers, and switches. The CMOS devices have the one big advantage over bipolar devices of essentially no gate current. This makes CMOS devices ideal in circuits where it is required to measure the charge on capacitors such as A/D converters. Bipolar devices would drain the charge during the measurement.

Development of CMOS devices for digital logic purposes is mainly driven by shrinking the device dimensions, thinning the gate oxide, and lowering supply voltages to achieve faster performance, increased density and lower power consumption. The smaller device lengths lower the parasitics and increase f_T, but also necessitate complex designs, including halo implants to minimize short channel effects and control punch-through. These implants have negative effects on important analog characteristics such as self-gain (g_m/g_0). Also, the thinner gate oxides in the advanced logic devices cannot support the higher voltages required in analog circuits. The solution to this is a dual oxide technology. The analog devices are designed with thicker oxides, longer channel lengths and unique source drain extensions. This added process complexity allows high performance logic devices and high voltage analog devices on the same chip.

Some specific parameters must be considered when designing analog CMOS devices such as noise and V_t matching. Noise is not a large concern in digital CMOS circuits, and some processes like nitrided gate oxide actually increase noise in a trade-off for decreased dopant penetration of the gate oxide and improved hot carrier degradation. In analog circuits V_t matching is much more critical than in logic circuits, and all variables that introduce mismatch, including process and layout, must be minimized. Other parameters that are important for analog devices are gm, Ro, back bias sensitivity (body effect), and f_T.

In 2002, IBM offered several SiGe and CMOS RF technologies that had been qualified for high-volume production. Examples include SiGe5HP, SiGe5HPE, SiGe6HP, SiGe7HP and CMOS6SFRF. SiGe5HP is a single-gate oxide technology, while the rest have an optional dual gate oxide process. SiGe5HP contains 3.3-V CMOS devices designed specifically for logic support in the BiCMOS technology. SiGe5HPE has 120-A gate oxide, 5-V CMOS with an additional isolated NFET (IsoNFET) device. The IsoNFET is a standard NFET surrounded by an isolation tub, which allows the IsoNFET P-well to be biased independently from the substrate. Independent well biasing enables a circuit designer to handle dual logic levels on chip, for example, by biasing the substrate at –5-V and the P-well in the tub at 0 V. Higher-voltage analog signals can also be handled in this way by stacking 5-V FETs inside and outside the isolation, with a 10-V signal across the combination. IsoNFET devices also have better noise isolation due to the independently biased P-well and isolation tub.

Dual oxide technologies include SiGe6HP (0.25 μm), CMOS6SFRF (0.25 μm), and SiGe7HP (0.18 μm). SiGe6HP contains 2.5-V and 3.3-V CMOS devices, which have 50-A and 70-A gate oxides, respectively. The thin-oxide FETs are used for the high-speed logic, with 0.25 μm L_{min} (min drawn gate lengths), while the thick-oxide devices are 0.4-μm L_{min} NFETs and 0.34-μm L_{min} PFETs. The thick-oxide FETs enable 3.3-V input/output (I/O) compatibility as well as analog signal handling. CMOS6SFRF is based on the same 2.5-V/3.3-V devices with an additional 2.5-V IsoNFET and a process option of 6.5-V thick-oxide (140-A) devices in place of the 3.3-V devices. The 6.5-V devices have L_{min} = 0.7 μm to support the higher voltage. These devices can be used as low-frequency power amps, high-voltage analog switches and voltage regulators in battery chargers. SiGe7HP has 1.8-V and 3.3-V CMOS devices with 35-A and 68-A gate oxides, respectively. There are 1.8-V standard V_t FETs and optional 1.8-V high V_t FETs, 3.3-V FETs, and both 1.8-V and 3.3-V IsoNFETs. The 1.8-V FETs have L_{min} = 0.18 μm and the 3.3-V FETs have L_{min} = 0.4 μm.

The main purpose of the CMOS devices in a SiGe BiCMOS technology is to provide integrated logic functionality. The logic design can be expedited by using IBM application-specific integrated circuit (ASIC) library books or industry standard-cell libraries already developed for the base CMOS technology. This approach can only be used if the CMOS device characteristics in the BiCMOS technology closely match those of the base CMOS technology. There are many process differences that can lead to significant device differences. Adjusting the process minimizes many of these differences, but some cannot be corrected. To verify that the ASIC library elements function correctly and that the CMOS timing models are still valid for the BiCMOS process, ASIC library testsites are built in the BiCMOS process. The chips are tested for functionality, and hardware-to-model correlation is done to validate the timing models. Any library elements that do not function or do not match the timing models are not offered. Some devices are known not to function correctly due to process differences and therefore any library elements containing the devices are not tested. Figure 1.6 shows a typical correlation plot with both SiGe and CMOS data showing that the same model can accurately represent both technologies.

1.1.3 Summary

In this chapter, we have discussed many of the key aspects in building world-class NPN and FET devices for SiGe BiCMOS process technologies. The presented methodology is pragmatic, and is both cost and time-to-market sensitive.

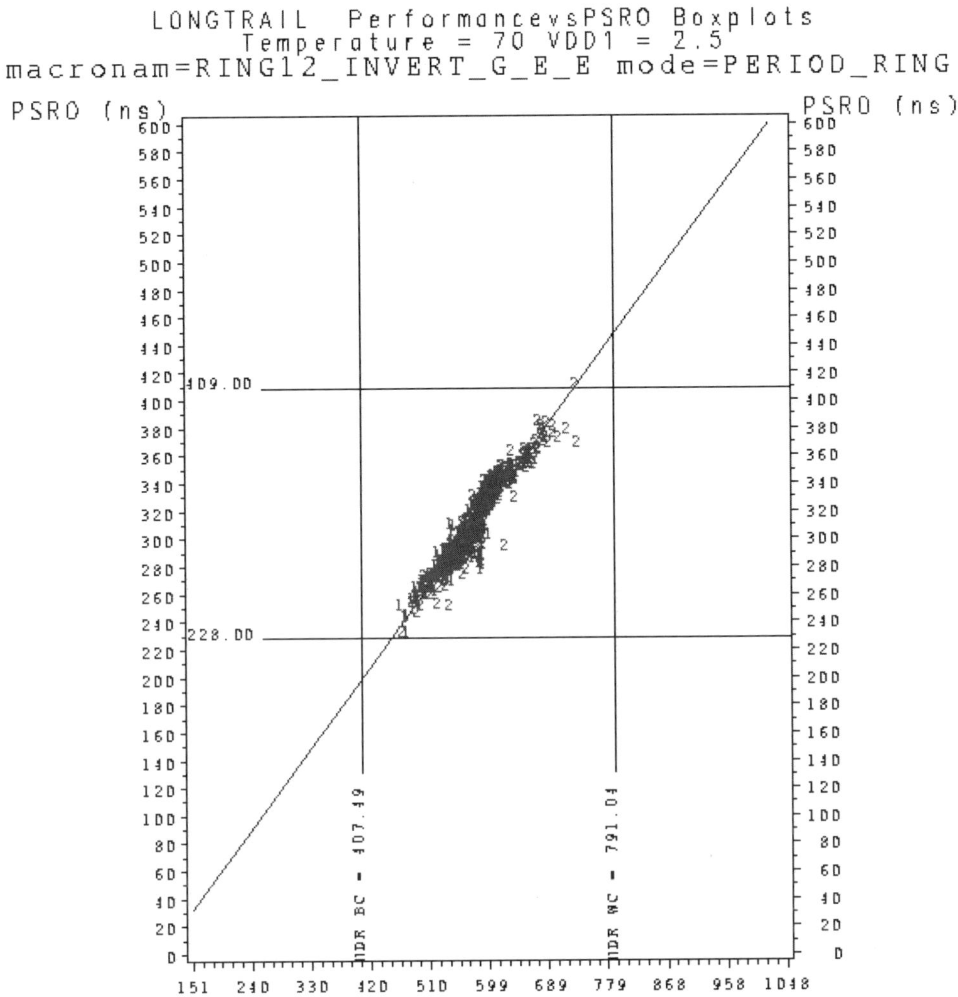

NDR = MODEL = Chips = 1108

BC NDR = 407.49 BC Hardware = 452.00

WC NDR = 791.04 WC Hardware = 730.58

BC Diff = (BC HW – BC NDR)/BC NDR * 100 = 10.92 WC Diff = (WC NDR – WC HW)/WC NDR * 100 = 7.64

NDR Pect difference = 94.13

Example Program Written by: Anthony Fazadxas 17MH02 Run by Anthony W. Fazadxas

Figure 1.6 Boxplot showing SiGe6SF CMOS performance set ring oscillator (PSRO) [2] identical to CMOS6SF Longtrail hardware [1].

1.1.4 References

1. M. Rickelt, H. M. Rein, and E. Rose, "Influence of Impact-Ionization-Induced Instabilities on the Maximum Usable Output Voltage of Si-Bipolar Transistors," *IEEE Trans. Electron Devices,* vol. 48, pp. 774–783, April 2001.
2. G. Niu, Q. Liang, J. D.Cressler, C. S. Webster, and D. L.Harame, "RF Linearity Characteristics of SiGe HBTs," *IEEE Trans. Microwave Theory Techniques,* vol. 49/9, pp. 1558–156, September 2001.
3. A. J. Joseph, J. D. Cressler, R. C. Jaeger, D. M. Richey, and D. L. Harame, "Neutral Base Recombination in Advanced SiGe HBTs and Its Impact on the Temperature Characteristics of Precision Analog Circuits," *IEDM Tech. Digest,* pp. 755–758, 1995.
4. M. D. R. Hashim, R. F. Lever, and P. Ashburn, "Two Dimensional Simulation of Transient Enhanced Boron Out-Diffusion from the Base of a SiGe HBT Due to an Extrinsic Base Implant," in *BCTM,* pp. 96–99, 1997.
5. J. B. Johnson, A. Stricker, A. Joseph, and J. A. Slinkman, "A Technology Simulation Methodology for AC-Performance Optimization of SiGe HBTs," *IEDM Tech. Digest,* pp. 489–492, December 2001.

1.2 TECHNOLOGY DEVELOPMENT: ADVANCED PASSIVES AND ESD PROTECTION

Passive devices, such as inductors, resistors, and capacitors, dominate the component count in modern wireless appliances. A passives-to-active device ratio of 20:1 is commonplace on a typical off-the-shelf cell phone [1]. To reduce form factors of handheld devices, traditional surface mounted passives are being integrated into the chip. In this section, we present details of the technology development of advanced passive devices in IBM's SiGe BiCMOS processes. Discussions cover resistors, capacitors, varactors, inductors, transformers, and back end of the line (BEOL) definition. In addition, we present details of ESD protection devices, with some comments on issues in RF-CMOS processes.

1.2.1 Passive Devices

BiCMOS technology development in IBM has been largely focused on the integration of high-performance SiGe HBTs in a base CMOS technology. Historically, passive devices are typically developed from existing processes used for these transistors, i.e., resistors are formed from CMOS FET source/drain implants, MOS capacitors from the reachthrough implants used for the HBT collector contact, and FET gate oxide/polysilicon gate, and inductors designed using last metal options for these technologies. The need for high integration and technology innovation in RF circuit design has changed the direction of passive development in the last several years, and has led to the developed of advanced process options. Examples are: analog (i.e., thick) metals used as last metal options for high-Q inductors, tantalum nitride (TaN) resistors integrated in the BEOL metalization for low parasitic capacitance/tolerance, and high-capacitance nitride metal insulator metal (MIM) capacitors.

Balancing the performance of passive devices with processing costs is a challenge. Some of the more critical parameters for the passive elements used in RF designs are resistor tolerance, varactor tunability (C_{max}/C_{min}) and linearity, MIM capacitance density and quality factor (Q), and inductor Q. In the following sections, the important passive elements offered in IBM's SiGe BiCMOS and RF CMOS technologies are described with focus on their RF application, key figures of merit, and reliability.

Resistors Resistors are used in all analog and mixed-signal circuit blocks. A wide variety of resistors are offered in IBM's SiGe BiCMOS and RF CMOS technologies to accommodate designer needs (see Table 1.1). Figures of merit for resistors are sheet resistance, tolerance, parasitic capacitance, voltage, and temperature coefficients. Three types of basic resistors are used in the SiGe BiCMOS process to achieve the desired properties and resistance ranges needed in analog circuit designs. These are P-doped polysilicon resistors, N- and P-type diffusion resistors, and BEOL TaN metal resistors.

Highly doped P-type polysilicon resistors are preferred in most cases for mixed-signal and analog applications due to their good matching, low parasitic capacitance to the substrate, and excellent temperature coefficient, as shown in Table 1.1. This resistor consists of gate or SiGe polysilicon doped with a high-dose boron implant, normally the PFET source/drain implant. Either shallow trench isolation or shallow/deep trench isolation is used under these resistors to reduce parasitic capacitance between the resistor and substrate. The ends of the resistor are silicided for low contact resistance to the BEOL wiring and the body of the resistor covered with silicon nitride to block the silicide. The P^+ polysilicon resistor has a sheet resistance of 270/sq and a 10–15% tolerance. Low tolerance is essential for efficient compact circuit designs. Table 1.1 shows that this resistor has a very low-temperature coefficient (TCR) of 21 ppm/°C, which is 2–3% of the TCR offered with other resistors. This makes the resistor most attractive for circuit applications due to low variation in resistance with changes in temperature over typical ranges of –40 to 125°C. The

Table 1.1 Electrical Parameters of Resistors Available in SiGe and RF-CMOS Technologies

Resistor	Sheet Resistance (Ω/Sq)	Tolerance (%)	TCR (ppm/C)	Parasitic Capacitance (fF/μm²)	Maximum Current (mA/μm)
P^+ polysilicon	270	10–15	21	0.11	0.6
P polysilicon	1,600	25	–1,105	0.09	0.1
N^+ diffn	72	10	1,751	1.00	1.0
N subcollector	8	15	1,460	0.12	1.0
Thin-film metal	142	10	–728	0.03	0.5

parasitic capacitance between the resistor and substrate is 10% of that of a diffusion resistor, but four times the value for a BEOL resistor, due to the distance of these devices from the substrate. A low-doped P-type polysilicon resistor is offered in these technologies as well. This provides a higher sheet resistance at 1600 Ω/Sq for applications requiring high resistance while maintaining good parasitic capacitance. This resistor is more difficult to control in the process, resulting in a 25% tolerance.

The N^+ diffusion resistor is formed with the NFET source/drain implant in single-crystal silicon. The ends of the resistor are silicided. With a sheet resistance of 72 Ω/Sq, the N^+ diffusion resistor is used in current source/biasing circuits where resistors in the 50–100-Ω range are needed. Since this resistor is made from the FET source/drain, it has a high capacitance, which limits its use. This resistor is typically controlled to a 10% tolerance.

The n-subcollector (NS) resistor is made from the low resistance NPN subcollector and contacted with the collector contact. This device has a low sheet resistance at 8 Ω/sq and a tolerance of 10–15%. Typical of diffusion resistors, this device has a high temperature coefficient at 1460 ppm/°C. This resistor is ideally used as a ballast resistor in applications such as power amplifiers.

A thin-film BEOL resistor has several attractive features, such as low tolerance, low parasitics, and the ability to make design changes with short lead times. A TaN resistor is offered in several of our SiGe BiCMOS technologies at metal levels M1 and M5. This device consists of a TaN film contacted by metal vias. The tolerance is low at 10%, and due to its distance from the substrate, it has very low parasitic capacitance to the substrate.

All resistors offered in IBM's SiGe BICMOS technologies meet stringent reliability requirements for 100K power-on hours (POHs). Reliability tests are performed by measuring the shift in resistance over a fixed period of time under constant current. By varying the bias conditions used to stress these devices, the amount the resistor will shift in 100K hours can be projected. Typically, resistance changes of less than several tenths of a percent are projected over the life of the resistor for the curent limits specifed in Table 1.1. Current limits of 1 mA/θm of width for diffusion resistors are normal. Low-resistance polysilicon and the BEOL resistors have a current limit of 0.5–0.6 mA/μm.

Capacitors Three types of capacitors have been developed in SiGe technologies to meet customer requirements for reduced board-level components. MOS (polygated capacitors on single crystal silicon), polysilcon–insulator–polysilicon (PIP) and MIM capacitors each have their own sweet spot for use in different application spaces, depending on capacitance desired and performance at the application frequency. An overview of process details for optimization, electrical performance, and reliability is presented for each device.

The simplest MOS capacitors are formed without additional masks from the FET elements in all SiGe BiCMOS generations using silicided gate polysilicon, thin gate oxide, and FET well-doped silicon. Although these devices have a very high capacitance/area owing to the ultrathin oxide, they are not particularly useful for RF applications because of the high resistance of the well doping (~250 Ω/sq) and poor voltage coefficient. A more optimized capacitor has been developed by heavily doping the silicon substrate to reduce parasitic resistance [2]. This is accomplished by using a high-dose phosphorus reachthrough implant (~25 Ω/sq) to dope the bottom plate of the capacitor. During gate oxidation, the insulator grown over high-dose phosphorus implants can result in a 50–100% increase in thickness relative to oxides grown over intrinsic silicon due to enhanced oxidation. Shallow reachthrough implants can cause unreliable oxides due to very high growth rates driven by high surface dopant concentrations. The quality of the oxide grown over the diffusion region increases significantly with implant depth. A comparison of the applied field in depletion mode at a 1-nA leakage current found that a shallow implant causes premature oxide leakage (at 5.2 MV/cm) relative to the deeper implant (6 to 7 MV/cm), which meets leakage requirements. Therefore, an optimized MOS capacitor will have a high-dose reachthrough implant at a moderate depth for low resistance and a reliable oxide.

PIP capacitors are fabricated in double polysilicon BiCMOS processes [3]. The unit capacitance is a product of the integration methodology, and typically the device comes for free in the process. Fabricated in SiGe 5HPE, the capacitor structure is formed using p^+ doped gate polysilicon as the bottom plate, a deposited oxide layer for the capacitor dielectric, and silicided extrinsic-base polysilicon as the top electrode. To minimize the bottom-plate capacitance to substrate, the doped gate polysilicon is patterned over shallow-trench isolation and/or deep-trench maze, an advantage gained over the MOS structure. The dielectric quality is critical to ensure high reliability and robust breakdown strength. Thermal oxides are rarely used, because during oxidation the polysilicon roughens along grain boundaries, yielding high field points that reduce the strength. An obvious alternative is plasma-enhanced chemical vapor deposition (PECVD) dielectric, but these typically have poor uniformity, poor conformality, and pinholes in thin films. For this application, a thin deposited hot thermal oxide (HTO) was developed. The HTO breaks down in the range of 9–10 MV/cm, and is very conformal yielding full thickness coverage at the gate poly corners where premature breakdown can occur under high electric fields. Optimizing linearity for the PIP capacitor was an important aspect of its development. To understand capacitance–voltage *CV* linearity as a function of dopant type and dose, a design of experiments was executed that varied dopant type and concentration for each of the capacitor electrodes. The total capacitance and linearity are a function of the polysilicon depletion capacitance in series with the dielectric capacitance. The optimal electrode configuration is when both plates of the device are doped n-type. At a given bias, one plate is in depletion, while the other plate is in accumulation, causing a small change in the net capacitance. In contrast, when one plate is doped n-type and the other p-type, both plates are either in depletion or accumulation simultaneously, causing a larger change in the net capacitance, resulting in reduced linearity.

While acceptable for use in low-frequency applications, MOS and PIP capacitors suffer from low-quality factors due to high resistance plates and capacitive losses in the 2–10-GHz range, rendering them nonideal or limiting their use. A novel MIM capacitor (MIMCAP) was developed that takes advantage of low-resistivity metal wiring, thick interlevel dielectric that physically distances the devices from the relatively low-resistivity substrate, and the planar BEOL topology to build in high reliability [4]. The SiGe BiCMOS planar MIM (Fig. 1.7) is fabricated by depositing a 50-nm PECVD oxide and a 200-nm metal stack on top of any metal wiring level except the first and last. A mask is applied and the top plate is etched, stopping at the capacitor dielectric. The metal-layer mask is then applied, which defines the metal wiring as well as the base capacitor plate. An interlevel dielectric is deposited and planarized, and vias added to connect to the next metal layer. The dielectric of the MIM capacitor is thicker than either PIP or MOS capacitors, because lower temperature dielectrics (compatible with BEOL processing) are generally of poorer quality than higher temperature chemical vapor deposition (CVD) or thermal oxides.

Designers prefer the MIM capacitor over the other two types, because of its advantageous performance (higher-Q) at higher frequencies. Thick metal plates offer lower resistance than doped and/or silicided polysilicon, and placement in the interconnect levels significantly reduces parasitic capacitance of the substrate. Finally, the ability to resize the MIM in a BEOL redesign reduces cycles of learning not afforded by silicon-based capacitors. The penalty paid for these benefits is that a large area of the chip is consumed, impacting the ability to reduce the chip's form factor. Capacitors may take up to as much as 50% of the chip area, depending on the application. In order to decrease the capacitor footprint, there are several options to increase the unit capacitance, as capacitance is a direct function of the insulator's dielectric constant and inversely proportional to film thickness.

The list of requirements that a high dielectric constant material must meet in order to address manufacturing, yield, design, and reliability concerns is a long and demanding one. Table 1.2 contains a short list of critical parameters. From a fabrication aspect, the deposition process must meet manufacturability targets, have

Figure 1.7 Cross section of SEM micrograph of MIM capacitor in SiGe BICMOS technology.

Table 1.2 Critical Parameters for High-k Dielectrics Required for MIM Capacitors

Property	Value
Deposition Temp	<400°C
Tks Uniformity	<5% (3 sigma)
Deposition Rate	30–50 nm/min
Thermal coefficient of capacitance (TCC)	<30 ppm
Voltage coefficient of capacitance (VCC)	<100 ppm
Operating voltage	± 5 V
Reliabiltiy	<10 ppm @ 100K POH
Leakage	10^{-6} A/cm^2 @ ± 5 V
Dielectric constant	7–25

slow and controllable deposition rates, have high throughput, and have no impact to BEOL yield or parametrics. From a design aspect, the new dielectric should not significantly change alternating-current (AC) device models established from prior generations. Finally, the film must have low defect density, negligible fixed or mobile charges, and be stable under thermal stress in order to pass stringent reliability qualification. The candidates under consideration, along with their area capacitance as a function of inverse thickness, are shown in Figure 1.8. A processing limitation for all new dielectrics will be deposition temperatures of ~400°C or less to prevent the shorting of narrowly spaced large metal lines due to metal extrusions. PECVD nitrides are known for their relatively high hydrogen content, bonded to both silicon and nitrogen. Recently, a PECVD nitride was qualified, increasing the area density

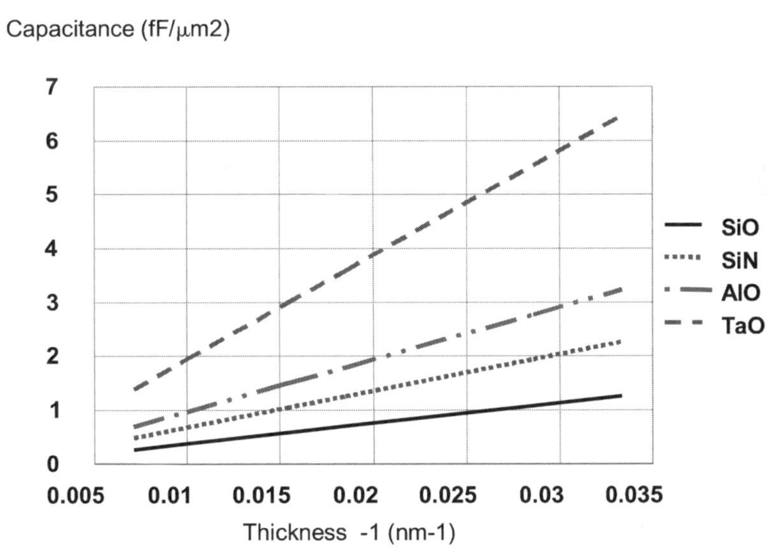

Figure 1.8 Capacitance density vs. thickness for various dielectrics.

by 43% over prior oxide dielectrics. Aluminum and tantalum oxide processes are coming on-line, primarily driven by next-generation CMOS gate oxide replacement and dynamic random-access memory (DRAM) node capacitors. These deposition processes cover the spectrum from atomic layer to bulk CVD, and often require clustering with rapid thermal-process modules to fully oxidize the unreacted carbon compounds derived from the source material. Very high-k and ferroelectric films, such as composites of barium, strontium, bismuth, and titanium, are still two to three generations away from implementation [5].

Scaling, or thinning the dielectric has not yielded the reliability performance necessary for qualification. Leakage currents through the dielectric are directly proportional to the film thickness to levels such that the reliability of the device is seriously compromised.

Finally, by taking advantage of the planarity and modularity of integration, wiring MIM capacitors in parallel on two or more levels can yield an equivalent multiple increase in unit capacitance, albeit at the cost of a mask level for each. Similarly, in SiGe 5HPE, the metal-oxide semiconductor capacitor (MOSCAP) and PIP capacitors can be integrated into one structure and wired in parallel to yield a total unit capacitance of 2.8 fF/μm^2.

The planar MIMCAP concept and fabrication techniques have been successfully transferred to all SiGe generations, as well as CMOS 0.35-μm and 0.25-μm technologies. The fabrication strategy has also been used to integrate the device in copper BEOL (CMOS 0.18 μm and beyond) [6]. Integration into a copper BEOL of more advanced CMOS technologies presented many challenges that ultimately resulted in a process similar to the aluminum MIMCAP. Approaches using the copper level as the bottom plate in an analogous fashion to the aluminum approach were not successful due to issues associated with copper metallurgy (formation of hillocks or extrusions), the requirement that large copper areas needed oxide plugs for chemical-mechanical polishing (CMP) to achieve planarity, and the oxidation of copper during the deposition of dielectric requiring oxygen precluded the direct deposition of all oxides. The so-called MIM over dielectric (MOD) provides a planar starting surface, relief from copper exposure to oxidizing and chlorine etch ambient, and capacitor size constraints. Plate materials and thickness are designed to be able to fit within the dual damascene stack while not impacting wiring and via parametrics and yield.

Table 1.3 summarizes key parameters for the optimized capacitor structures, comparing areal capacitance, temperature linearity, and voltage linearity. The PIP capacitor has improved parasitics relative to the MOS capacitor, because it is over shallow-trench isolation/deep-trench (STI/DT) isolation. Area capacitances of the bottom plate to substrate are one-tenth the parasitics for the bottom plate of the MOS capacitor. The PIP capacitor has a 15% reduction in bottom plate to substrate parasitics for a device over STI/DT relative to a device over STI only. Because the MIM capacitor is produced in the BEOL, it has the lowest parasitics of the devices described. The area capacitance of the bottom plate to substrate is 5% of that for the PIP capacitor over DT when the MIM is at M5. However, the MIM perimeter parasitics are much higher. The MIM has the highest Q of the devices described. For a

Table 1.3 Summary of Capacitor Offering Across Several SiGe Technologies

Capacitor	Parameter	SiGe 5HP/AM/DM	SiGe 5HPE	BiCMOS 6HP	BiCMOS 7HP
MOSCAP	C_0 (fF/μm2)	1.5	1.2	3.1	2.5
	Tol. (%)	10	15	15	15
	T_{CC} ppm/°C	48	21	20	
	V_{CC} +5/–5 V ppm/V	1740/1740	–990/–450	–7500/–1500	–5480/–1240
Poly-Poly	C_0 (fF/μm2)	—	1.6	—	—
	Tol. (%)	—	25	—	—
	T_{CC}	—	21	—	—
	V_{CC} +5/–5 V ppm/V	—	–3525/–2475	—	—
MIM	C_0 (fF/μm2)	0.7	1.35/2.7	0.7/1.4	1.0
	Stacking	Single	2 Stack	2 Stack	Single
	Tol. (%)	15	15	15	15
	T_{CC} ppm/°C	–57	–44	–44	–15
	V_{CC} +5/–5V ppm/V	<25	<25	<25	<25

20×20-μm device, the MIM has a Q of 90, and the MOS capacitor has a Q of 20. A comparison of the MOS and PIP capacitors for a 10×50-μm device shows Q values of 3 and 6, respectively.

Reliability is central in the design and development of capacitors for BiCMOS technologies. The goal is to maximize capacitance by reducing dielectric thickness while still maintaining acceptable reliability at the rated use voltage. Capacitor reliability is determined using a time-dependent dielectric breakdown (TDDB) test. The devices are biased at high electric fields to accelerate the time to failure. The failures are plotted as a Weibull distribution plot at multiple stress voltages, and then converted to a lifetime plot (Fig. 1.9) to extrapolate to 100,000 power-on hours to determine the electric field the capacitor will survive for this period of time. The MIM capacitor is less reliable than the MOS or PIP capacitors. This result is expected since the MIM oxide is a PECVD oxide, while the MOS and PIP capacitors are higher quality oxides. Therefore, the MIM has to have a thicker oxide and lower capacitance to meet the same reliability requirements. Also, the PIP capacitor shows equivalent reliability results to the MOS capacitor, indicating that a high-quality CVD oxide can be as robust as a thermally grown oxide over n^+ type diffusion.

Capacitors are used in a variety of low- and high-frequency applications, ranging from power supply bypass and decoupling to resonators, filters and tank circuits. MIM capacitors with Q values of 90 at 2 GHz are preferred in narrow-band applications such as resonators, filters, and tank circuits where high-Qs and low parasitic capacitance are required [7]. In contrast, PIP and MOS are typically used for power-supply bypass and decoupling. With an optimized PIP capacitor using an n^+/n^+ polysilicon stack, the device may have Qs approaching 30 and would be appropriate for low frequency RF applications.

The device parameters, process, and reliability issues have been reviewed for the MOS, PIP, and MIM capacitors. The MIM has the lowest series resistance and par-

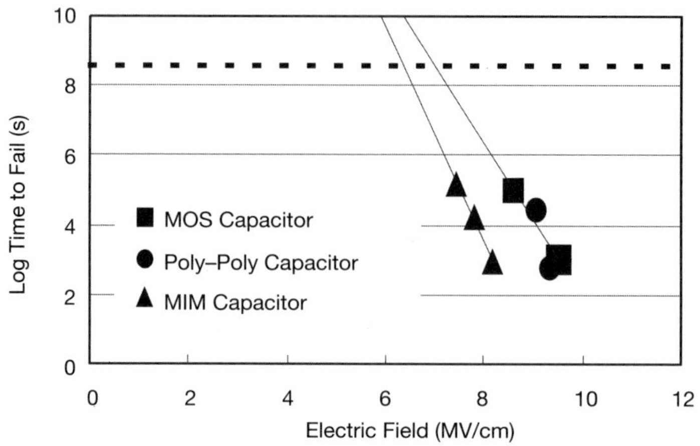

Figure 1.9 TDDB-accelerated stress results for MOS, PIP, and MIM capacitors.

asitics, resulting in the highest Q of the three devices by taking advantage of low-resistivity metal plates and their presence in the interconnect levels. However, a trade-off is made, since it has a lower capacitance to compensate for its lower reliability. The MOS capacitor is on the other side of the spectrum, with high capacitance and reliability, but poor linearity and parasitics. The PIP capacitor is a higher-capacitor device and when designed over DT isolation, can afford very good parasitics as well. By selecting the proper dopant polarities, the linearity of the PIP can approach that of the MIM. The device has improved linearity over the MOS capacitor, and if both polysilicon plates are similarly doped, very good voltage coefficients can be achieved.

Varactors Varactors are essential elements of some key RF circuits, such as voltage-controlled-oscillator (VCOs), phase shifters and frequency multipliers. The key figures of merit for varactors are (1) tunability (C_{max}/C_{min}), (2) CV linearity for VCO gain variation, (3) quality factor Q, (4) tolerance, and (5) capacitance density. Traditionally, the varactor offered in BiCMOS technologies is the standard collector-base (CB) junction varactor. Three varactors have been developed for the SiGe BiCMOS and RF CMOS processes to augment this offering. These are a "quasi-hyperabrupt" enhancement of the CB diode, a MOS accumulation mode capacitor, and a CMOS diode.

Because it relies on the doping profile of the existing SiGe HBT, the standard CB diode is not optimum for RF circuit performance. This is evidenced by a tunability of 1.7:1 between 0 and 3 V and a low degree of linearity, as shown in Fig. 1.10. To overcome these problems, the quasi-hyperabrupt varactor utilizing a retrograde implant to modify the CB doping profile has been developed. To ensure independent optimization of the varactor and the HBT, an extra mask is used to introduce the retrograde implant. The quasi-hyperabrupt device has excel-

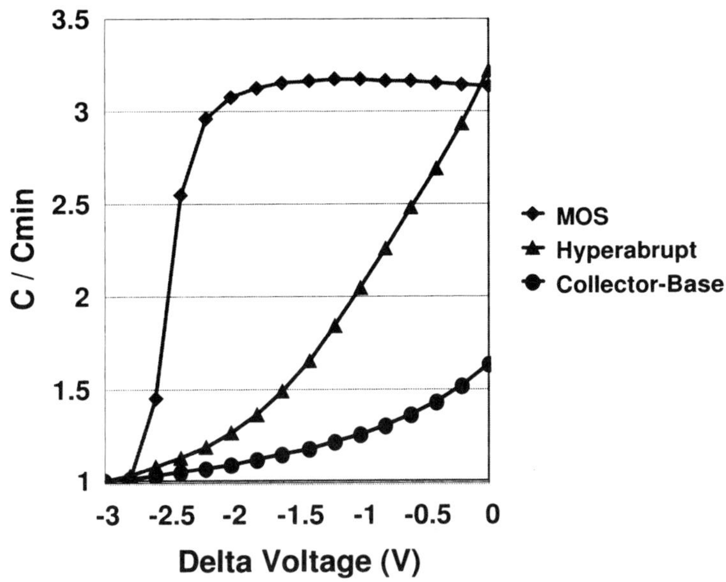

Figure 1.10 Normalized CV curves for hyperabrupt and MOS varactor compared to a (standard) collector-base junction varactor.

lent VCO circuit performance due to an outstanding tunability of 3.4 (0, 3 V) and a high degree of linearity, specifically in the 0- to 2-V range (see Fig. 1.10). The extra mask is compatible with an interdigitated layout, thereby preserving the device's high-quality factor, Q. Figure 1.11, which is a plot of Q versus frequency at various biases, demonstrates that the device has a Q in excess of 100 at 2 GHz.

The MOS varactor is an n-channel MOSFET fabricated in an N-well. The capacitance of this varactor is high in accumulation and decreases sharply as the device enters and goes further into depletion. This results in excellent tunability. The tunability is further enhanced with technology scaling, since the gate oxide capacitance increases as the gate oxide becomes thinner. The MOS varactor, however, can have less than optimum Q due to the series resistances associated with the N-well. To maximize Q, our MOS varactor features an interdigitated layout. Interdigitating the source/drain regions minimizes the N-well series resistance, resulting in a high-quality factor. The MOS varactor is built in IBM RF-CMOS processes, as well as BiCMOS technologies, without any extra masks.

The CMOS diode utilizes the halo typically present in state-of-the-art PFETs to achieve the desired hyperabrupt doping profile. The varactor is basically a P^+/N-well diode. The tunability is optimized by maximizing the gate-bounded perimeter and thus the overall area of the halo. The CMOS diode exhibits a tunability of about 1.7:1 over a 3-V range. This varactor can be useful for RF CMOS technologies that require better varactor linearity than a MOS varactor.

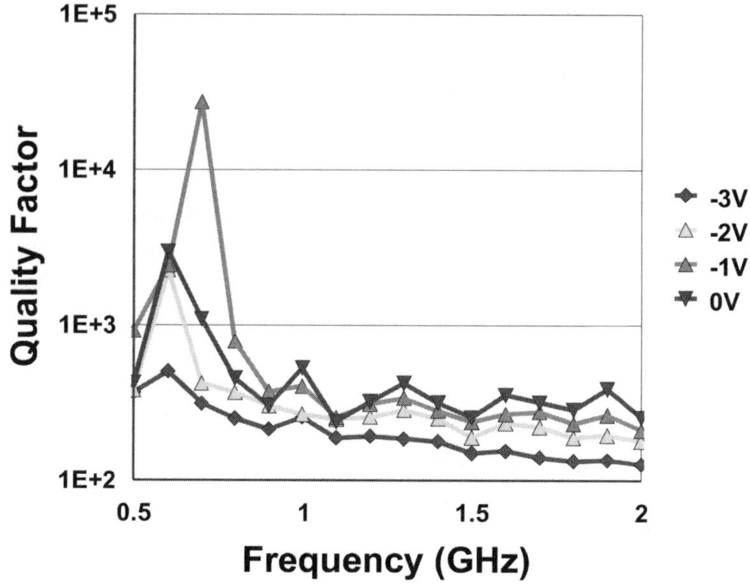

Figure 1.11 Quality factor Q vs. frequency for hyperabrupt varactor, Q is lower for increased reverse bias as the depletion layer, and parasitic resistance, increases.

Back End of the Line (BEOL), Inductors, and Transformers The need for high-performance passive elements in SiGe technologies has become increasingly important recently due to the technology and integration requirements of high-functionality/low cost RF circuit applications. On-chip passive elements, specifically monolithic spiral and multilevel spiral inductors, are a key component of these passive offerings. On-chip integration of spiral inductors poses numerous challenges. Issues related to quality factor (Q), parasitics, manufacturability, design, and cost must be balanced in order to provide a competitve inductor offering [8]. For high-performance inductors, it is necessary to provide thick metallization with low via resistance for reduced series resistance [9]. The typical interconnect scaling associated with digital circuit technologies (thinning metal and dielectric levels to increase wiring density) is inconsistent with the need to keep series resistive losses low for high-quality inductors. Additionally, minimizing substrate losses due to eddy currents and capacitive coupling is desired to increase Q. Integrating the inductors in the last, thick metal levels of the BEOL and using large vias to provide a thick dielectric between the inductor and substrate helps reduce these effects [10].

Three basic metallization options are provided for spirals inductors in IBM's SiGe BiCMOS (and RF CMOS) technologies: a 2-μm-thick aluminum level, a 4-μm-thick aluminum level, and a dual metal stack of 4-μm aluminum and 3-μm of copper. Figure 1.12 shows the integration of these levels in the BEOL. These inductor offerings have evolved since around 1997 to provide high-performance in-

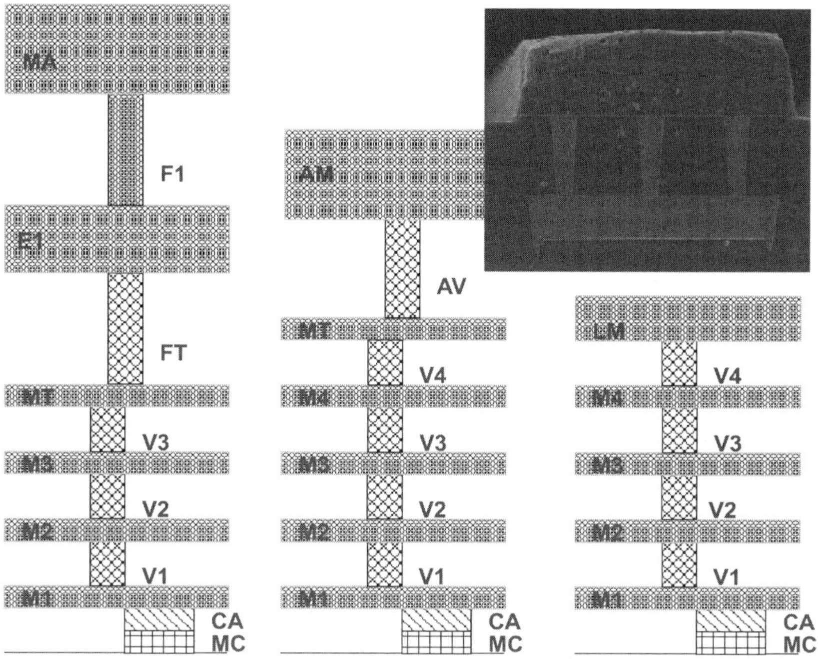

Figure 1.12 Schematic cross-sectional view of metalization (5HP, 5AM, 5DM). The inset shows a SEM micrograph of the top (two) thick metal layers in 5DM. The difference in these three technologies is the thick metallization for passive improvement and analog scaling.

ductors as RF circuit function and integration needs have evolved. Initially, the SiGe 5HP BiCMOS technology was qualified in 1998 with 2-μm Al as the last metal [11]. As can be seen from Table 1.4, this aluminum layer has a relatively high sheet resistance of 14 mΩ/sq, and therefore a peak Q for a 1-nH inductor at 2–4 GHz in the 5–9 range. The need for higher-Q inductors drove the qualification of SiGe 5AM, an upgrade to the 5HP process with a 4-μm Al layer for improved inductor performance [12]. This layout has a 4-μm-high via below it to allow for an additional thick interlayer dielectric (ILD) layer between the inductor and substrate. For a 1-nH inductor, a peak Q of 18 can be realized with the 4-μm

Table 1.4 Metal Options Available for Fabricating High-Quality Passives such as Inductors or Transmission Lines

Metal	Rs (mΩ/sq)	Peak Q–1 nH/2–4 GHz
2 μm Al	14	5–9
4 μm Al	7	18
4 μm Al/3 μm Cu	3.2	28

Al inductors in SiGe 5AM at a frequency of about 5 GHz. The underpass, a 0.8-μm-thick aluminum layer has a high sheet resistance relative to the thick aluminum layer. This fixed underpass resistance shows up in series with the relatively low resistance of the 4-μm aluminum spiral metallization limiting the achievable peak Q.

Numerous IBM SiGe technologies (e.g., 5DM, 6DM, 7WL) include additional two thick, low-resistivity metal levels above the standard BEOL stack (called "dual metal"). This stack was qualified for production in November 2001. The uppermost level is thick aluminum compatible with wire bonding or C4 interconnections with a second thick level composed of copper. Each of these levels has an additional 4 μm of oxide below it to increase the dielectric spacing from level to level and between the two levels and the substrate (see Fig. 1.12). This high-performance offering allows not only high-Q inductors (achieved by paralleling the two thick metals for extremely low series resistance), but also high inductance density (achieved by connecting a spiral from the thick aluminum level in series with a spiral stacked below it on the thick copper level). Figure 1.13 shows a comparison between the single-level, parallel-stacked, and series stacked spiral inductance and Q results. The results show the clear advantage of using series-stacked inductors for higher inductance and parallel-stacked inductors for higher Q. Efficiently coupled structures such as transformers (impedance matching and power splitting) and baluns (bal-

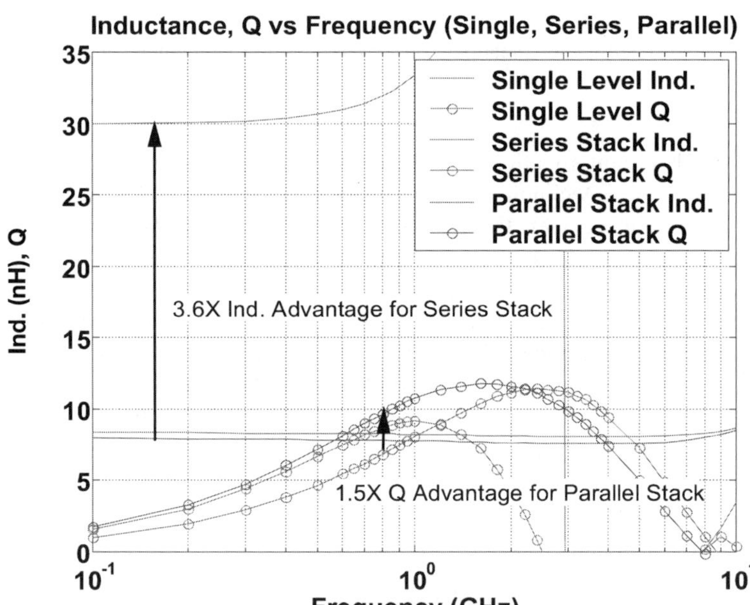

Figure 1.13 Comparison of inductance and quality factor Q for single-level, series- and parallel-stacked spirals in 5DM. The results show the clear advantage of using series stacked inductors for higher inductance and parallel stacked inductors for higher-Q.

anced-unbalanced transformers used to convert differential signals to/from single-ended signals and achieved by vertically stacking windings) are also possible, with a 1:1 balun achieving 3.5-dB untuned insertion loss (Fig. 1.14). In addition to the two thick levels, an optional polysilicon shield is offered that can increase the peak Q by as much as 30% for certain geometry spirals in addition to reducing substrate noise coupling from the spiral.

The high-quality inductors achievable with IBM's Dual Metal Technology enable designers to integrate high-Q resonant circuits in support of low-phase noise VCOs, narrow-band filters, low-loss impedance matching, etc. The added ability to achieve high-quality integrated magnetically coupled structures permits the realization of on-chip baluns and transformers. Another unique advantage of using two thick metal levels with a large separation distance is the ability to create nearly ideal microstrip and/or coplanar waveguide (CPW) transmission lines with very low-loss. Typical CMOS/BiCMOS/SiGe technologies are not able to achieve low loss, ideal transmission lines due to the excessive series losses and high capacitive coupling inherent in CMOS metallization schemes.

1.2.2 ESD Protection Devices

ESD protection of RF products is important, as application frequencies exceed 1-GHz. Below 1 GHz application frequency, the ability to simultaneously achieve excellent ESD protection and performance objectives was possible in most CMOS, BiCMOS, and silicon-on-insulator (SOI) applications. Since semiconductor appli-

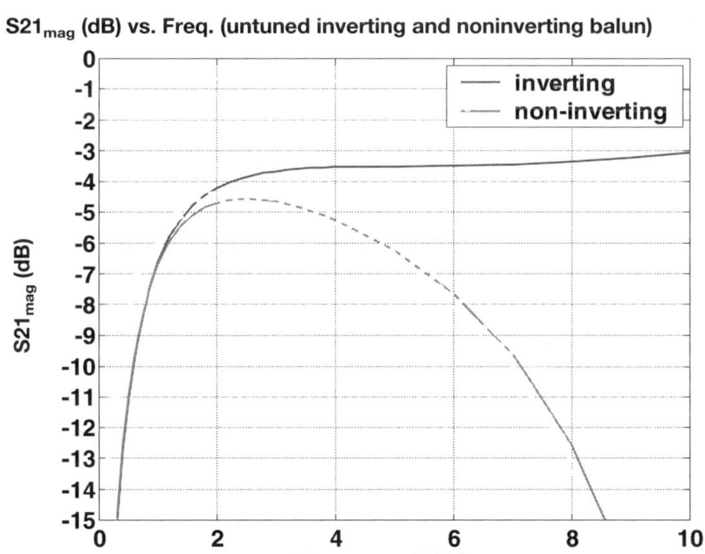

Figure 1.14 Balun insertion loss—inverting and noninverting—for IBM's SiGe 5DM process.

cations extend beyond 1 GHz, providing ESD protection and satisfying performance goals will increase in difficulty. Today, high-speed data-rate wired, wireless, test equipment, and disk-drive applications are extending well above 1 GHz. These high data rate telecommunication applications will contain RF CMOS, SiGe gallium arsenide (GaAs), and indium phosphide (InP) devices. With transistors whose unity current gain cutoff frequency f_T exceeds 100 GHz, the ability to provide ESD protection without impacting performance will be a significant challenge. For RF applications, ESD elements must have low capacitance, a high-quality factor (Q), linearity, and low noise.

RF CMOS can utilize some of the traditional ESD solutions that are common in the industry, although many ESD solutions are unacceptable in order to achieve low capacitance, high-Q, and low noise. MOSFETs have significant $1/f$ noise and capacitance loading making diode-based and diode configured ESD implementations the preferred solutions for RF applications [13]. The integration of STI has allowed for both optimization and scaling of the STI-bound p^+/n-well diode, the STI-bound n^+/substrate diode, and n-well-substrate diode structures from 0.5-μm to 0.1-μm CMOS technologies. To maintain a constant ESD robustness, to counter the impact of dimensional scaling from MOSFET constant electric-field scaling theory, constant ESD scaling theory indicates that ESD robustness can be preserved by increasing the n-well retrograde dose with successive technology generations. With the introduction of high retrograde well dose, the p^+/n-well diode capacitance can be maintained by adjustment of the n-well implant energy. For RF technologies, the introduction of low-doped p/substrates allows for lower n^+/substrate and n-well-to-substrate diode capacitance as well as noise isolation of the ESD elements on adjacent circuitry. *RC*-Discriminator networks, whose *RC* time constant is tuned to the ESD pulse rise time, are also utilized for triggering large MOSFETs located between power supplies.

RF BiCMOS SiGe technologies offer even more opportunity to introduce low-capacitance, high-Q, low-resistance robust ESD elements for ESD protection of RF circuitry. First, for mixed-signal chips that contain digital, analog, and RF circuitry, different ESD solutions can be applied to different functional circuit blocks. The aforementioned CMOS ESD diode elements and *RC*-triggered MOSFET ESD power clamps can be utilized for the digital functional block. Additionally, with the myriad additional elements that BiCMOS technology offers, new ESD elements and circuits can be utilized to provide ESD protection of digital, analog, or RF networks. This is possible by taking advantage of the SiGe library elements, and the SiGe NPN base, subcollector, and isolation structures. SiGe passive elements, such as SiGe varactors, SiGe Schottky diodes, and active SiGe HBT NPN devices and SiGe HBT PNP transistors can be utilized in either diodic- or bipolar-configurations for networks. Low-capacitance emitter-base and base-collector junctions provide well-controlled high-Q junctions for ESD protection networks. Low diode–anode series resistance is achievable in SiGe heterojunction bipolar transistor devices because of the high base doping concentration utilized in heterojunction transistors. Heterojunctions decouple the emitter-base junction capacitance from the base doping concentration design point. For ESD structures, this allows for improved current uniformity in multifinger base-collector diode structures. Additionally, low diode series cathode resistance, significantly lower than the CMOS well design point, is achieved using the heavily doped subcollector, reachthrough, and collector implants. Additionally, removal of the Kirk-effect limiting pedestal implant provides a 2× lower capacitance base-collector junction for usage of SiGe varactors in the forward-bias mode with no ESD robustness degradation. Deep-trench isolation also provides for usage of DT-bound subcollector-substrate, DT-bound n-well-to-substrate, and DT-bound p^+/n-well diode elements. The DT-bound structures can provide low capacitance, improved latchup immunity, noise injection reduction, as well as higher density.

SiGe technology also allows for the introduction of scalable ESD power clamp for analog and RF functional blocks. To provide an ESD solution that naturally scales with the BiCMOS technology, and utilizes the limitation of bipolar transistors, ESD power clamps were designed that take advantage of the Johnson limit of SiGe HBT devices. The Johnson limit in its power formulation is given as

$$(P_m X_c)^{1/2} f_T = E_m v_s / 2\pi \tag{1.3}$$

where P_m is the maximum power, X_C is the reactance $X_c = ½\Pi f_T C_{bc}$, f_T is the unity current gain cutoff frequency, E_m is the maximum electric field, and v_s is the electron saturation velocity. This can also be expressed in terms of maximum voltage, V_m, where

$$V_m f_T = E_m v_s / 2\pi \tag{1.4}$$

Hence from the Johnson limit equation,

$$V_m^* f_T^* = V_m f_T = E_m v_s / 2\pi \tag{1.5}$$

where $V_m^* f_T^*$ is associated with the first transistor and $V_m f_T$ is associated with the second transistor. The ratio of breakdown voltages can be determined as

$$\frac{V_m^*}{V_m} = \frac{f_T}{f_T^*} \tag{1.6}$$

Using this Johnson relationship, an ESD power clamp can be synthesized where a trigger device with the lowest breakdown voltage can be created by using the highest cutoff frequency (f_T) transistor, and a clamp device with the highest breakdown device will have the lowest cutoff frequency. Figure 1.15A shows an example of a Darlington configured bipolar power clamp with a 47 GHz/4 V BV_{CEO} trigger device that supplies the 27-GHz/6 V BV_{CEO} clamp device. A 7-Ω ballast resistor was used for each leg of the clamp device. A 7-kΩ bias resistor was used below the trigger device to limit the current. In this power clamp, the trigger device had an open base configuration, allowing early breakdown of the trigger circuit. Figure 1.15B shows the human-body model (HBM) ESD results. In

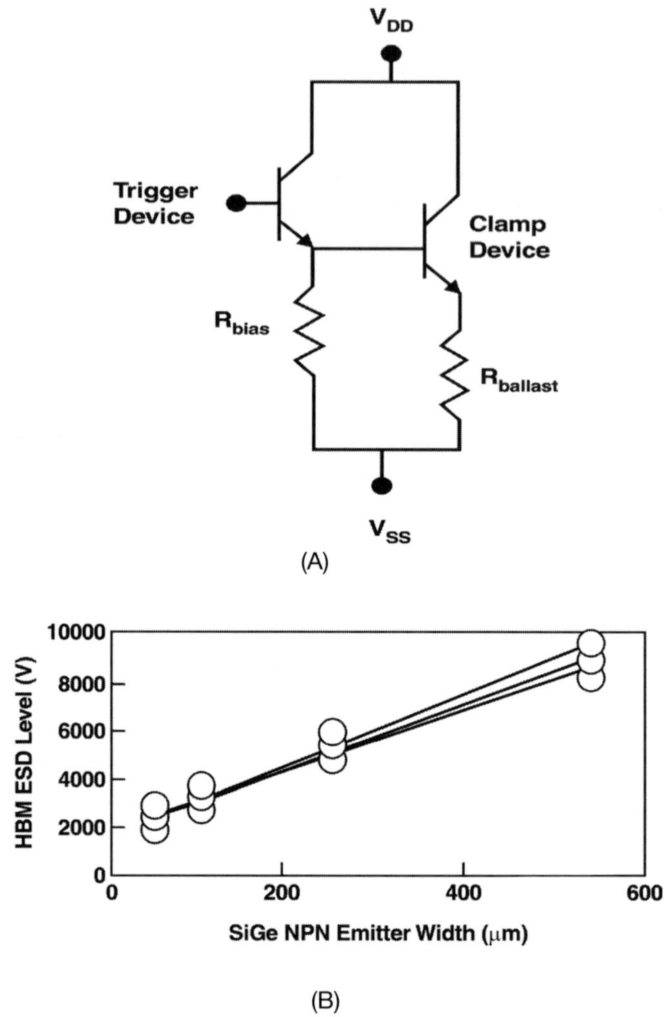

Figure 1.15 (A) SiGe HBT ESD power clamp utilizing a high-f_T/low-BV_{CEO} trigger element and low-f_T/high BV_{CEO} clamp element. (B) Human body model (HBM) ESD robustness as a function of SiGe HBT power clamp emitter width.

the given BiCMOS SiGe technology, a plurality of different breakdown voltage SiGe HBT is available to establish different trigger conditions for different power supply applications.

There are many advantages of the SiGe ESD clamp compared to *RC*-triggered MOSFET ESD power clamps. First, SiGe ESD power clamps provide a much higher robustness per unit micron. *RC*-Triggered MOSFET ESD power clamp achieves less than 2.5 V/μm, while the SiGe HBT NPN achieves 15–30 V/μm of NPN width. Second, it can be used for negative power-supply voltage products. A disadvantage of the MOSFET RC-triggered clamp in a single- or dual-well technology is that it cannot be used for between V_{DD} and negative power supply, V_{EE}, due to MOSFET overstress. Third, in contrast to MOSFET ESD networks, the ESD robustness per unit micron (width) does not decrease with successive RF technology generations. Experimental results from three successive SiGe technologies do not show ESD robustness degradation.

To evaluate the ESD scaling of a SiGe HBT device for a dimensionless group can be established, explaining the relationship between thermal conduction, thermal capacity, failure temperature, pulse width, saturation velocity, maximum electric-field condition, and the unity-current-gain cutoff frequency. Defining a dimensionless group, the power-to-failure can be compared to the Johnson limit maximum power condition:

$$V_o = \frac{P_f}{P_m} = \frac{A\sqrt{\dfrac{\pi K C_p \rho (T - T_o)}{\tau}}}{\dfrac{\left(\dfrac{E_m^2 v_s^2}{(2\pi)^2}\right)}{X_c f_T^2}} \tag{1.7}$$

From the dimensionless group, as the frequency of the SiGe HBT increases, the maximum power decreases. Because the device dimensions are scaled to achieve these objectives, the power-to-failure will decrease, unless doping and material changes are addressed. To produce future high f_T devices, dimensional scaling, doping concentration, and material changes will be needed. This will entail optimization of the base doping concentration, Ge concentration, and the introduction of carbon. In the evolution of these HBT devices, the choices made to achieve the BV_{CEO} and the f_T will influence both the maximum power and ESD robustness.

1.2.3 Summary

In this section, we have presented details of the technology development of passive devices in IBM's SiGe BiCMOS processes. The discussion has included numerous devices, including resistors, capacitors, varactors, inductors, transformers, and BEOL definition. In addition, we have presented details of ESD protection devices, with some comments on issues in RF-CMOS processes.

1.2.4 References

1. R. K. Ulrich, W. D. Brown, S. S. Ang, F. D. Barlow, A. Elshabini, T. Lenihan, H. A. Naseem, D. M. Nelms, J. Parkerson, L. W. Schaper, G. Morcan, "Getting aggressive with passive devices," *Circuits and Devices Magazine,* IEEE, Vol. 16, No. 5, pp. 16–25, September 2000.

2. S. A. St. Onge, D. L. Harame, J. S. Dunn, S. Subbanna, D. C. Ahlgren, G. Freeman, B. Jagannathan, J. Jeng, K. Schonenberg, K. Stein, R. Groves, D. Coolbaugh, N. Feilchenfeld, P. Geiss, M. Gordon, P. Gray, D. Hershberger, S. Kilpatrick, R. Johnson, A. Joseph, L. Lanzerotti, J. Malinowski, B. Orner, and M. Zierak, "A 0.24 um SiGe BiCMOS Mixed-Signal RF Production Technology Featuring a 47 GHz f_t HBT and 0.18 μm L_{EFF} CMOS," in *Proceedings of 1999 BCTM,* pp. 117–120, 1999.
3. A. Pruijmboom, D. Szmyd, R. Brock, R. Wall, N. Morris, Keng Fong, F. Jovenin, "QUBiC3: A 0.5 μm BiCMOS Production Technology with f_T = 30GHz, f_{max} = 60 GHz and High-Quality Passive Components for Wireless Telecommunication Applications," in *Proceedings of 1988 BCTM,* p. 120, 1998.
4. K. Stein, J. Kocis, G. Heuckel, E. Eld, T. Bartush, R. Groves, N. Greco, D. Haramo, T. Tewksbury, "High Reliability Metal Insulator Metal Capacitors for Silicon Germanium Analog Applications," in *Proceedings of 1997 BCTM,* p. 191, 1997.
5. B. Hendrix and G. Stauf, "Low Temperature Process for High Density Thin Film Integrated Capacitors," *Proceedings of the 2000 International Conference on High-Density Interconnect and Systems Packaging,* pp. 342–345, 2000.
6. M. Armacost, A. Augustin, P. Felsner, Y. Feng, G. Friese, J. Heidenreich, G. Hueckel, O. Prigge, K. Stein, "A High Reliability Metal Insulator Metal Capacitor for 0.18 μm Copper Technology," *IEDM Tech. Digest,* pp. 157–160, December 2000.
7. J. N. Burghartz, M. Soyuer, K. A. Jenkins, M. Kies, M. Dolan, K. J. Stein, J. Malinowski, D. L. Harame, "Integrated RF Components in a SiGe Bipolar Technology," *IEEE Trans. Solid-State Circuits,* vol. 32, no. 9, p. 1440–1445, 1997.
8. A. M., Niknejad and R. G. Meyer, *Design, Simulation and Applications of Inductors and Transformers for Si RF IC's,* Kluwer Academic Publishers, Boston, 2000.
9. J. N. Burghartz, D. C. Edelstein, K. A. Jenkins, C. Jahnes, C. Uzoh, E. J. O'Sullivan, K. K. Chan, M. Soyuer, P. Roper, S. Cordes, "Monolithic Spiral Inductors Fabricated Using a VLSI Cu-Damascene Interconnect Technology and Low-Loss Substrates," *IEDM Tech. Digest,* 1996, pp. 96–102, December 1996.
10. D. L. Harame, D. C. Ahlgren, D. D. Coolbaugh, J. S. Dunn, G. G. Freeman, J. D. Gillis, R. A. Groves, G. N. Hendersen, R. A. Johnson, A. J. Joseph, S. Subbanna, A. M. Victor, K. M. Watson, C. S. Webster, P. J. Zampardi, "Current Status and Future Trends of SiGe BiCMOS Technology," *IEEE Trans. Electron Devices,* vol. 48, no. 11, pp. 2575–2594, November 2001.
11. D. C. Ahlgren, G. Freeman, S. Subbanna, R. Groves, D. Greenberg, J. Malinowski, D. Nguyen-Ngoc, S. J. Jeng, K. Stein, K. Schonenburg, D. Kiesling, B. Martin, S. Wu, D. Harame, B. Meyerson, A SiGe HBT BiCMOS Technology for Mixed Signal RF Applications," in *Proceedings of 1997 BCTM,* 195–197, 1997.
12. S. A. St. Onge, D. L. Harame, J. S. Dunn, S. Subbanna, D. C. Ahlgren, G. Freeman, B. Jagannathan, J. Jeng, K. Schonenberg, K. Stein, R. Groves, D. Coolbaugh, N. Feilchenfeld, P. Geiss, M. Gordon, P. Gray, D. Hershberger, S. Kilpatrick, R. Johnson, A. Joseph, L. Lanzerotti, J. Malinowski, B. Orner, and M. Zierak, "A 0.24 um SiGe BiCMOS Mixed-Signal RF Production Technology Featuring a 47 GHz f_t HBT and 0.18 μm L_{EFF} CMOS," in *Proceedings of 1999 BCTM,* pp. 117–120, 1999.
13. S. Voldman and V. Gross, "Scaling, Optimization and Design of Considerations of Electrostatic Discharge Protection Networks in CMOS Technology," in *EOS/ESD Symposium Proceedings,* pp. 251–261, 1993.

1.3 PROCESS DEVELOPMENT

As described in the Introduction, IBM has demonstrated leadership in development of world-class analog bipolar and BiCMOS process technologies. In this section, we briefly overview some of the process integration issues addressed by IBM in building SiGe processes. Importantly, we discuss some of the key manufacturability issues.

1.3.1 Process Integration

The basis for development of an integrated SiGe BiCMOS technology starts with an understanding of semiconductor requirements of the marketplace. The wireless customer envisioned for SiGe 5HP could be serviced with a cost-competitive CMOS technology (ASIC CMOS 5×) and analog capabilities achieved with a 50-GHz HBT capable of driving speeds for applications in the 2–5-GHz range. HBT and CMOS technologies were successfully merged by sharing a common polysilicon film for both bipolar extrinsic-base and CMOS gate structures. This integration method, referred to as "base during gate," (Fig. 1.16) allowed cost-efficient use of films and masking levels, and was made possible by the thermal compatibility of early CMOS structures with the rather deep emitter profile developed in the early HBT. It was fully appreciated that Ge doping profiles had a profound effect on both HBT performance and strain that could result in decreased circuit yields, so Ge doping was intentionally kept low to facilitate early production learning. Acceptability by the analog marketplace also required the addition of high-quality passive elements (resistors, capacitors, inductors, etc.) discussed in detail in Section 1.2.

SiGe 6HP was developed primarily for applications used in storage products that require a very large circuit count of high-performance CMOS. For these applications, it was necessary to replace the 0.5-μm CMOS of SiGe 5HP with 0.25-μm CMOS 6SF. The reduced thermal cycle used in this generation CMOS also required a modification of the integration from base-equals-gate to an integration methodology, which fabricated the critical emitter/base region of the HBT after CMOS structures were in place (Figure 1.16). While this integration requires careful protection of the CMOS gate conductor during the processing of the HBT, and subsequent removal of these protective films, this base-after-gate integration was successful in decoupling of thermal cycles required for narrow HBT profiles and the CMOS device activation.

While the first two generations of SiGe BiCMOS could successfully service applications up to 10 Gbps (OC-192 synchronous optical network (SONET)), 40 Gbps (OC-768) optical networking products require a substantially faster bipolar transistor. SiGe 7HP was developed for this switching-speed-driven application space. Since designers request bipolar devices that switch at approximately three times the bit rate of multiplexing and demultiplexing circuits, the SiGe 7HP target was established at a maximum oscillation frequency of 120 GHz. To successfully achieve these speeds, previously widely believed only to be achievable with III–V

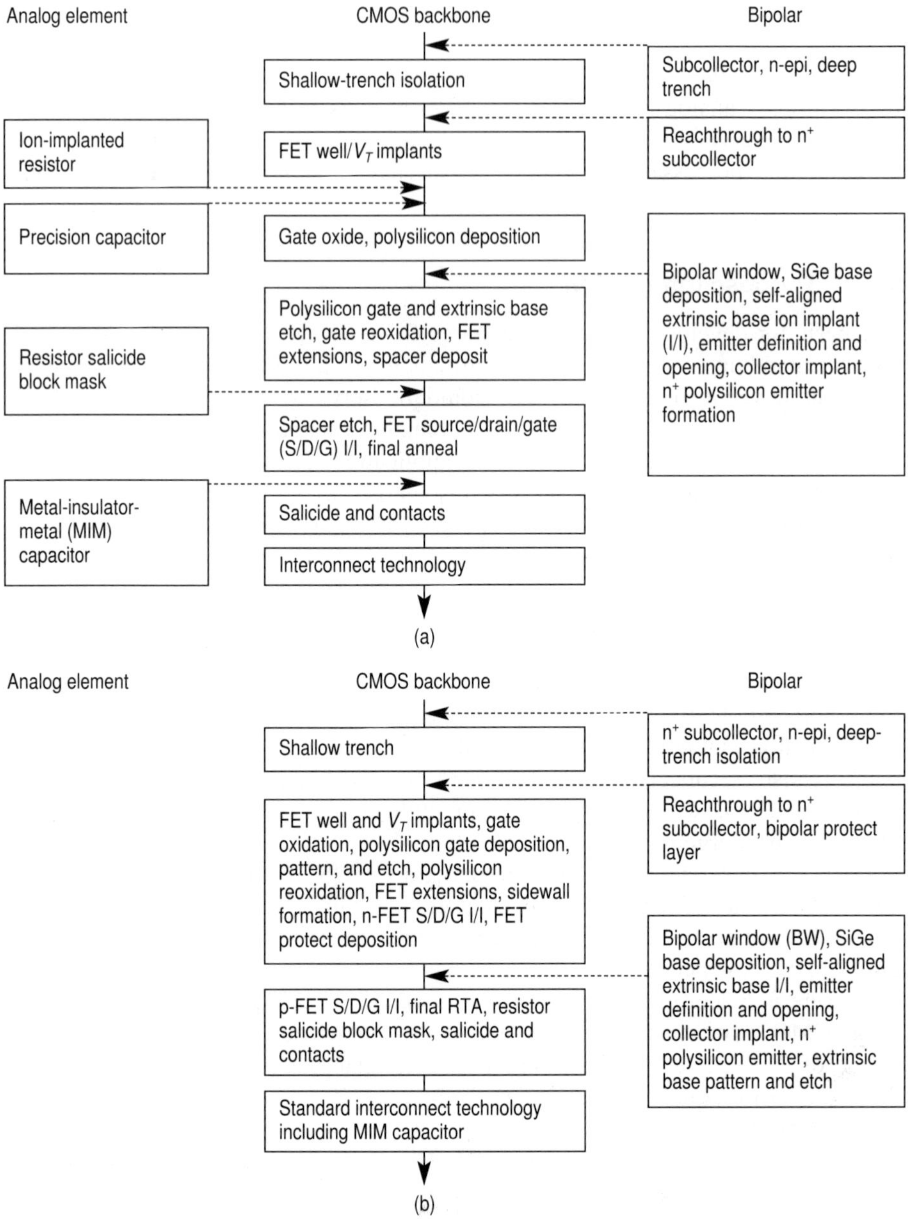

Figure 1.16 Schematic process flow diagram for (A) 0.5-μm base-during-gate production technology. (The CMOS backbone represents the base CMOS process. Insertion points are shown for analog and bipolar process modules.) (B) 0.25-μm and 0.18-μm base-after-gate production technology. (The CMOS backbone represents the base CMOS process. Insertion points are shown for analog and bipolar process modules.) Note the reduced number of analog and bipolar blocks compared to the base-after-gate process, because the CMOS now contains many analog modules.

compound semiconductors (notably InP-based HBTs), vertical scaling (reduction of eptitaxial (epi) base and collector profiles) as well as horizontal scaling of the HBT device was necessary. The reduction of vertical profiles necessitated careful control of process thermal cycles as well as the use of carbon doping [1], which had not been necessary in IBM's early generations of SiGe BiCMOS. In addition to continued improvements in passives, the integration of this technology with CMOS7SF (0.18-μm gate length) marked the introduction of Cu interconnection technology, which further enhanced current carrying capability of high-performance circuits.

The self-aligned HBT transistor utilized in generations 5, 6, and 7HP required evolutionary modifications of the ETX transistor [2], which featured the ion implantation of the heavily doped extrinsic-base region of the device. Further scaling of the ETX device would require extensive reduction in intrinsic-base doping (high pinch-base sheet resistance) which would lead to a nonoptimal design point for a technology based on high manufacturability. Necessary improvements to ETX in base resistance (and collector-base capacitance), which determined the maximum available power-gain frequency (f_{MAX}), were limited by this device structure. This shortcoming therefore led to the development of a new integration scheme. The RXB bipolar was introduced to allow further vertical and horizontal scaling without the limitations imposed by the ETX device. By contacting the lightly doped intrinsic base with an additional boron-doped polysilicon film, these constraints were lifted and the HBT introduced for SiGe 8HP integration has demonstrated f_T and f_{MAX} values in excess of 200 GHz [3].

In addition to the flagship 5HP, 6HP, and 7HP SiGe technologies, several derivative SiGe technologies have been developed to capture smaller markets with more specific requirements. These derivative technologies include SiGe 5PA, SiGe 5MR, and BiCMOS 5HPE, which are targeted at RF power-amplifier, disk drive preamplifier, and wireless applications, respectively.

The SiGe 5PA technology was developed as a derivative of the SiGe 5HP technology targeted at the RF power-amplifier market. Power amplifiers for wireless handsets have historically been built using GaAs, due to its ability to achieve the competing breakdown and linearity requirements that were previously unattainable with silicon technologies. SiGe 5PA meets these requirements with a high breakdown SiGe HBT that has a BV_{CEO} of 7 volts and BV_{CBO} of 20 volts. The increased breakdown voltages are achieved, while minimizing the impact to the device linearity and f_T, through modification and optimization of the collector doping profile. The technology is defined such that the remaining device set is common with those in the parent SiGe 5HP technology.

The SiGe 5MR technology was developed for disk drive preamplifier applications, which have a conflicting combination of performance and voltage technology requirements between the read and write portions of the chips. The read-amplifier portion of the chip must amplify extremely small signals from the magneto-resistive read element while not significantly degrading the signal-to-noise ratio (SNR). Conversely the write portion of the chip requires high-voltage devices to drive high-speed signals through the inductive write head. The SiGe 5MR technology includes a SiGe HBT capable of achieving the performance and noise requirements of the read amplifier, yet having a BV_{CEO} of 10 volts to support the voltage requirements of driving the inductive write head. The CMOS devices included in the technology for logic control circuitry supports the industry standard disk drive supply voltage of 5 volts. The technology was developed by combining the 5-V CMOS devices from CMOS 5SF, the proven SiGe HBT from SiGe 5HP, breakdown voltage enhancements similar to SiGe 5PA, and the base after the gate integration scheme from BiCMOS 6HP. Additional development efforts focused on adding features such as a lateral PNP, an isolated NFET, and a poly-poly capacitor, as well as process simplifications targeted at cost and cycle-time reduction.

These features of the SiGe 5MR technology are all key elements included in its derivative technology, BiCMOS 5HPE [4]. This technology's combination of a rich feature set and simplified process make it well suited for the competitive wireless marketplace. The BiCMOS 5HPE technology includes the features of SiGe 5MR plus an additional higher-performance SiGe HBT and enhanced passives such as the 1.35-fF/μm2 MIM, high-Q inductors, and MOS varactor. Process simplifications, such as replacing long furnace anneals with rapid thermal anneals (RTA) significantly reduces the production cycle time. This cycle time reduction is extremely valuable for wireless product developers, who depend on multiple design passes to debug sensitive RF circuits.

1.3.2 Manufacturing Issues

In order to introduce semiconductor technology into a production environment, it is necessary to demonstrate a rigorous set of manufacturability standards. These requirements can be loosely categorized as repeatability, reliability, and yieldability. Device models are developed to describe devices fabricated using a large statistical database collected on a substantial quantity (100–300 typically) of wafers fabricated during this assessment. These models begin by analysis of device parameters, as illustrated in Fig. 1.17. The models describe nominal values and distributions of manufactured devices from probability plots. Similar data are collected on in-line measurements of physical parameters such as lithographic-critical dimensions and overlay, film thicknesses, and layer resistances and physical properties. Both device electrical data measured on test structures and in-line parametric measurements are collected during the qualification in order to demonstrate statistical process control (SPC) [5]. SPC methods compare measured data to well-defined process specifications and must meet rigorous standards in order to proceed with volume production. SPC measurement of C_p (a gauge of process centering) and C_{pk} (process capability index) are illustrated in Fig. 1.18. A C_{pk} value of 1.0 represents the point at which 3-sigma of the population distribution is within the specified manufacturing control limits, with values in excess of 1.0 indicative of much tighter production control.

SiGe HBT reliability stressing is conducted in both forward-bias and reverse-bias modes, using the same accelerating stress conditions and analysis techniques developed for previous implanted-base silicon bipolar technologies [6]. In forward bias, a less than 5% change in NPN current gain, β, over a 500-hour stress at 140°C and 1.3 mA/μm^2 (near peak f_T) was found for the 0.5-μm generation SiGe HBTs.

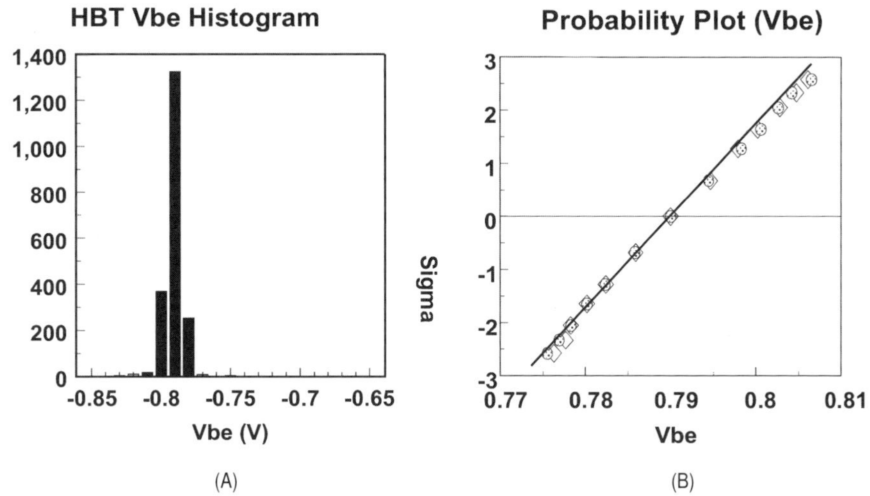

Figure 1.17 Example of statistical-device models showing (A) histogram of V_{BE} for $0.5 \times 1.0\ \mu m^2$ HBT at 10-μA emitter current; (B) illustrates a probability plot for V_{BE} comparing measured hardware (diamonds) on approximately 2000 samples and device model (circles) to a normal distribution (line).

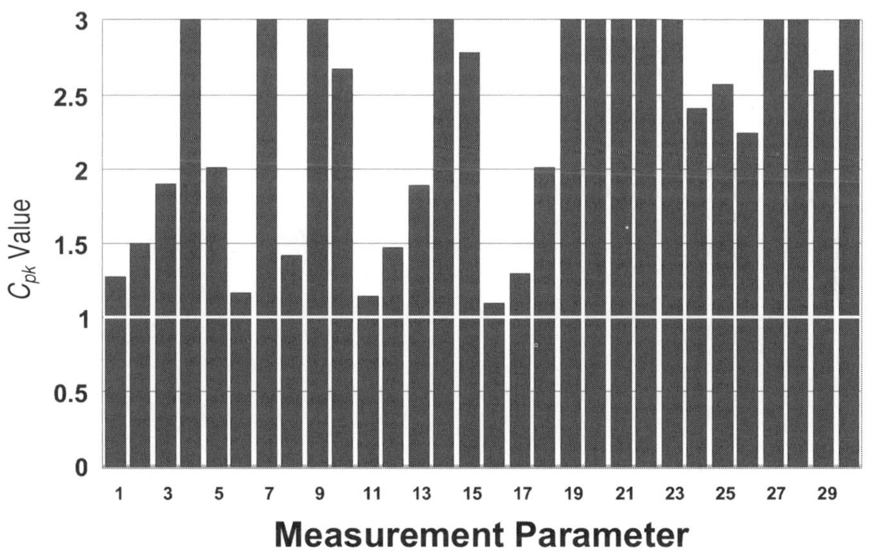

Figure 1.18 Statistical process-control chart showing achievement of good process control over a wide range of in-line process measurements. Values of the process capability index (C_{pk}) of 1.0 indicate process measurements at the three-sigma control point. Values of 1.5 are considered under six-sigma control, while values above 2.0 are not of statistical validity.

Most GaAs HBT technologies are stressed at currents far less than the peak f_T current [7], which does not represent realistic use conditions. Using empirically determined acceleration factors, the SiGe HBT forward-bias result is equivalent to less than 10% β degradation under typical-use conditions (1.25 mA/μm^2 at 100°C) after 100K POH. The change in β has been attributed to electromigration-induced pressure on the emitter contact, causing decreased collector current with stress condition [8]. Reliability at high current densities with SiGe 7HP and 8HP vertical scaling also require careful look at the choice of interconnection technology and device self-heating considerations [9]. With reverse bias of 2.0 V or less, no measurable change in base or collector currents is observed at a worst-case condition of –40°C. Hot-electron degradation with reverse-bias stress is also typical of bipolar junction transistors, and is substantially lower for the epi-base devices when compared to ion-implanted base double-polysilicon BJTs [8]. This result is attributed to the reduction of the electric field at the emitter-base junction due to a base doping setback that can be achieved with epitaxial-base growth and is expected to be similar with HBT vertical and horizontal scaling.

A major focus for final production qualification decisions is a technology's capability of achieving high manufacturing yields. FET yield assessment and control in SiGe BiCMOS technologies is adopted from the parent CMOS technologies, with an added focus on ensuring that the SiGe process steps do not degrade the FET yield. For example, with the base-after-gate integration scheme, several bipolar films are deposited over the FETs and then subsequently etched away. Defect-free removal of these films over the FET regions is critical to minimize yield loss due to shorting. Successful removal of the bipolar films from the FET regions is monitored with both large-area defect monitors and CMOS static random-access memory (SRAM) yields. A BiCMOS 6HP 30-lot yield trend for large-area monitor struc-

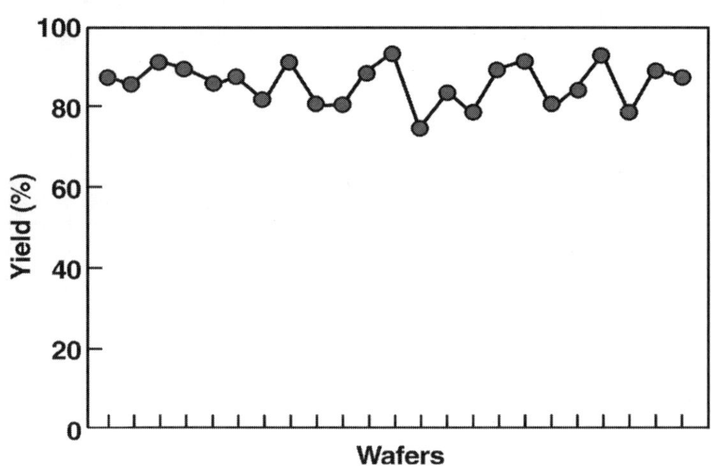

Figure 1.19 Static RAM functional perfect (zero bit fails) yield data from a 154K RAM (12-μm^2 cell size, 1 million FETs) by wafer.

tures [3] demonstrates that BiCMOS yields are equivalent to the typical yield of the parent CMOS technology. CMOS SRAM yields were also shown to be within the parent CMOS target range as indicated by the BiCMOS 6HP CMOS SRAM yield trend shown in Fig. 1.19.

Silicon bipolar transistor device yield has historically been limited by dislocations in the silicon crystal that cause junction shorting [10]. One would expect SiGe HBTs to have a higher density of these dislocations due to defect formation during epi growth and strain in the germanium layer. However, the low-temperature epitaxy (LTE) process technology produces a relatively defect-free epi layer and does not significantly contribute to yield loss. Also, minimizing the thermal cycles post-epi growth helps reduce the possibility of forming defects due to the strained layer. The primary cause for defect generation has been found to be related to stress due to the STI. As a result, the STI formation process has been optimized in order to minimize the stress and increase the HBT yield. Figure 1.20 shows a SiGe 5HP 20 month all good yield trend of an array of 4000 HBTs. Initially, the yield was depressed, averaging in the 60–70% range. Through careful optimization of the HBT process, the yield has been improved substantially to above 90% [11].

1.3.3 Summary

In this section, we have overviewed some of the process integration issues seen in IBM's SiGe processes. We have also discussed the key manufacturability issues in building SiGe IC products.

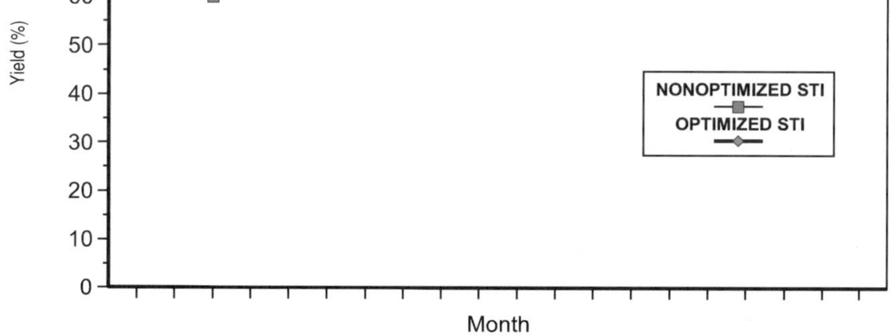

Figure 1.20 An illustrative example of yield learning, shows HBT yield improvement over time made through modification of process controls.

1.3.4 References

1. L. Lanzerotti, J. C. Sturm, E. Stach, R. Hull, T. Buyuklimanli, and C. Magee, "Suppression of Boron Outdiffusion in SiGe HBTs with Carbon Incorporation," *IEDM Tech. Digest,* p. 96, December 1996
2. D. L. Harame, J. H. Comfort, J. D. Cressler, E. F. Crabbe, J. Y. C.-Sun, B. S. Meyerson, and T. Tice, "Si/SiGe Epitaxial-Base Transistors, Part I: Materials, Physics and Circuits," *IEEE Trans. Electron Devices,* pp. 455–468, March 1995.
3. B. Jaganathan, unpublished communication.
4. N. B. Feilchenfeld, S. J. Astravas, M. D. Gallagher, S. Gibson, and J. J. Twomey, "BiCMOS 5HPE: A New Silicon Germanium Technology for Hi-Frequency RF Applications" *MicroNews,* vol. 7, no. 4, pp. 32–36, 2001.
5. V. Kane, *J. Qual. Tech.,* vol. 18, p. 41, 1986.
6. D. C. Ahlgren, G. Hueckel, G. Freeman, K. Walter, R. Groves, T. H. Ting, A. Vayshenker, and E. Hostetter, "Device Reliability and Repeatability of a High Performance Si/SiGe HBT BiCMOS Technology," in *Proceedings of 28th ESSDERC Conf.,* pp. 452–455, 1998.
7. T. S. Low, C. P. Hutchinson, P. C. Canfield, T. S. Shirley, R. E. Yeats, J. S. C. Chang, G. K. Essilfie, M. K. Culver, W. C. Whiteley, D. C. D'Avanzo, N. Pan, J. Elliot, and C. Lutz, "Migration from an AlGaAs to InGaP emitter HBT IC process for improved reliability," in *Technical Digest of the 1998 Gallium Arsenide Integrated Circuit (GaAs IC) Symposium,* pp. 153–156, 1998.
8. R. Hemmert, G. Prokop, J. Lloyd, P. Smith, and G. Calabrese, "The Relationship among Electromigration, Passivation Thickness and Common-Emitter Current Gain Degradation within Shallow Junction NPN Bipolar Transistors," *J. Appl. Phys,* vol. 53, pp. 4456–4462, 1982.
9. J.-S. Rieh, D. Greenberg, B. Jagannathan, G. Freeman, and S. Subbanna, "Measurement and Modeling of Thermal Resistance of High Speed SiGe Heterojunction Bipolar Transistors," in *Proceedings of the Topical Meeting on Silicon Monolithic Integrated Circuits in RF Systems,* pp. 110–113, 2001.
10. S. A. St. Onge, D. L. Harame, J. S. Dunn, S. Subbanna, D. C. Ahlgren, G. Freeman, B. Jagannathan, S. J. Jeng, K. Schonenberg, K. Stein, R. Groves, D. Coolbaugh, N. Feilchenfeld, P. Geiss, M. Gordon, P. Gray, D. Hershberger, S. Kilpatrick, R. Johnson, A. Joseph, L. Lanzerotti, J. Malinowski, B. Orner, and M. Zierak, "A 0.24 μm SiGe BiCMOS Mixed-Signal RF production Technology Featuring a 47 GHz f_T HBT and 0.18 μm Leff CMOS," in *Proceedings of 1999 BCTM Conference,* pp. 117–120, 1999.
11. D. C. Ahlgren, A. G. Domenicucci, R. Karcher, S. R. Mader, and M. R. Poponiak, "Defect Induced Leakage from Bipolar Transistor Isolation Leakage," in *Proceedings of Symposium on Defects in Silicon,* vol. 83–9, pp. 472–481, 1983.

1.4 TECHNOLOGY IMPLICATIONS IN SiGe DESIGN

In this section, we discuss the issues of device noise and reliability. Understanding these issues is key to successful product yield, and complements the previous sections in this chapter.

1.4.1 Device Noise

In SiGe BiCMOS technology, designers enjoy the option of using BJT and/or FET transistors as active devices in analog and RF circuits. However, as discussed in earlier sections, the SiGe bipolar device features several intrinsic properties that make it particularly attractive for low-noise circuits such as low-noise amplifiers (LNAs).

The first source of noise is thermal noise generated in proportion to the temperature and value of the base resistance, R_B. The design of the SiGe bipolar achieves a low value for this resistance. Germanium narrows the bandgap of the base with respect to the emitter, decreasing the amount of hole injection into the emitter for a given electron injection into the base. This effect compensates for the loss of beta that would otherwise occur with high base doping, allowing a reduction in the sheet resistance of the intrinsic base. Further, the lateral distance that base current must travel through the intrinsic base from edge to center can be made quite small through the use of advanced lithography to define a very narrow emitter stripe width. Such a narrow width is made possible by a polysilicon emitter construction that achieves uniform dopant distribution even at tenth-micron dimensions.

A second source of high-frequency noise in the bipolar transistor is the fundamental operation of the emitter-base p-n junction itself, which contributes shot noise in proportion to both the base (I_B) and collector (I_C) currents. The SiGe bipolar device keeps these noise components under control by two means. The high beta created by the presence of Ge in the base keeps I_B low for any given I_C bias, reducing I_B shot noise. Further, the low parasitic resistances and capacitances that contribute to high f_{MAX} (e.g., R_B and C_{CB}) result in high power gain at typical frequencies of interest. This high gain allows an input signal to be amplified to much larger levels before the addition of shot noise, reducing the impact of this shot noise on the output SNR and thus improving the noise figure.

Although these generic SiGe bipolar features favor a low-noise figure for any given generation, there is also a generation-to-generation scaling trend toward a lower-noise figure. This improvement is due to distinct lateral and vertical scaling trends in device design. Lateral scaling reduces R_B and improves gain by reducing parasitic capacitances. Vertical scaling leads to reductions in transit time that further improve power gain. Both R_B and gain can also improve as the result of fundamental changes in device architecture aimed at reducing parasitics.

Fig. 1.21 displays both the minimum noise figure F_{MIN} and associated gain G_A as a function of I_C for the 0.5-μm generation (45 GHz f_T, 65 GHz f_{MAX}) device with an emitter area (A_E) of 40 μm^2. The 0.5-μm technologies, including IBM SiGe 5HP and 5AM, are the generations most widely used at present for 1–5-GHz wireless telephony and data networking due to the attractive cost-performance balance. The figure shows data for two frequency bands. At 2 GHz, near the band used by several standards of wireless telephony, we observe that the device achieves a minimum noise figure of 0.8 dB at a current of 1 mA. At the higher bias of 5 mA, the device maintains F_{MIN} below 1 dB with an increased associated gain of 20 dB. Looking to 5 GHz, a band used for high bit-rate wireless data networking, F_{MIN} remains below

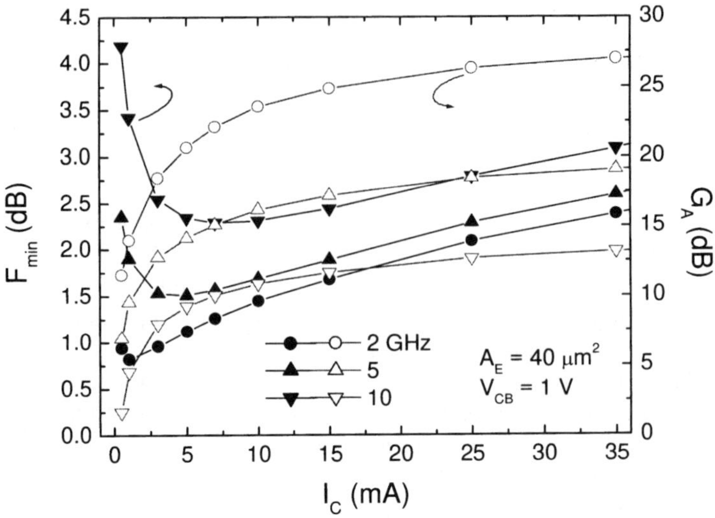

Figure 1.21 Minimum noise figure and associated gain for a 0.5-μm-generation SiGe bipolar transistor with an emitter area of 40 μm², measured at 2, 5, and 10 GHz (V_{CB} = 1 V).

1.5 dB with an associated gain of 14.2 dB at the same 5-mA bias. While there were no high-volume wireless applications in the 10-GHz regime in 2002, we note that an IBM device shows F_{MIN} of about 2.3 dB in this band, with associated gains of 10–12 dB. As we note below, a 40-μm² device area is nonoptimal for 10 GHz operation due to source-matching difficulties. Indeed, smaller devices (e.g., A_E = 2.5 μm²) have demonstrated 10-GHz F_{MIN} values of < 1.8 dB.

High-frequency noise performance improves moving to 0.25-μm SiGe BiCMOS generation, such as 6HP. While the basic structure of the 0.25-μm bipolar transistor is similar to that of the 0.5-μm generation, the lithographic improvements allow for a somewhat narrower emitter stripe, achieving a reduced R_B. The result is an improvement in F_{MIN} of about 0.1 dB in the 1–10-GHz band. Combined with the ability to integrate a more aggressive CMOS, this generation is also attractive for use in the 1–5-GHz band, while adding the ability to mix analog/RF circuits with large gate counts of high-speed logic.

Both the 0.5-μm and 0.25-μm SiGe generations offer a high-breakdown voltage variant of the bipolar transistor, featuring a modified collector profile that also results in reduced R_B and C_{CB}. Although this transistor suffers f_T falloff from the Kirk effect at much lower current densities compared with the high-f_T variant, this current threshold is much higher than the biases resulting in best noise performance, and is thus not a limitation. A designer can achieve an additional improvement in F_{MIN} of about 0.1 dB by using this device, although the poorer linearity characteristics of the device may limit suitability for many applications.

Although the first applications of the 0.18-μm SiGe 7HP technology have been 40-Gbit/s optical networking [serialize–deserialize (SERDES)], the technology also represents a significant improvement in noise performance and may therefore be attractive for wireless design as well. The benefits result from both lateral and vertical scaling, which improve R_B, beta, and power gain in the device. The improvements in gain are particularly important for reducing the noise figure at higher frequencies (e.g., 10 GHz and above). As an example, Fig. 1.22 illustrates $F_{\rm MIN}$ and G_A behavior vs. I_C for a 25.6-μm^2 device at 3, 10, and 20 GHz. Three GHz is close to the unlicensed 2.4-GHz band commonly used for wireless data networking. At this frequency, the 7HP device achieves an $F_{\rm MIN}$ value of only 0.4 dB and maintains this value for associated gains exceeding 20 dB. At 10 GHz, $F_{\rm MIN}$ climbs to only 0.6 dB, demonstrating the possibly greater suitability of the 0.18-μm generation for higher-frequency wireless work compared to 0.5 and 0.25 μm. Associated gains exceed 12 dB. In addition, $F_{\rm MIN}$ values of 1.7 dB are achievable out to 20 GHz, approximately double the frequency of the earlier generations for equivalent performance and consistent with the 2.5× (1.5×) increase in f_T ($f_{\rm max}$).

Tuning for a desired combination of noise figure and associated gain requires a designer to create a network to match the source impedance (typically 50 Ω) to the optimum noise impedance of the device. This task is greatly simplified if the real part of the desired match is already 50 Ω, allowing the imaginary part to be matched with a simple inductor in series with the base. This condition may be realized by scaling the device area appropriately. For example, for a 0.5-μm generation, 40-μm^2 bipolar transistor biased at an I_C of 5 mA, $F_{\rm MIN}$ at 2 GHz is achieved at a

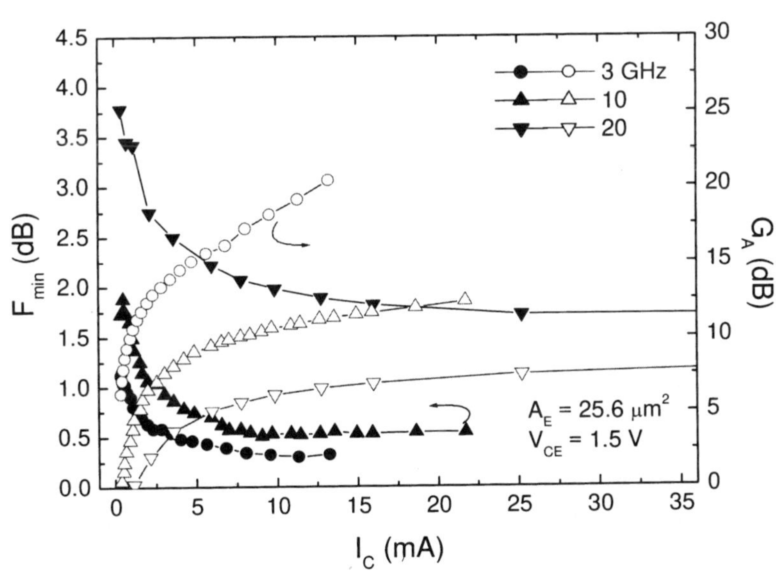

Figure 1.22 Minimum noise figure and associated gain for a 0.18-μm-generation SiGe bipolar transistor with an emitter area of 25.6 μm^2, measured at 3, 10, and 20 GHz (V_{CE} = 1.5 V).

matching impedance of $(68 + j37)\ \Omega$. Scaling the device up to the slightly larger area of 54 μm^2 would achieve the desired 50-Ω real match, requiring an approximately 3-nH series inductor to complete to match. In comparison, a 0.18-μm generation, 25.6-μm^2 bipolar biased at 10 mA achieves an optimal match at 10 GHz with a source impedance of $(7.7 + j13.6)\ \Omega$, pointing out the need to scale the device down to a more suitable emitter area of 3.9 μm^2.

High-frequency, or broadband, noise is not the only category of noise pertinent to analog and RF design. In addition to low-defect-density single-crystal silicon, the structures of both bipolar and FET transistors contain a number of interfaces to surface films, particularly to silicon dioxide (SiO_2). These interfaces contain a relatively large number of defects that act as electron and hole traps. Exchange of carriers between these traps and the active regions of the device lead to the phenomenon of $1/f$ noise. This mechanism manifests as a high noise energy content near direct current (DC), tapering off with frequency approximately as $(1/f)^n$ before merging with the broadband noise spectrum at a higher frequency approximated by a figure of merit known as the corner frequency. As initially introduced into the signal by the device, this low-frequency noise is generally out of band. However, subsequent signal processing by a mixer can shift the noise spectrum to straddle the carrier frequency, potentially causing jitter as well as interfering with adjacent channels. Thus, designers will seek a device with good $1/f$ noise properties when creating circuits such as LNAs and VCOs that eventually send their output to a mixer.

Bipolar and FET devices differ greatly in their $1/f$ noise properties. Because the electrical characteristics of the bipolar transistor are determined by mechanisms that occur below the surface of silicon substrate, away from oxide interfaces, these devices show significantly better $1/f$ noise results compared with FETs. Indeed, generation-to-generation scaling in FETs moves oxide interfaces even closer to the active channel, resulting in worsening of $1/f$ performance. The bipolar device is largely immune from this, maintaining a low corner frequency across generations. Figure 1.23 compares the $1/f$ performance of four devices from the 0.18-μm SiGe 7HP technology: a 3.3-V NFET, a 3.3-V PFET, a high-f_T SiGe bipolar, and a high-breakdown SiGe bipolar. We focus on the 3.3-V FET variants here, since these devices support the higher supply voltages commonly used to achieve signal headroom in analog circuit design. The proximity of surfaces contributes significant noise signal in the NFET, extending out well beyond 100 kHz. The PFET is an order of magnitude better, partly due to the lower drain current of the PFET at the same gate bias.

Compared with both FETs, the SiGe bipolar transistor shows significantly better $1/f$ noise performance, with an output noise signal that is lower than with an NFET signal by a factor of 50 and with a resulting corner frequency of less than 1 kHz. This bipolar $1/f$ noise performance is similar to that achieved by prior generations, emphasizing the device's tolerance to scaling. The modified collector profile of the high-breakdown bipolar results in reduced avalanche current even below breakdown, resulting in yet lower noise at higher frequencies compared with the high-f_T variant.

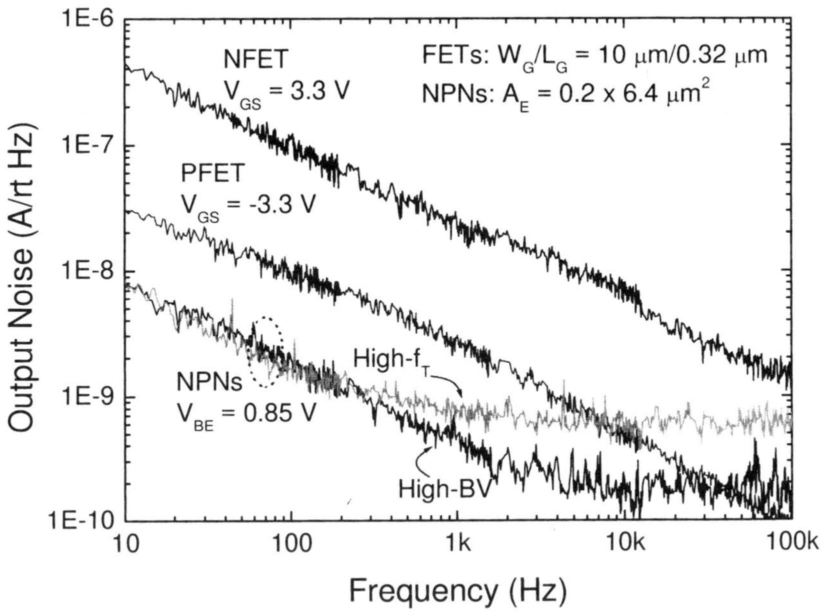

Figure 1.23 $1/f$ output noise vs. frequency for a 3.3 V NFET, 3.3 V PFET, high-f_T SiGe bipolar and high-breakdown SiGe bipolar.

1.4.2 HBT Device Reliability and Operating Issues

HBT Thermal Effects The operating temperature of semiconductor devices is of great importance, since it modulates most of the device characteristics and strongly affects the time dependence of those characteristics, or long-term reliability, as well. The actual temperature inside the device under operation, called *junction temperature,* is generally higher than the ambient temperature due to the self-heating effect. The internal temperature rise is determined by the thermal resistance of the device when power is dissipated. Therefore, accurate estimation of the thermal resistance is critical for the correct prediction of the device performance and its degradation under thermal/electrical stress.

Thermal resistance is commonly extracted by exploiting the temperature-sensitive electrical parameters (TSEPs), the most popular of which includes base-emitter voltage V_{BE} [1,2] and current gain β [3,4]. The thermal resistances of SiGe HBTs from IBM SiGe7HP technology have been extracted utilizing V_{BE} as a TSEP [5]. Figure 1.24 shows the measured thermal resistances with various emitter dimensions, plotted as a function of reciprocal emitter area as a convenient comparison of different emitter widths. As expected, the thermal resistance increases as device size shrinks (both emitter length and width), since the heat path downward to the substrate, particularly the cross-section area surrounded by the DT, decreases with device dimension. Interestingly, however, it turned out that the junction tempera-

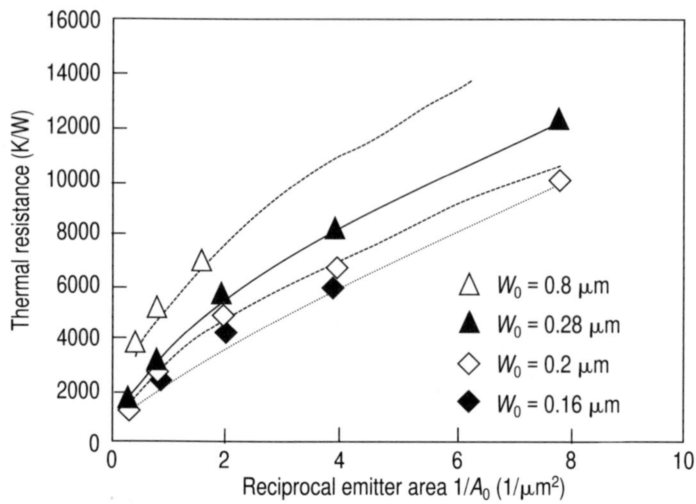

Figure 1.24 Measured thermal resistance for various device dimensions (symbols). Compared are the calculated thermal resistances based on the analytical model developed (lines).

ture rise due to self-heating decreases with reduced device size for a fixed power density level. This can be attributed to the fact that the cross section surrounded by the DT per emitter area is larger for smaller devices, thereby providing a wider heat path for a given current density (thus the power density for an identical voltage condition).

An accurate modeling of the thermal resistance would provide a reliable prediction of the thermal behavior of devices with arbitrary structural dimensions. The thermal resistance of DT-isolated bipolar transistors has been modeled assuming 45° cone-shaped heat flow to the substrate [5], this model exhibits an excellent agreement with measurement data for a wide range of emitter width and length (Fig. 1.24). The model suggests the thermal resistance strongly depends on the geometry of the DT and a substantial decrease of the thermal resistance is expected with a reduction of the DT depth and an increase of the space between the emitter finger and the DT.

HBT Safe Operation Area (SOA) The bias condition of a transistor is often chosen in a way to optimize the device performance for a given application. However, there exists a boundary for bias points beyond which the device may experience a serious performance drop or electrical and/or thermal instability. As these may lead to a catastrophic device failure or device lifetime reduction as well as the malfunction of the device, it is recommended that the device be operated within such a boundary, inside of which is called the *safe operation area* (SOA). The boundary is usually composed of a combination of two different operation limits: the voltage operation limit and the current operation limit.

The voltage operation limit is generally dictated by the avalanche multiplication process within the device, and thus can be represented by the avalanche breakdown voltages. Historically, BV_{CEO} (base-open collector to emitter breakdown voltage) has been perceived widely as a universal voltage limit beyond which bipolar transistors are inoperative. However, the voltage limit of a device cannot be represented by one single parameter, because its avalanche behavior strongly depends on the external bias configurations and bias levels. In fact, BV_{CEO} represents just one of those cases, which happens to be the worst, as will be clear in the following brief review of the voltage limits given for a few widely employed bias configurations. The common emitter forced-V_{BE} configuration, dominantly found in practical bipolar circuits, exhibits a collector-to-emitter breakdown voltage that is generally larger than BV_{CEO}, as the positive feedback to the avalanche multiplication is reduced compared to the open-base situation for which BV_{CEO} is defined. The actual breakdown for this configuration is determined by the effective external resistance seen from the base as it modulates the strength of the positive feedback. Another frequently employed bias configuration is the common base forced-I_E case. For this type of configuration, the collector-to-emitter breakdown voltage is virtually identical to BV_{CBO} (emitter-open collector to base breakdown voltage), because the positive feedback is completely suppressed due to the firmly fixed emitter current level. It has been observed, however, that a lateral current instability stemming from the "pinch-in" effect [6] can result in lowered voltage operation limits. The common emitter forced-I_B configuration is rarely found in practical circuit applications, although transistor I-V characteristics are routinely plotted with this configuration. The voltage limit with this configuration is smallest and roughly falls around BV_{CEO}, with slight variation over actual base current level. This confirms that BV_{CEO} represents the worst case of the voltage limit and bears only minor practical importance. Typical voltage limits measured from each of these configurations are collected and illustrated in Fig. 1.25 with a device size of $A_E = 0.2 \times 6.4\ \mu m^2$ from 7HP technology as an example.

The current operation limit of the SiGe HBTs is mostly determined by the electromigration effect of metal lines and/or vias connected to the devices, since the device reliability is closely related to this effect. With the electromigration as a limiting factor, the maximum allowed current is given as an exponential function of the temperature, making it highly sensitive to temperature variations. (For example, a 5°C change leads to ~35% change in the current limit.) Hence, accurate temperature information at the metal interconnects becomes critical for the current limit estimation, causing the self-heating effect crucial for this purpose. Considering the power dissipation–dependence of the junction temperature rise from the self-heating, it is appropriate to express the current limit as a function of the device bias condition. The bias-dependent current limit has been calculated for SiGe HBTs from 7HP technology with various emitter lengths and ambient temperature of 85°C (Fig. 1.25). The limit decreases with increasing voltage applied (larger power dissipation) and with increasing emitter length (larger thermal resistance). The current limit can be relaxed significantly with modified metal interconnect schemes. Also note that the limit depends on assumed values for chip lifetime, number of metal lines

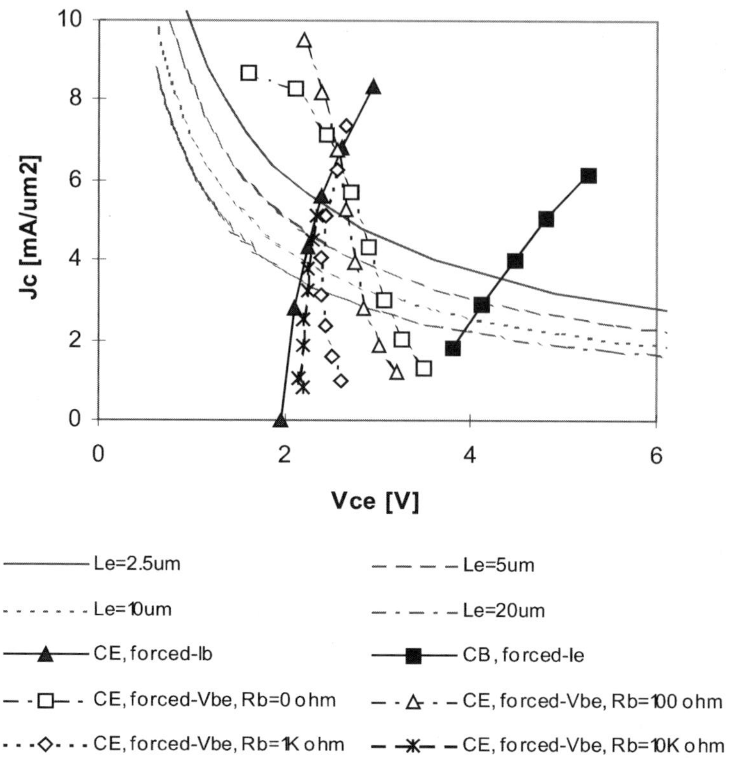

Figure 1.25 Measured voltage boundaries for various bias configurations (lines + symbols). For common the emitter forced-V_{BE} case, the voltage limits are given with four different resistance values connected to the base. The device size was $A_E = 0.2 \times 6.4\ \mu m^2$. Also shown are the calculated current boundary for four different emitter lengths (lines only). Emitter width was fixed at 0.2 μm. $T_{amb} = 85°C$.

per chip, and signal duty cycle, which were chosen conservatively in the present calculation. Further relaxation of the limit can be achieved with a new set of these assumptions.

HBT Reliability The only structural difference of SiGe HBTs from conventional epi-base Si bipolar transistors is the addition of a small amount of Ge in the base region. In terms of reliability, this implies that a SiGe HBT should be perceived as an extension of a Si bipolar transistor, except for any effect of Ge on the reliability, if such an effect exists. After three generations of production-qualified IBM SiGe BiCMOS technology (0.5 μm, 0.25 μm, 0.18 μm), with Ge composition in 2002 reaching up to 25%, we have not identified any reliability issue that can be directed to the addition of Ge. This is consistent with the observation from a similar non-IBM SiGe technology [7]. Instead, considerable suppression of boron out-diffusion from the base layer with the addition of Ge has been observed, which would result in the improvement of the base-width control and stability. Therefore, it is legitimate to say that SiGe HBTs inherit the advantages of Si bipolar transistors over HBTs based on III–V systems in terms of reliability, such as minimal exposed surface area with planar structure, better surface control with oxidation, and stable metal contact with silicidation, all contributing to a better reliability. The relatively high current density of SiGe HBTs, which results from the efforts to improve the operation speed, is occasionally brought up as a potential reliability issue. However, it comes from experience in III–V systems where current densities in those semiconductor systems have historically been a reliability issue. Dopant diffusion in silicon systems is significantly lower than in III–V systems at operating conditions due to the highly stable Si bonding. For example, base dopant diffusion coefficients in Si are approximately 18 orders of magnitude lower than in InP systems at device operating temperatures. Thus the effect of high current density on dopant diffusion is not a serious concern for Si systems as opposed to III–V systems.

Typical degradation of the SiGe HBT with forward-bias long-term stress can be characterized as a gradual rolloff of the current gain, as illustrated in Fig. 1.26. The degradation in the current gain usually exhibits saturation at less than ~10% with nominal operating temperatures and bias conditions, and no catastrophic failure is normally observed. The change in the current gain is attributed to electromigration-induced pressure on the emitter contact [5], which is consistent with the observation from conventional Si bipolar transistors. The electromigration effect does not belong to the intrinsic characteristics of the device, and can be alleviated with the modification of structural and material properties of the metal interconnect.

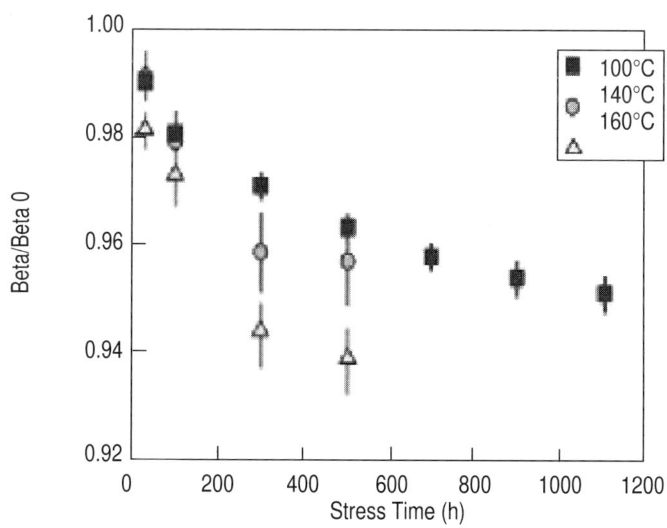

Figure 1.26 Typical HBT current-gain degradation pattern with forward-bias stress. The current density is 1.3 mA/μm^2 (near peak f_T) with device size of $A_E = 0.5 \times 2.5$ μm^2.

1.4.3 FET Device Reliability and Operating Issues

As part of the SiGe BiCMOS device suite the FET devices must also pass a full reliability qualification. To do this they must pass the standard digital reliability tests as well as additional ones for RF/analog applications. It has been shown [8] that the reliability of FET devices in SiGe integrated technologies is comparable to that of the base technology for the major reliability mechanisms of hot carrier, gate dielectric, and negative-bias temperature instability (NBTI). The reliability models generated for the base technologies will adequately predict integrated SiGe parameter shifts and estimated lifetimes. In this section, we present the major FET reliability concerns that need to be addressed for analog and RF applications.

Implications of Analog and RF operating point to Hot Carrier Degradation For logic applications it has been the standard industry practice to project hot carrier lifetimes by performing a set of stresses for a particular channel length using 3 or 4 V_{DS} stress voltages and a single V_{GS} stress voltage condition. At each stress V_{DS} the V_{GS} is found that gives the peak *Ib* for the NFET and the peak I_G for the pFET. The V_{GS} stress conditions are typically $V_{GS} = V_{DD}/2$ to achieve peak *Ib* for the nFET and $V_{GS} = V_T$ to achieve peak I_G for the PFET. The standard definition of lifetime (τ) corresponds to the length of time required to achieve a given degradation level for a given device parameter. The most commonly used definition is the time it takes an FET to reach an arbitrary level of degradation, typically 5% or 10% in I_{ON}. Notice that the degradation corresponds to the measurement of I_D at the nominal power-supply voltage for the technology and is located in the high V_{GS} regions shown in Figs. 1.27 and 1.28. At this point a semilogarithmic plot of Log (τ) vs. $(V_D)^{-1}$ is constructed. Lifetime estimates are generated by extrapolating the measured lifetimes across several orders of magnitude, thus generating predictions at the use condition supply voltage, as illustrated in Fig. 1.29. This technique has provided reasonable estimates for various generations of logic CMOS applications, but it cannot be directly utilized for analog and RF applications, as it relies on the measurement of I_D at the $V_G = V_D$ (ion) operating point which, as we point out in the following discussion, could produce inaccurate lifetime predictions.

The operating conditions imposed by analog and RF applications require careful consideration per the reliability implications of the gate-voltage biasing point. It can easily be shown that the V_G voltage, selected to measure the drain current I_D, constitutes a significant source for the large variations in the observed degradation. This degradation greatly influences the lifetime projected to achieve a given percentage of degradation. Figure 1.27 shows the decrease in drain current for an NFET transistor under hot carrier stress. Notice that the percentage of the reduction in I_D changes drastically across the V_G voltage value selected for the measurement. The drastic reduction in the projected lifetime is shown in Fig. 1.30, normalized for a nominal power supply voltage of 1.8 V. The data show a two order-of-magnitude reduction in lifetime for a gate-voltage operating point of 0.6 V. Notice that using I_D at the nominal power-supply voltage as a metric for hot carrier damage will yield

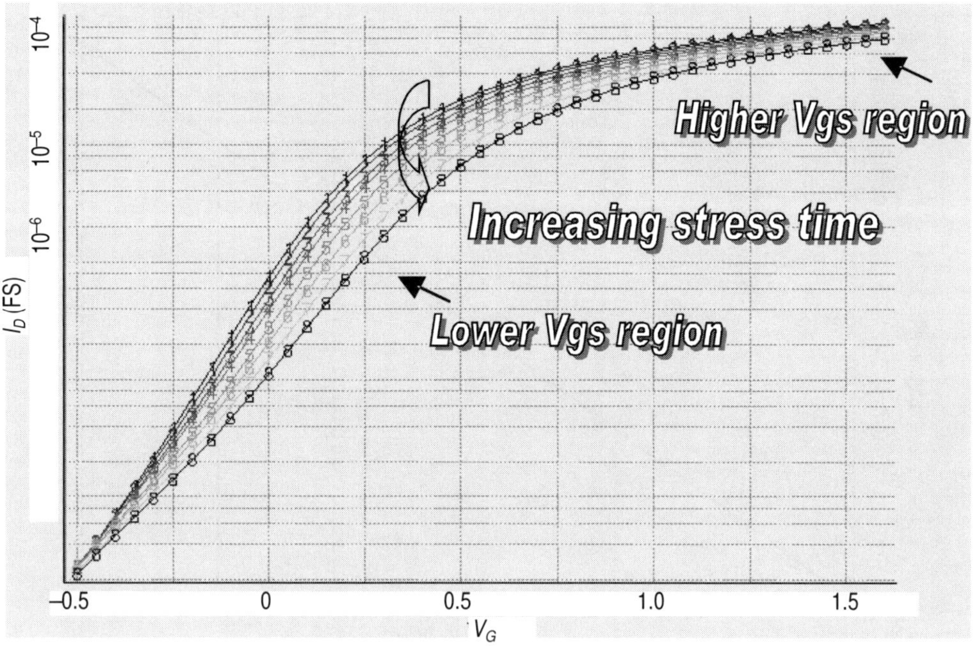

Figure 1.27 I_D vs. V_G characteristics for the various readouts performed under hot carrier stress.

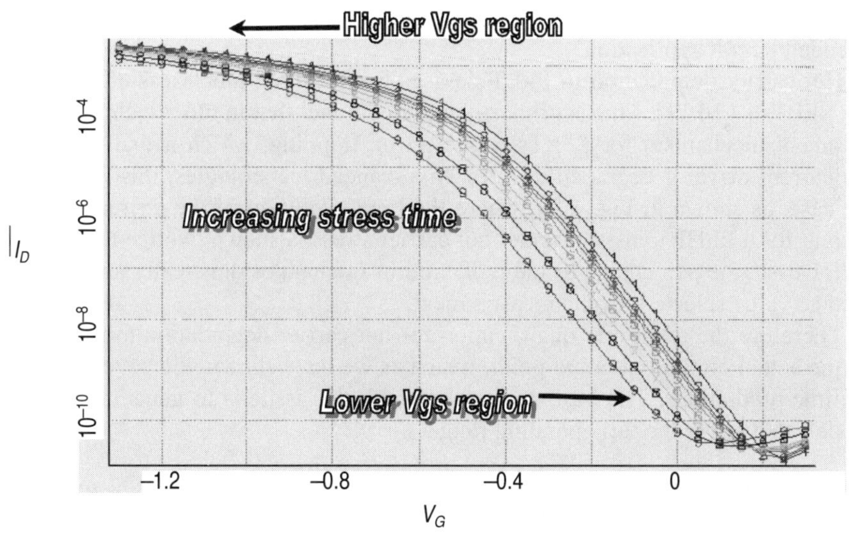

Figure 1.28 Shift in I_D vs. V_G characteristics for the various readouts performed under hot carrier stress.

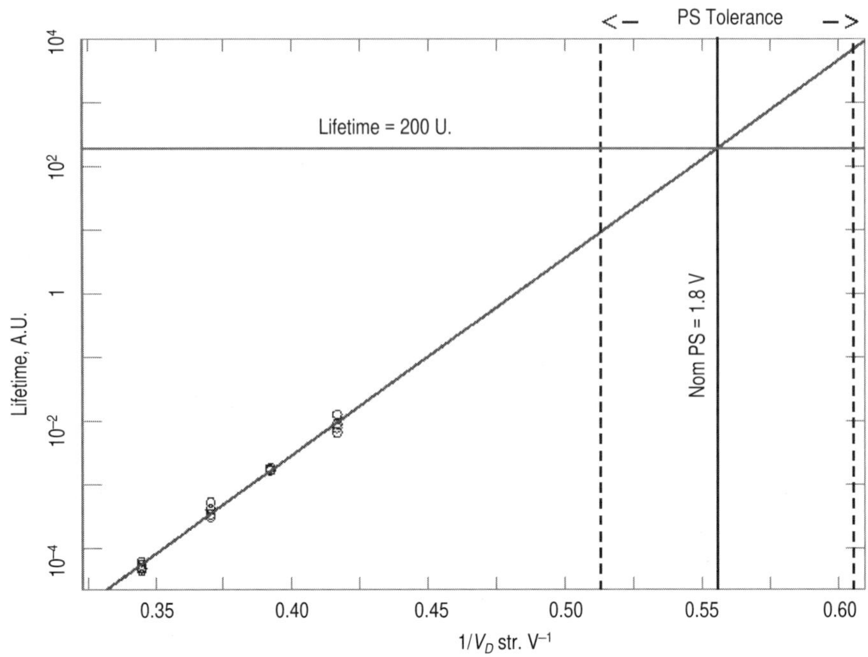

Figure 1.29 Lifetime prediction by extrapolation from V_D stress conditions to nominal power-supply condition.

overly optimistic estimates and should be carefully reviewed for adequacy with the intended circuit application.

Hot carrier degradation in PFET devices has received much less attention than for NFETs [9,10,11]. This is primarily due to the fact that in older technologies the dominant mechanism for PFETs was electron trapping, which actually increased the current drive of degraded devices. In advanced technologies, this is no longer the case, as shown in Fig. 1.29, where the time evolution of the decrease in drain current for a PFET transistor under hot carrier stress is shown. Notice that as in the NFET case, the percentage of the reduction in I_D changes drastically across the V_G voltage value selected for the measurement.

Therefore the reliability qualification for hot carrier degradation must be based on the actual circuit operating point, which as we have shown will severely impact lifetime predictions. The best way to deal with this issue is to generate reliability models that cover the full operating range.

Negative-Bias Temperature Instability (NBTI) NBTI [12] is a PMOSFET wearout mechanism that results in a positive-charge buildup and interface states generation at the Si/SiO_2 interface under the influence of applied negative gate voltages. This mechanism exhibits a strong temperature dependence and will very like-

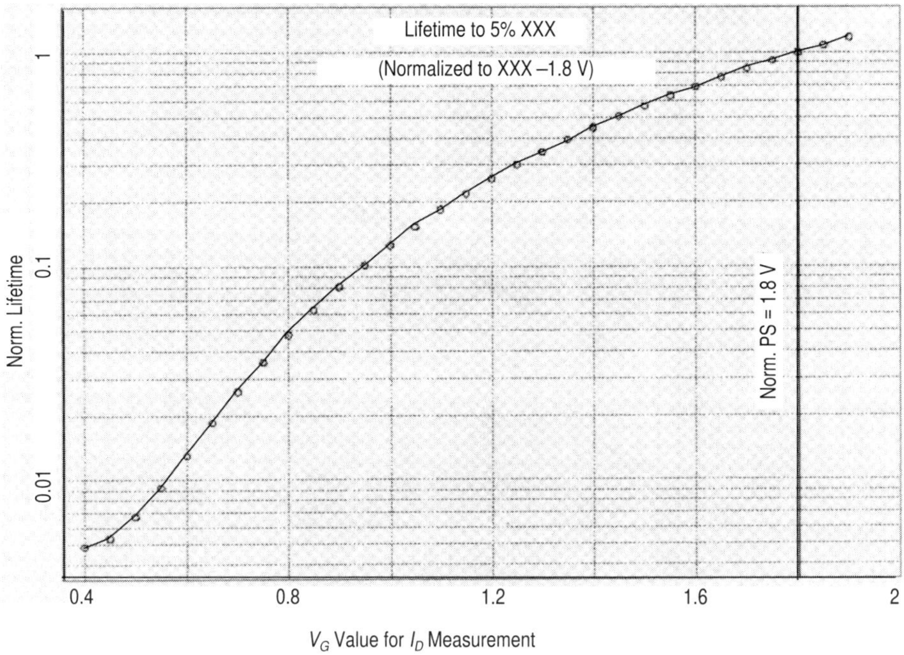

Figure 1.30 Influence of the FET gate-voltage selection for I_D measurement on lifetime reductions.

ly be the dominant degradation mechanism during burn-in. It also exhibits a strong electric-field dependence, and will play an increasingly important role as oxides are made thinner. ($E_{OX} < 6.0$ MV/cm – lower than *F/N* injection.) This mechanism has traditionally been observed with no directly measurable gate currents.

NBTI competes with channel hot carrier damage to PMOSFET degradation during circuit operations in advanced CMOS technologies. The hot carrier and NBTI contributions can be separated by stressing at different temperatures, as shown in Fig. 1.31. Where the NBTI is shown to be dominant at 125°C, while the hot carrier contribution can be seen at 0°C, notice that for short-channel devices (higher I_{NW}) the hot carrier dominates the degradation at both temperatures. The NBTI degradation is typically predicted using V_t shift models that depend primarily on T_{ox}, V_G, temperature as well as other geometrical, process, and second-order effects. This mechanism will play a significant role and must be taken into account for RF/analog applications..

Other Reliability Considerations Some of the additional parameters that would require attention for their possible impacts to the reliability of RF and analog applications are: device matching, S-parameter stability, V_t, dV_t/DV_x (body effect), gm, gds and $1/f$ noise.

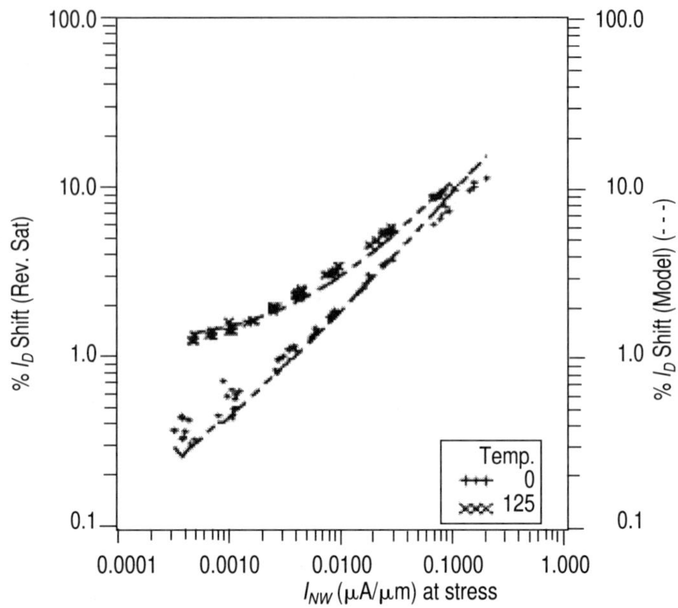

Figure 1.31 Separating PMOSFET hot carrier and NBTI degradation in advanced CMOS technologies by stressing at different temperatures.

1.4.4 Electromigration

Electromigration is the biased atom diffusion in interconnects during electrical current loading [13]. This redistribution of material can induce voiding near the cathode end of interconnects, which results in resistance increase, or even electrical open, in circuits. Therefore, this phenomenon imposes a major reliability concern in IC operation. In particular, electromigration plays a critical role in determining the current operation limit, and thus the performance, of HBT devices in their high current applications. Also, the self-heating of HBTs at high-current further aggravates the electromigration degradation.

Due to stringent reliability requirements, traditional Al(Cu) metallurgy used in BiCMOS of five and six generations is no longer suitable for more advanced HBT circuits. Starting from BiCMOS 7HP technology, Cu interconnects are introduced for better electrical performance and significantly improved electromigration reliability. As clearly demonstrated in [14] and [15], Cu interconnects enjoy up to ~1000× improvement in electromigration lifetime over that of traditional Al(Cu) interconnects at various elevated temperatures.

Experimentally, the electromigration current limit is determined from accelerated tests using high DC current at elevated temperatures. Based on the test results, the DC current limit allowed for reliability operation under use condition can then be projected, with the acceleration factor described by

$$\text{Acceleration factor} = \left(\frac{j_{\text{stress}}}{j_{\text{use}}}\right)^n \exp\left(\left(\frac{\Delta H}{k}\right)\left(\frac{1}{T_{\text{use}}} - \frac{1}{T_{\text{stress}}}\right)\right) \tag{1.8}$$

where j_{stress} and j_{use} are current densities under stress and use conditions, respectively; n is the current exponent; ΔH is the activation energy associated with metal diffusion; and T_{stress} and T_{use} are the interconnect temperatures under stress and use conditions, respectively. As suggested in this equation, the electromigration current limit is very sensitive to the interconnect temperature, dictated by the activation energy ΔH. Figure 1.32 compares the temperature dependence of electromigration current limit between copper and aluminum, using reported ΔH and n values of 0.9 electron volts (eV) and 1.1 for Cu and 0.85 eVand 1.7 for Al(Cu), respectively. Note that the current limits shown in Fig. 1.32 are normalized by their respective limits at the 125°C reference temperature. As can be observed, the current limits increase with decreasing temperature in both cases, and Cu benefits much more significantly from reduced temperature. Therefore, it is crucial to know the interconnect temperature under operating condition in design for high-current applications.

Another factor in determining the current operation limit is the failure-rate tolerance, considering the total number of malfunctioned chips due to electromigration wearout within the expected product lifetime. For example, for a chip containing a million interconnects, a failure rate tolerance of $1E - 12$ allows only one electromigration failure out of a million chips during the expected lifetime. Figure 1.33 illustrates the effect of failure-rate tolerance on the current limit, calculated for a typical range of sigma values (or the breadths of failure distributions). These results indicate that the current limit can be adjusted according to the total number of interconnects in a particular product chip, as well as the product-reliability tolerance. Fur-

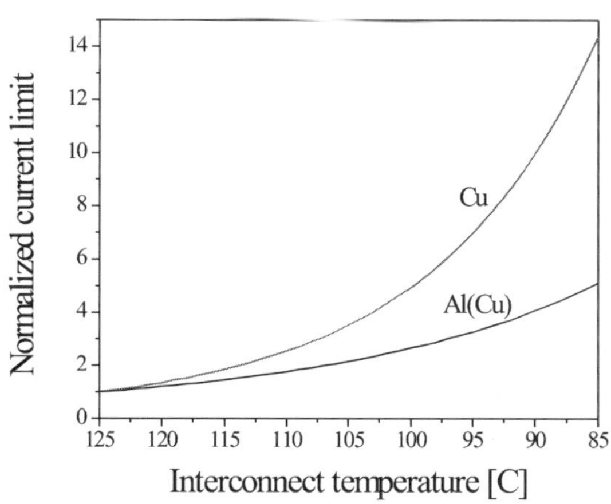

Figure 1.32 Normalized current operation limit of Cu and Al(Cu) over a range of temperature under use condition. Note that the current limits are normalized by their respective values at 125°C as a reference temperature.

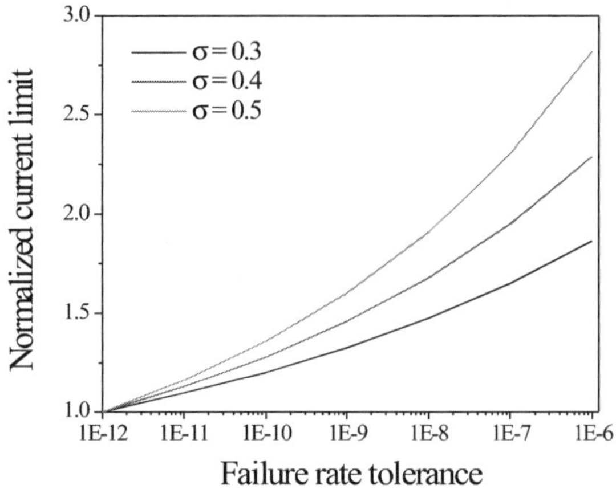

Figure 1.33 Normalized current operation limit as a function of failure-rate tolerance. Note that the current limits are normalized by the value at $1E - 12$ failure rate.

thermore, the current operation limit can also be adjusted for individual interconnects by considering their duty cycle during circuit operation.

1.4.5 Summary

In this section, we discussed many of the key aspects of the technology implications in SiGe (and RF-CMOS) circuit design. Specifically, the topics of device noise and reliability were covered.

1.4.6 References

1. M. G. Adlerstein and M. P. Zaitlin, "Thermal Resistance Measurements for AlGaAs/GaAs Heterojunction Bipolar Transistors," *IEEE Trans. Electron Devices,* vol. 38, no. 6, pp. 1533–1554, 1991.
2. B. M. Cain, P. A. Goud, and C. G. Englefield, "Electrical Measurement of the Junction Temperature of an RF Power Transistor," *IEEE Trans. Inst. Meas.,* vol. 41, no. 5, pp. 663–665, 1992.
3. J. R. Waldrop, K. C. Wang, and P. M. Asbeck, "Determination of Junction Temperature in AlGaAs/GaAs Heterojunction Bipolar Transistors by Electrical Measurement," *IEEE Trans. Electron Devices,* vol. 39, no. 5, pp. 1248–1250, 1992.
4. D. E. Dawson, A. K. Gupta, and M. L. Salib, "CW Measurement of HBT Thermal Resistance," *IEEE Trans. Electron Devices,* vol. 39, no. 10, pp. 2235–2239, October 1992.
5. J.-S. Rieh, D. Greenberg, B. Jagannathan, G. Freeman, and S. Subbanna, "Measurement and Modeling of Thermal Resistance of High Speed SiGe Heterojunction Bipolar Transistors," in *Proceedings of Topical Meeting on Silicon Monolithic Integrated Circuits in RF Systems,* pp. 110–113, 2001.

6. M. Rickelt, H.-M. Rein, and E. Rose, "Influence of Impact-Ionization-Induced Instabilities on the Maximum Usable Output Voltage of Si-Bipolar Transistors," *IEEE Trans. Electron Devices,* vol. 48, no. 4, pp. 774–783, April 2001.
7. A. Neugroschel, C.-T. Sah, J. M. Ford, J. Steele, R. Tang, and C. Stein, "Comparison of Time-to-Failure of GeSi and Si Bipolar Transistors," *IEEE Electron Devices Lett.,* vol. 17, no. 5, pp. 211–213, May 1996.
8. G. Hueckel, P. Wang, K. Watson, K. Brelsford, and W. Ansley, "BTV SiGe 7HP BiCMOS Reliability Report T1 Qualification," September 2001.
9. B. S. Doyle, and K. R. Mistry, "A lifetime prediction method for hot-carrier degradation in surface-channel p-MOS devices," *IEEE Trans. Electron Devices,* vol. 37, p. 1301–1307, 1990.
10. T. Tsuchiya, Y. Okazaki, M. Miyake, and T. Kobayashi, "New hot-carrier degradation mode and lifetime prediction method in quarter-micrometer PMOSFET," *IEEE Trans. Electron Devices,* vol. 39, p. 404, 1992.
11. R. Woltjer and G. M. Paulzen, "Modeling of oxide-charge generation during hot-carrier degradation of PMOSFETs," *IEEE Trans. Electron Devices,* vol. 41, p. 1639–1644, 1994.
12. G. Larosa, S. Rauch, and F. Guarin, "NBTI-Channel Hot Carrier Effects in Advanced Sub-micron PFET Technologies," *IRPS Proceedings,* pp. 282–286, April 1997.
13. C.-K. Hu, K. P. Rodbell, T. D. Sullivan, K. Y. Lee, and D. P. Bouldin, "Electromigration and stress-induced voiding in fine Al and Al-alloy thin film lines," *IBM J. Res. Develop.,* vol. 39, p. 465, 1995.
14. D. Edelstein, J. Heidenreich, R. Goldblatt, W. Cote, C. Uzoh, N. Lustig, P. Roper, T. McDevitt, W. Motsiff, A. Simon, J. Dukovic, R. Wachnik, H. Rathore, R. Schulz, L. Su, S. Luce, and J. Slattery, "Full copper wiring in a sub-0.25 μm CMOS ULSI Technology," *IEDM Tech. Digest,* p. 773–776, 1997.
15. C.-K. Hu, R. Rosenberg, H. S. Rathore, D. B. Nguyen, and B. Agarwala, "Scaling effect on electromigration in on-chip Cu wiring," *Proceedings of IEEE Int. Interconnect Technology Conf.* p. 267–269, 1999.

2

MODELING AND CHARACTERIZATION

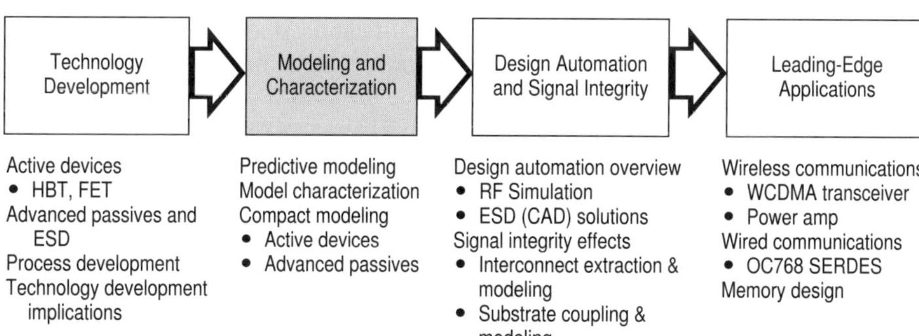

OVERVIEW

In Chapter 1, we described the details behind the development and manufacture of the SiGe technologies. This section describes the next step in IBM's enablement, namely, modeling and characterization. These methodologies have been matured over a decade of world-class development, and are a critical part of IBM's ability to provide first-pass enablement for designers.

This section reviews the predictive modeling work, presents the model charac-

Silicon Germanium: Technology, Modeling, and Design. By Singh, Harame, and Oprysko
ISBN 0-471-44653-X

terization work, and discusses in detail the compact modeling of a broad range of devices, which were discussed in Chapter 1.

- Section 2.1 discusses the background and methodology of IBM's predictive modeling work.
- Section 2.2 discusses IBM's model characterization development activities.
- Section 2.3 presents the compact modeling work for the active HBT and FET devices.
- Section 2.4 discusses compact modeling methodology for the advanced passive devices.

2.1 PREDICTIVE MODELING

A fundamental difference between digital-technology development and analog/RF-technology development is the sensitivity of analog/RF circuits to many manufacturability/performance trade-offs that are an inevitable part of the technology-development process. Thus, while feedback between IC designers and technologists during technology definition is critical for timely product development, it is difficult to realize in practice, because to an IC designer the technology is enabled through the design kit, including a complete and accurate set of compact models for both active and passive devices as well as interconnects. It is expected to reflect a stable and well-characterized process, while to the technologist the technology is a process recipe that is roughly characterized, often in flux, and, in practice, often not actually realized until the final technology qualification stage. The role of predictive modeling is to utilize detailed process and device simulation, or technology CAD (TCAD), in place of hardware to facilitate the feedback loop between circuit designer and process technologist until definitive hardware data are finally incorporated into the design kit. This feedback path provides timely notification to the technologist of potential shortcomings in the targeted technology design point, thus optimizing the use of the available experimental wafer budget. For the circuit designer, it provides more design turns with the technology and thus a greater likelihood of meeting circuit performance targets.

In this section, we present the concepts of TCAD and how IBM has aggressively used this capability to predict active- and passive-device performance.

2.1.1 Technology CAD

Semiconductor TCAD originated in the early 1960s [1] with efforts to understand and optimize bipolar transistors. This effort continues today, with the increases in computing power available leveraged to understand and engineer devices with higher operating speeds and fabricated with more complex processes. The TCAD paradigm is applied to all conceivable types of active and passive devices, and intensive TCAD studies are part of all modern technology development efforts.

The TCAD paradigm is depicted in Fig. 2.1. Detailed process simulation creates one-, two-, or three-dimensional device representations, consisting of structure (film thicknesses and shapes) and impurity concentrations used as input for device simulation. Device simulation produces the DC and AC characteristics of interest, which are in turn used to define compact models for use in a prototype design kit for the technology. In this way, process options can receive circuit-design feedback before time and budget are expended to define these options in silicon.

2.1.2 Process Simulation

Physical process simulation is the critical component in a predictive TCAD capability. Research and development of existing process simulation capabilities are due to the last decade's worth of logic CMOS scaling. Ever more sophisticated process simulation capabilities are being developed as semiconductor processing capabilities, driven by an extremely competitive microelectronics industry, continually progress. However, despite intensive efforts to bring higher-level modeling capabilities such as molecular dynamics and kinetic Monte Carlo codes into practical use, continuum codes based on silicon process physics are still the primary platform for semiconductor process simulation and will be focused on here. The critical silicon process operations are impurity implantation and diffusion, oxidation, and material deposition and etching. Silicon and SiGe epitaxy, an increasingly critical silicon process step, are treated via a series of material depositions and diffusions.

Impurity implantation is typically addressed by analytic distribution functions, fitted to both experimentally determined distributions and distributions based on sophisticated simulation approaches, such as Monte Carlo methods. A given implant dose and energy specify a tabular search that determines the implant profile para-

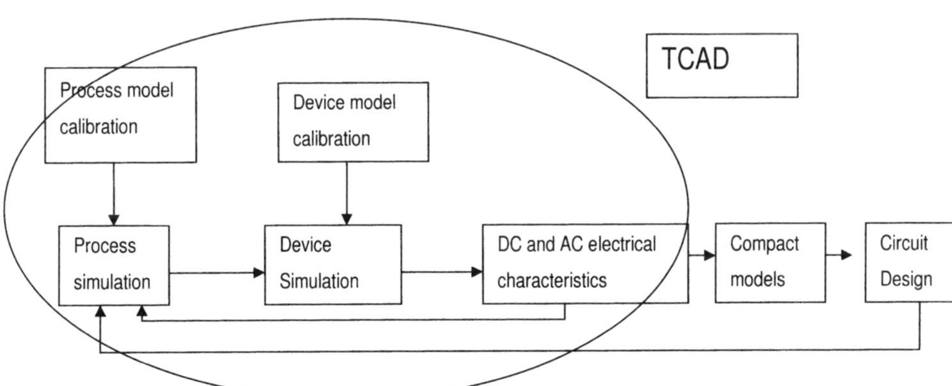

Figure 2.1 TCAD is typically used to assist technology development, but is leveraged to its fullest extent when combined with compact model development to provide early technology access for circuit design.

meters—range, standard deviation, etc.—of the resulting multidimensional distribution, considering the materials through which the implanted ions pass and conservation of the total impurity dose. Particle-based (Monte Carlo) simulation approaches are finding increasing use in industrial environments, however, as device scaling drives impurity implants to increasingly higher doses, lower energies, and smaller (in terms of surface area) targets, and the resulting profile become too complex to be characterized by compact parameterizable functions.

In continuum-process modeling codes, dopants are treated as impurity concentrations redistributed via the classic diffusion equation derived from Fick's laws. A significant amount of experimental and theoretical work has been done over the last three decades to understand and model and impurity diffusion in silicon, largely driven by the need to understand and engineer CMOS short-channel effects [2]. The resulting general and accurate approach to silicon impurity diffusion, in which both dopant impurities and silicon point defects (interstitials and vacancies) codiffuse, has found widespread utility in all aspects of silicon process development. Ancillary models to handle point-defect kinetics in such silicon process phenomena as pre- and self-amorphization, oxidation-enhanced diffusion and extended silicon defects and dislocations continue to be developed and characterized. Diffusion modeling continues to be addressed by continuum methods; more sophisticated approaches have not yet found extensive use in industrial environments.

Oxidation of silicon is based on the 1965 work of Deal and Grove [3], which considers the flux of oxidant from the ambient to the oxide surface, its diffusion across the oxide, and finally its reaction with the silicon surface. The Deal–Grove model has been well characterized and modified extensively to treat, for example, growth of very thin oxides and the effect of dopant impurities in both the silicon and oxide on the oxidation process. By the appropriate treatment of how the oxide flows during growth, including the effects of stress through interaction with surrounding materials on the oxidation process, accurate simulation of the effects of oxidation on complex silicon structures can be expected.

Etching and deposition simulation approaches span a wide range of approaches, from simple addition and subtraction of polygonal regions to more sophisticated approaches based on evolving etching or deposition fronts, including both continuum and particle-transport approaches for the material fluxes. Industrial TCAD depends primarily on simple etching and deposition models, including conformal material deposition of materials with a specified thickness and single-material etching with a given degree of lateral to vertical anisotropy.

Despite the intensive effort to understand, characterize, and model silicon process physics driven by the CMOS logic microelectronics industry, industrial use still requires continual calibration of model parameters, and the predictive range of any process modeling capability can vary significantly between process modules. The most effective approach to TCAD calibration has been found to be that depicted in Fig. 2.2. The calibration process in Fig. 2.2 begins with optimizing essential process-simulation models, such as impurity diffusion coefficients, implant models, and film thicknesses, to available physical data such as secondary ion mass spectrometry (SIMS) and structural cross sections. This initial calibration provides a

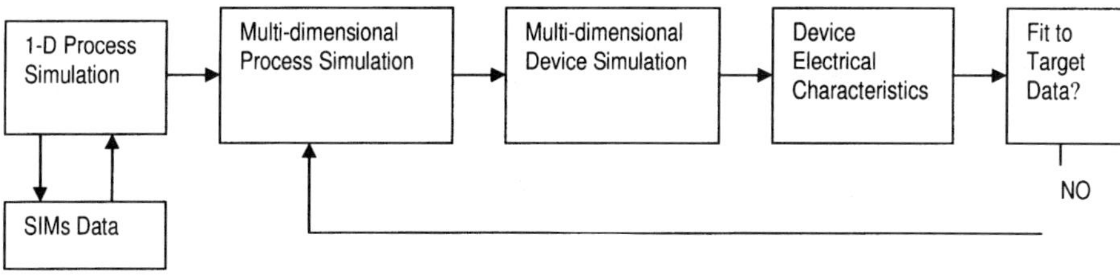

Figure 2.2 Self-consistent calibration of the process- and device-simulation models is critical for leveraging TCAD for technology development. A successful calibration links process development levers to device electrical performance, providing the ability to explore design space and optimize device performance outside of the lengthy and expensive silicon fabrication process.

good starting point for multidimensional process and device simulations, the goal of which is to fit electrical data considered critical for target use of the simulation capability. A feedback loop is then established and exercised, in which a few critical process and device model parameters are optimized to fit the target electrical data. A successful calibration effort provides a self-consistent TCAD capability that links process levers to electrical characteristics through simulation, which has been found to be a very powerful tool for technology development and optimization.

As an example of the simulation of a complex process module, simulation of the formation of a deep trench, a fundamental isolation structure for high-performance SiGe bipolar transistors, is shown in the Fig. 2.3. Figure 2.3A shows the silicon surface, composed of a light n-type epitaxial layer grown over a lightly p-type substrate, masked by oxide and nitride layers, which have been opened and the silicon deep trench etched. The buried subcollector was implanted and diffused prior to the etch. Following this etch, the interior of the trench is oxidized and boron implanted through the bottom of the trench to inhibit leakage under the trench (Fig. 2.3B). Additional oxide is deposited and the remainder of the trench filled with polysilicon. The structure is then recessed by the SDI process (Fig. 2.3C). Figure 2.3D shows the simulated DT isolation structure at the end of the process.

2.1.3 Device Simulation

Similar to process simulation, device simulation has been implemented at several levels of physical sophistication. The approach used in the pioneering continuum-based device simulation efforts [1,4,5] continues to be the foundation of most silicon TCAD efforts. This is particularly true in the predictive modeling mode, which for compact model generation demands simulation of the complete device structure, including all parasitic capacitances and resistances, and often represents a considerable expanse of silicon, over a wide range of bias conditions.

Continuum-based device simulation consists of the solution of Poisson's equation along with two or more equations accounting for the transport, generation, and

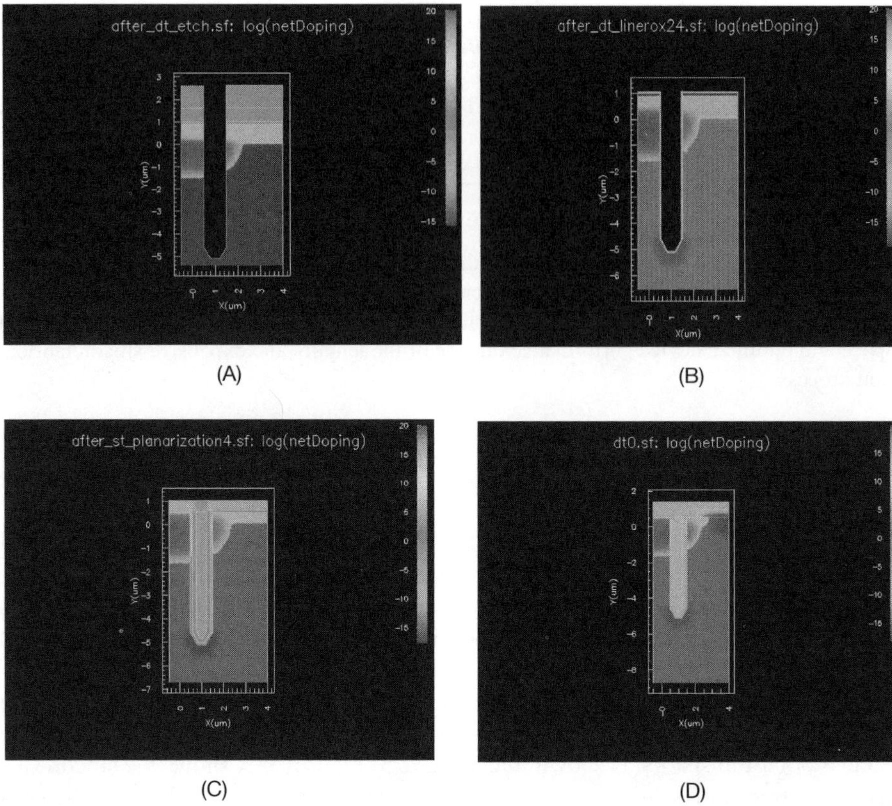

Figure 2.3 Snapshots of DT-isolation process simulation. (A) After etch; (B) after channel stop implant; (C) after DT fill and planarization; (D) at the end of the process.

recombination of holes and electrons in the semiconductor. Additional equations can be added to account for carrier energy exchange with the silicon lattice, generating average carrier temperatures, and thus address nonstationary transport effects such as velocity overshoot, and for device lattice self-heating. A further enhancement for simulation of field-effect devices is the incorporation of some form of Schrodinger's equation to account for carrier quantization effects. An immense amount of theoretical and experimental work has been done to formulate and calibrate the many physical models required in a typical continuum device simulation, such as carrier mobility as a function of electric field, doping concentration, lattice temperature, and surface roughness, and the effects of impurity concentration and species on the silicon bandgap [6].

Continuum-based device simulation is able to duplicate experimental AC measurements, such as S-parameter extraction, by application of the AC small-signal approximation to the Poisson and transport equations [7]. Implementation of the small-signal approximation completes the demand on the TCAD capability of supplying all necessary electrical data for building a compact model. When device-simulation is combined with process simulation and suitably calibrated, the system so formed provides the technology development effort and the ability to quickly, and with a useful range of self-consistent physical accuracy, connect process levers to all critical device electrical characteristics.

Semiconductor device scaling has driven future-generation devices into a physical regime in which typical operation is dominated by physical transport effects that are not well addressed by enhanced continuum approaches. A larger amount of physical accuracy and detail is provided by Monte Carlo particle transport codes [8], albeit on, typically, reduced computational regions and requiring significantly longer solution times.

2.1.4 Examples

An example of TCAD application in the technology development mode follows. A SiGe HBT with AC performance characterized by $f_T = 50$ GHz and $f_{MAX} = 60$ GHz was the technology starting point. It was desired to size the AC-performance consequences of simplifying the extrinsic base formation. The study began by calibrating the process and device simulation to the technology. The results of the concurrent two-dimensional process and device calibration that are shown in Fig. 2.4 compares the simulated and experimental summary AC parametrics of cutoff frequency, f_T, and maximum frequency of oscillation f_{MAX}. Both parameters are well matched by the their respective simulations, and as they involve all aspects of the device performance—transport in both low- and high-doped and depletion regions, across signal frequencies and bias conditions—a strong indication that the AC performance of the device is captured with an adequate degree of accuracy. Figure 2.5A plots simulated versus experimental base resistance for the target device. Base resistance has a strong three-dimensional component that cannot be fully captured by the two-di-

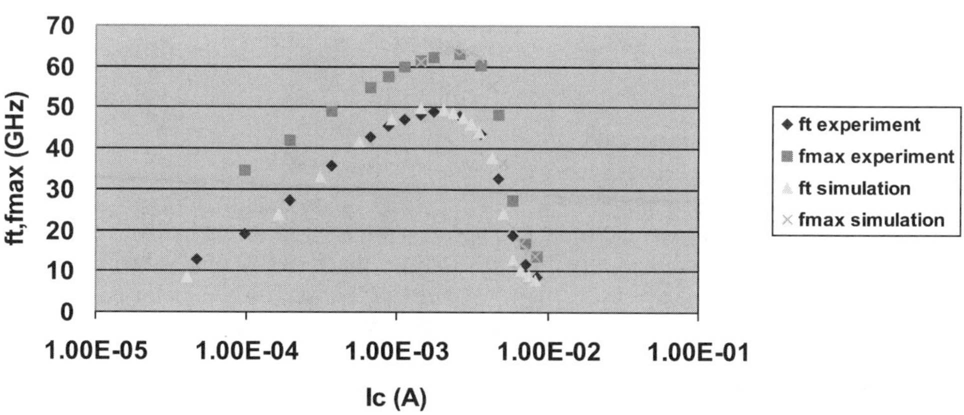

Figure 2.4 Results from TCAD calibration for 50-GHz/60-GHz f_T/f_{MAX} SiGe HBT.

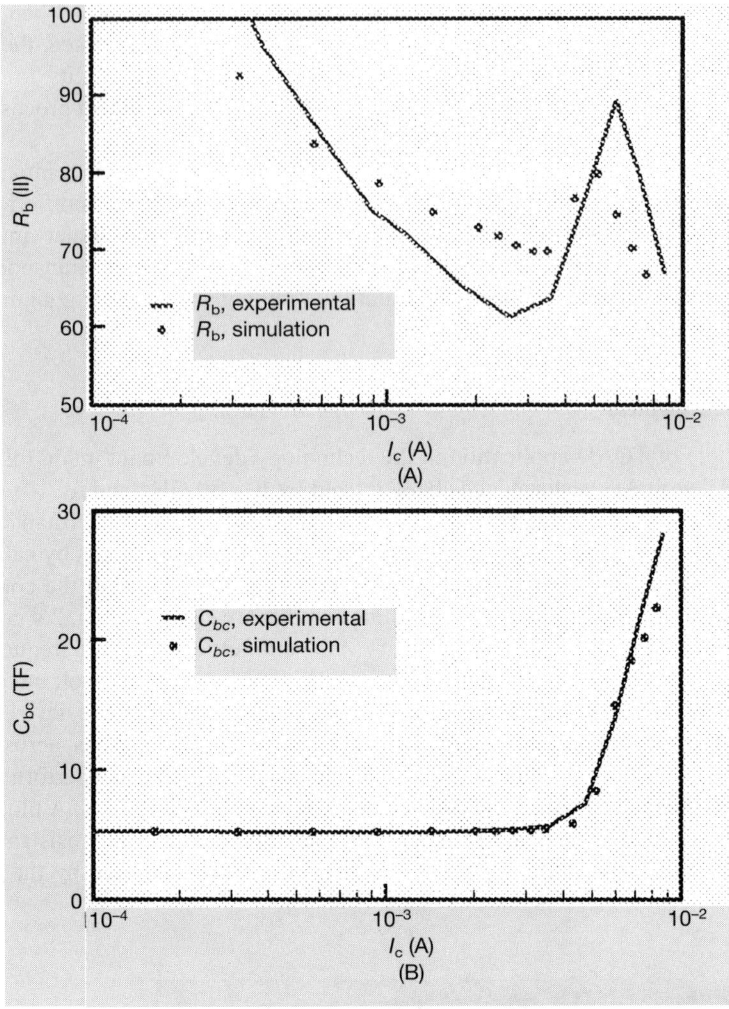

Figure 2.5 Comparison of simulated and experimental R_b (a) and C_{bc} for 50-GHz/60-GHz f_T/f_{MAX} SiGe HBT.

mensional simulations; however, the simulation shows the same qualitative characteristics as the hardware and is within 10–15% of the experimental values at the current densities at which peak AC performance (f_T) is observed. Collector-base capacitance (Fig. 2.5B) can be well modeled by the two-dimensional approach. Both simulation and experimental parameters are extracted from the low-frequency Z-parameters. From this calibrated starting point, the consequences for the device AC performance due to a simplified, NSA process for the extrinsic base was investigated.

Figure 2.6A depicts the process used to form the extrinsic-base in the device for which the process and device simulation calibration just discussed was performed. In this self-aligned process, a pedestal structure is formed to define the future position of the emitter polysilicon, and is surrounded by a sidewall spacer. The extrinsic base is then implanted, with the spacer controlling the separation, independent of the emitter width, between the emitter and extrinsic base. An alternative extrinsic-base formation method is shown in Fig. 2.6B. In this NSA method, the extrinsic-base implant occurs immediately after the emitter spacer is etched. This latter

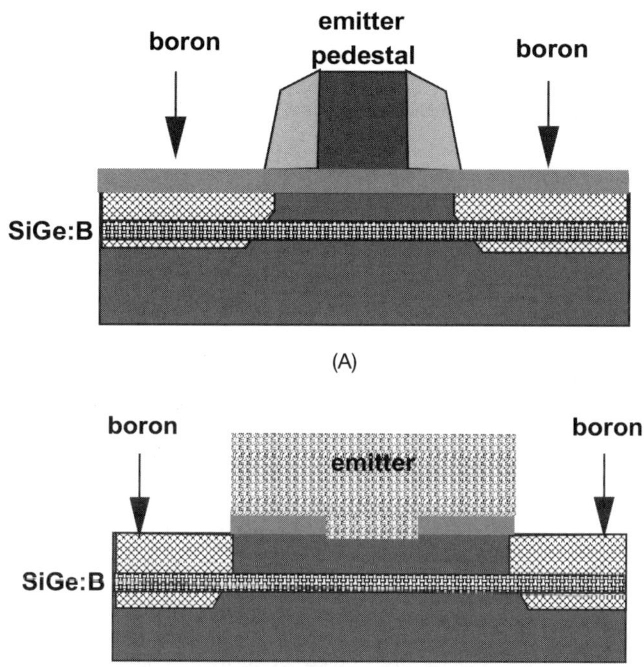

Figure 2.6 (A) Formation of a self-aligned extrinsic base for a 50-GHz/60-GHz f_T/f_{MAX} SiGe HBT. The emitter pedestal defines where the polysilicon emitter will be formed later in the process. The spacers on either side of the emitter pedestal provide control over the trade-off between reduced extrinsic-base resistance and deleterious effects of the point defects introduced by the extrinsic base implant. (B) Alternate extrinsic-base formation for 50-GHz/60-GHz f_T/f_{MAX} SiGe HBT. In this process the extrinsic-base implant is performed after the polysilicon emitter has been deposited and etched. This is simpler and cheaper than the self-aligned process depicted in (A), but is expected to result in significant degradation of the extrinsic-base resistance. How this base-resistance degradation will interact with possible changes in device capacitances and reductions in point-defect-enhanced diffusion in the intrinsic base to modify the critical AC performance metrics of f_T and f_{MAX} was the question addressed by the TCAD simulations. The results predicted by calibrated TCAD simulations and from the resulting hardware are shown in Table 2.1.

Table 2.1 Comparison of TCAD Evaluation of AC-Performance Consequences of NSA Extrinsic-Base Process

Parameter	Experiment	Simulation
ΔC_{BC}	~0	~0
ΔC_{BE}	~0	~0
ΔR_B	+125 Ω	+110 Ω
Δf_{MAX}	–21 GHz	–18 GHz
Δf_T	+7 GHz	+8 GHz

process is simpler and cheaper than the self-aligned process, but necessarily results in a higher base resistance, since process tolerances demand that the distance between the emitter out-diffusion and the extrinsic-base implant be larger. A more subtle issue is the effect of the extrinsic-base implant on the AC performance of the device. Both experimental results and understanding of silicon process physics suggest that point defects introduced by the extrinsic-base implant can induce unwanted additional diffusion in the intrinsic base, thus degrading the AC performance of the transistor [9]. It is expected that the calibrated process and device-simulation capability will accurately reflect those phenomena and provide a method to size the many trade-offs involved in this option.

Figure 2.7 shows the predictions of the calibrated process and device simulation and the subsequent experimental results. The simulation accurately predicted the increase in base resistance and simultaneous reduction in base-collector capacitance arising from the alternate extrinsic-base process. Further, it accurately captured the improvement in cutoff frequency f_T resulting from the extrinsic-base implant of the NSA process that differed in both dose and energy, as well as being spaced farther from the emitter-base junction. Finally, these modifications in device resistance, capacitance, and cutoff frequency were accurately reflected in the predicted maximum oscillation frequency, f_{MAX}.

Another example of the TCAD predictive technology characterization paradigm is taken from the 200-GHz SiGe HBT technology performance path [10]. Based on the previous-generation f_T = 120-GHz technology [11], exploration of the process and device modifications were performed with TCAD to identify promising ultra-high-performance SiGe HBT design points [12]. In particular, significant modifications to the extrinsic-base process module and LTE structure were suggested. Because the new process modules required significant process development effort, test structures based on a simple NSA structure were used to prototype the vertical device structure. Initial hardware showed promising AC performance results for a f_T = 200 GHz SiGe HBT. While process development continued apace, an early design kit was formulated based on calibrated TCAD models. The process to produce this early NSA hardware was used to calibrate the process and device simulation, following Fig. 2.2. The complete target process was then simulated and a scalable compact vertical bipolar intercompany (VBIC) [13] model was extracted from electrical simulations of these devices, following the methodology depicted in Fig. 2.1.

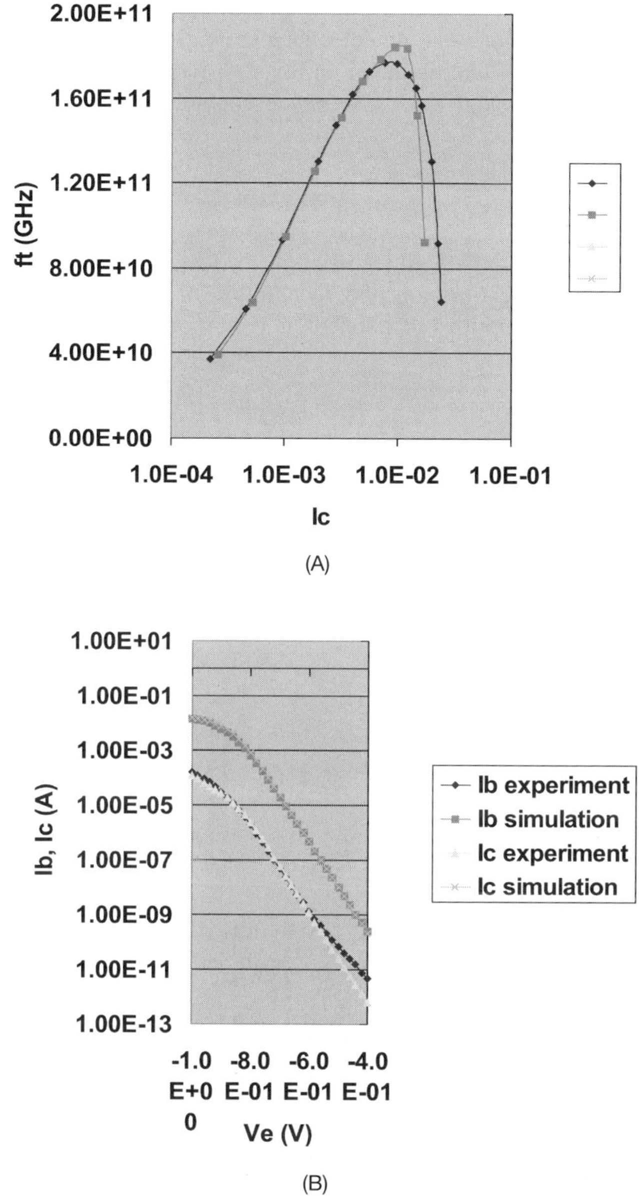

Figure 2.7 NSA TCAD calibration in preparation for TCAD definition and model extraction for f_T = 200-GHz SiGe HBT device development. The figures compare the process/device simulation vs. measured wafer characteristics. AC characteristic compaison is shown in (A), and DC characteristic comparison is shown in (B).

Basic AC and DC electrical results used in the calibration are shown in Fig. 2.7. A comparison with the compact model characteristics arising from the complete process fabricated a year later is shown in Fig. 2.8. Working ring-oscillator circuits, showing record silicon-stage delays, were subsequently designed using these models [14].

2.1.5 Summary

Technology CAD is becoming an increasingly critical part of RF technology development. With careful calibration and recognition of the predictive range of the many models and assumptions that constitute the process and device simulation ap-

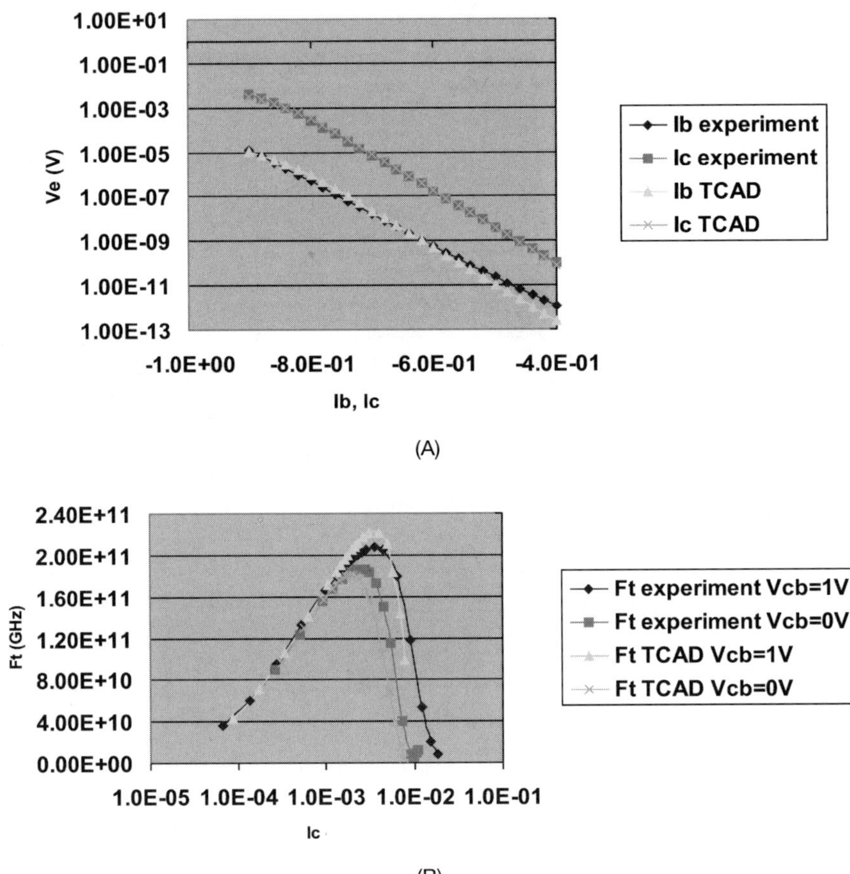

Figure 2.8 Comparison of DC (A) and AC (B) characteristics of $f_T = 200$-GHz SiGe HBT hardware and compact models based on TCAD simulations of target process produced nine months prior to hardware.

proaches relied upon for industrial technology development, TCAD can significantly improve the process learning and technology performance progress achieved with a given experimental wafer budget. When leveraged to its fullest extent via the predictive TCAD paradigm presented in Fig. 2.1, TCAD can assume an important strategic role in RF product development by providing an efficient and inexpensive link between circuit designers and technologists that allows tighter coupling between the technology and its target products, increasing the likelihood of meeting time-to-market goals. By providing early design-kit information prior to the appearance of experimental hardware, more design turns are available to circuit designers, further ensuring the likelihood of timely product development.

2.1.6 References

1. H. K. Gummel, "A Self-Consistent Iterative Scheme for One-Dimensional Steady Sate Transistor Calculations," *IEEE Trans. Electron Devices,* vol. ED-11, pp. 455–465, 1964.
2. C. Rafferty, "Progress in Predicting Transient Diffusion," in *Proceedings of 1997 Intl. Conference on Simulation of Semiconductor Processes and Devices, SISPAD'97,* pp. 1–4, 1997.
3. B. E. Deal and A. S. Grove, "General Relationship for the Thermal Oxidation of Silicon," *J. Appl. Phys.,* vol. 36, pp. 3770–3778, 1965.
4. D. L. Scharfetter and H. K. Gummel, "Large-Signal Analysis of a Silicon Read Diode Oscillator," *IEEE Trans. Electron Devices,* vol. 16, pp. 64–77, 1969.
5. E. M. Buturla and P. E. Cottrell, "Two-Dimensional Finite Element Analysis of Semiconductor Steady-State Transport Equations," paper presented at the International Conference on Computational Methods in Nonlinear Mechanics, Austin, TX, 1974.
6. S. Selberherr, *Analysis and Simulation of Semiconductor Devices,* Springer-Verlag, New York, 1984.
7. S. E. Laux, "Techniques for Small-Signal Analysis of Semiconductor Devices," *IEEE Trans. Electron Devices,* vol. 38, pp. 2028–2037, 1985.
8. S. E. Laux and M. V. Fischetti, "Full Band Monte Carlo Simulations of Small MOSFETS," in *Proceedings of Intl. Conference Solid State Devices and Materials, Tokyo, Japan,* 1999.
9. J. B. Johnson, A. Stricker, A. J. Joseph, and J. A. Slinkmanet, "A Technology CAD Methodology for AC Performance Optimization of SiGe HBTs," *IEDM Tech. Digest,* pp. 234–238, December, 2001.
10. B. Jagannathan, M. Khater, F. Pagette, J.-S. Rieh, D. Angell, H. Chen, J. Florkey, F. Golan, D. R. Greenberg, R. Groves, S. J. Jeng, J. Johnson, E. Mengistu, K. T. Schonenberg, C. M. Schnabel, P. Smith, A. Stricker, D. Ahlgren, G. Freeman, K. Stein, and S. Subbanna, "Self-Aligned SiGe NPN Transistor with 285GHz fmax and 207GHz ft in a Manufacturable Technology," *IEEE Elec. Device Lett.,* vol. 23, no. 6, pp. 268–260, May 2002.
11. A. Joseph, D. Coolbaugh, M. Zierak, R. Wuthrich, P. Geiss, Z. He, X. Liu, B. Orner, J. Johnson, G. Freeman, D. Ahlgren, B. Jagannathan, L. Lanzerotti, V. Ramachandran, J. Malinowski, H. Chen, J. Chu, P. Gray, R. Johnson, J. Dunn, S. Subbanna, K. Schonenberg, D. Harame, R. Groves, K. Watson, D. Jadus, M. Meghelli, and A. Rylyakov, "A 0.18 mm BiCMOS Technology Featuring 120/100 GHz (f_T/f_{MAX}) HBT and ASIC-Com-

patible CMOS Using Copper Interconnect," in *Proceedings of 2001 BCTM,* pp. 143–146, 2001.

12. S. Subbanna, J. Johnson, G. Freeman, R. Volant, R. Groves, D. Herman, B. Meyerson, "Prospects of Silicon-Germanium Technology for Very High-Speed Circuits," *IEEE MTT-S Microwave Symposium Digest,* vol. 1, pp. 31–364, 2000.
13. C. C. McAndrew, J. A. Seitchik, D. F. Bowers, M. Dunn, M. Foisy, I. Getren, M. McSwain, S. Moinan, J. Parker, D. J. Roulston, M. Schroter, P. van Wijnen, L. F. Wagner, "VBIC95, The Vertical Bipolar Inter Company Model," *IEEE J. Solid-State Circuits,* vol. 31, 1996.
14. J. S. Dunn, D. C. Ahlgren, D. D. Coolbaugh, N. B. Feilchenfeld, G. Freeman, D. R. Greenberg, R. A. Groves, F. J. Guarin, Y. Hammad, A. J. Joseph, L. D. Lanzerotti, S. A. St. Onge, B. A. Omer, J.-S. Rieh, K. J. Stein, S. H. Voldman, P.-C. Wang, M. J. Zierak, S. Subbanna, D. L. Harame, D. A. Harame Jr., B. S. Meyerson, "Communications Technologies—SiGe BiCMOS and RF CMOS," *IBM J. Res. Develop.*

2.2 CHARACTERIZATION

In this chapter, we present the importance of characterization as a step in the development of compact-device models. While the model provides the circuit designer with the ability to predict the behavior of the device, and thus the circuit, it is the characterization effort that allows for the device model to be developed. Compact-model development and parameter extraction involve both DC and AC data. AC data are usually in the form of two-port S-parameter measurements. Test sites are designed and wafers fabricated to provide a full complement of device test structures, both with DC and AC padsets, to allow for on-wafer measurements in support of model development. Test structures are designed to target the measurement of specific process parameters, specific model parameters, and to characterize the complete device.

Additionally, it is also essential to make other types of measurements to verify model functionality and circuit simulation accuracy. These additional measurements should include both low- and high-frequency noise measurements, large signal measurements, and two-tone intermodulation distortion measurements.

2.2.1 In-Line Wafer Measurements

A complete set of DC measurements is taken in line during wafer fabrication, and this provides an accurate method for determining the process parameters for each individual wafer. Data are collected for each wafer run through the fabrication line, so this provides a large statistical database for many process and modeling parameters. A detailed description of the problems associated with collecting and storing in-line data and the software programs needed to analyze the data, was previously presented in [1]. Electrical parameter data analysis benefits enormously from object-oriented techniques that make it easy to create comparison charts. An example of in-line data collection and analysis is shown in Fig. 2.9. These data are used later in the selection

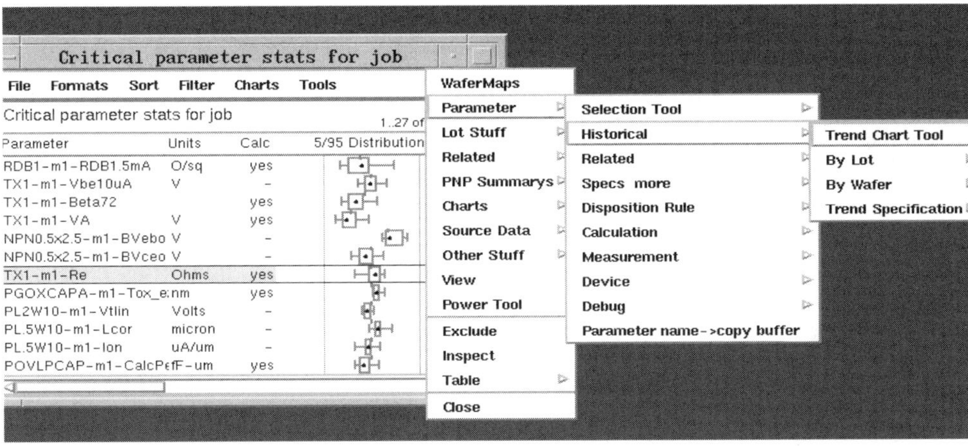

Figure 2.9 Example of in-line data collection and analysis. These data provide a statistical database for process and model parameters as well as allowing an accurate selection of a Plan of Record (POR) wafer. (The data were taken from [1].)

of modeling wafers to be slated for characterization. Since it is essential that process variations are fully determined, wafer characterization from the in-line data is an important first step in the data acquisition and device characterization process.

2.2.2 DC, *CV*, and Matching Measurements

DC and capacitance measurements are taken on a full set of on-wafer test structures for each device type and a statistical database is built. Numerous wafers from several lots are tested to generate a large sample size to allow for the employment of statistical modeling techniques. DC measurements are performed on devices to characterize and model device performance, whereas *CV* measurements, in addition to allowing capacitors themselves to be characterized, are also performed to determine specific device model circuit parameters. For example, capacitance measurements on specially designed structures allow for the gate oxide and overlap capacitances to be determined and included in the circuit model for FETs. Matching measurements are important to fully determine the variation of device parametrics between adjacent devices.

Measurements are taken on each type of device across operating conditions and at numerous bias conditions to fully represent all possible operating modes. The full set of test structures includes a complete range of device sizes. This enables geometry dependencies to be incorporated into the models. Data are also collected at a full range of temperatures (–50 to +145°C) to enable device temperature dependencies to be fully determined.

All measurements should be taken with a flexible instrument control and data-acquisition software program that allows the data to be easily accessible among numerous characterization labs and engineers. At IBM, a program has been developed in object-oriented language, referred to as device measurement and characterization system (DMACS), which has been described in detail elsewhere [2]. Some key attributes of the software program are an ability to store large quantities of different kinds of data, store automated testing programs, drive a large variety of instruments under program control, display the raw data in numerous types of plots and charts as well as manipulating the data with various types of calculations, and finally, an ability to output data into a wide variety of formats so that it can be ported to other modeling software tools. The program serves two functions: an engineering development tool, and a manufacturing characterization tool.

2.2.3 AC S-Parameter Measurements

Small-signal analysis and model development of devices require extensive AC measurements. Small-signal equivalent-circuit models were developed to enable devices to be represented by lumped elements rather than complicated, nonlinear equations. High-frequency two-port parameters such as S-parameters are measured to enable the development of these small-signal-equivalent models and determine their associated model parameters. As with DC measurements, S-parameter measurements are taken on both active and passive devices, across both a full range of geometries and temperatures, and, for active devices, at different operating conditions and across numerous biases. For passives, S-parameters are also converted to Y-parameters and used to determine the input and output characteristics as a function of frequency. In addition, S-parameter measurements are used to calculate two important figures of merit for transistors that are the cutoff frequency of the AC-current gain, f_T, and the cutoff frequency of the maximum power gain, also called the maximum oscillation frequency, f_{MAX}.

It is currently common to measure device AC characteristics up to as high as 110 GHz. Special steps must be taken at these frequencies to ensure data integrity and quality. The reference plane must be firmly established and its position known. In addition, at these frequencies, padset parasitics become an issue. A probe-tip calibration is done first to move the measurement reference plane to the probe tips. Then a two-step de-embedding procedure is used to subtract out both the series and parallel parasitics of the padsets, thus obtaining the high-frequency characteristics of the device itself. This procedure is described in detail elsewhere [3]. The two-port S-parameters are measured using a vector network analyzer and ground–signal–ground (GSG) probes. For active devices, special care must be taken regarding the power levels to prevent gain compression [4] and to ensure the device remains in the small-signal regime.

To calculate f_T, the S-parameters are converted to H-parameters and, in graphical terms, the AC current gain, $|H_{21}|$ (in dB), is plotted on a linear scale versus frequency on a log scale. The f_T of the transistor is the point at which $|H_{21}|$ crosses the x-axis. To facilitate the determination of f_T IBM's instrument control and data acquisition software (DMACS) calculates this parameter in the following manner. The $|H_{21}|$ curve is assumed to have perfect single-pole, 20-dB/decade, rolloff characteristics. The f_T is calculated using the base transit time τ_T, as defined in Equations (2.1) and (2.2).

$$\tau_T = \left| \frac{\sin \angle H_{21}}{|H_{21}|\omega} \right| \tag{2.1}$$

Then f_T is given by

$$f_T = \frac{1}{2\pi\tau_T} \tag{2.2}$$

It can be shown from circuit theory that these equations are valid at any frequency, though care must be taken to do the calculation at a high enough frequency to avoid inaccuracies due to phase limitations in the instrumentation. The maximum oscillation frequency, f_{MAX}, is determined as the frequency at which the maximum-available power gain (MAG), also a quantity exhibiting single-pole transfer function characteristics, is unity (see Equation (2.3):

$$\mathrm{MAG} = \left| \frac{S_{21}}{S_{12}} \right| (k - \sqrt{k^2 - 1}) \tag{2.3}$$

where k is Kurokawa's stability factor and is given by Equation (2.4):

$$k = \frac{1 - |S_{11}|^2 - |S_{22}|^2 + |S_{11}S_{22} - S_{12}S_{21}|^2}{2|S_{12}||S_{21}|} [4] \tag{2.4}$$

Using the calculation of MAG across the frequency range, DMACS then uses a linear regression routine to determine the x-intercept, and thus, f_{MAX}. Again, care must be taken to ensure that an appropriate frequency range is used when performing the linear regression.

2.2.4 Noise Characterization

Both low-frequency flicker noise ($1/f$ noise) and high-frequency, broadband noise parameters, including noise figure, associated gain, optimum reflection coefficient, and noise resistance, must be measured to facilitate the modeling efforts and the design of integrated telecommunication circuits in BiCMOS technologies. Flicker noise measurements are made to determine several subcircuit parameters in the modeling of the SiGe BiCMOS technologies. In addition to model verification, the broadband noise performance characterization gives the circuit designer an idea of the signal-to-noise level degradation that will result by adding the device to the circuit, an important consideration in telecommunications systems, which typically process very low-level signals.

Noise figure describes the degradation of the SNR ratio between the input and output of the device. The circuit designer is faced with wanting both minimum noise figure and maximum gain from the device when placed in the circuit. Since both cannot be achieved simultaneously for a given source impedance, the designer must determine what trade-offs must be made. Gain characterization is provided by S-parameter measurements, but noise parameters require separate measurements. For a linear two-port device, the dependence of noise figure on source reflection coefficient is described by Equation (2.5):

$$F(\Gamma_S) = F_{\text{MIN}} + 4r_n \frac{|\Gamma_S - \Gamma_{\text{opt}}|^2}{(1 - |\Gamma_S|^2)|1 + \Gamma_{\text{opt}}|^2} \tag{2.5}$$

where Γ_s is the source reflection coefficient, F_{MIN} is the minimum noise figure, r_n is the normalized noise resistance (the sensitivity of noise figure to changes in the source reflection coefficient), and Γ_{opt} is the optimum source reflection coefficient that gives minimum noise figure [5].

A graphical representation of the noise parameter equation is shown in Fig. 2.10, indicating that a value of source impedance can be determined for which a minimum noise figure is achieved. Thus, high-frequency broadband noise characterization, together with the active device's gain characterization, can provide the circuit designer with the information needed to determine what kind of trade-off must be made when impedance matching to the device for minimum noise figure versus maximum gain. This trade-off is illustrated in Fig. 2.11A and 2.11B. Figure 2.11A shows the constant gain circles for a FET, and Fig. 2.11B shows the corresponding constant noise circles for the same device. The center of the smallest circle indicates the optimum source impedance for each parameter, and the spacing between the cir-

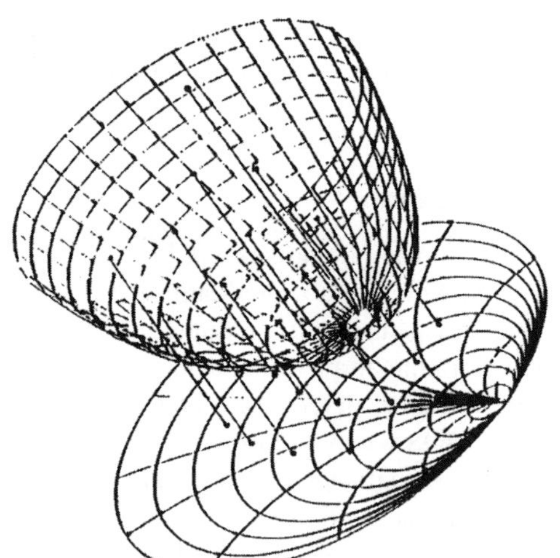

Figure 2.10 The graphical representation of the noise equation parabola showing that a source impedance exists at whichever minimum noise figure is achieved [5].

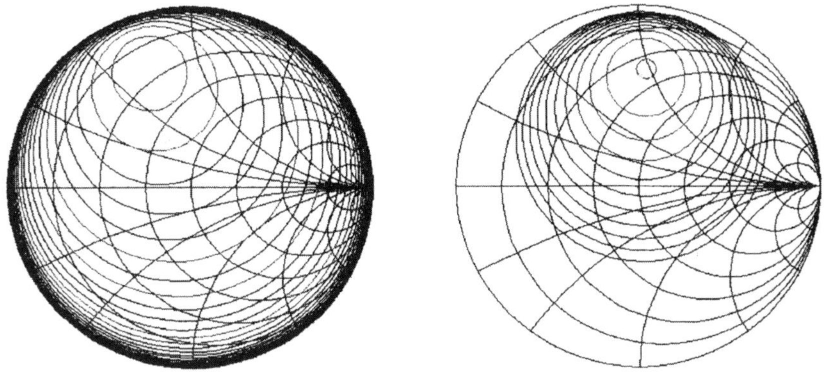

(A) (B)

Figure 2.11 Constant noise and constant gain circles vs. source impedance for a MOSFET. Illustrating the trade-0ffs that must be made between minimum noise and maximum gain when designing the circuit. (A) Constant gain circules vs. source impedance for a MOSFET. Contour start: 7 dB; contour step: 1 dB. (B) Constant noise circles vs. source impedance for a MOSFET. Contour start: 5 dB; contour step: 0.5 dB.

cles is an indication of how much the designer must give away when moving away from that optimum point.

In addition to its contribution to the model, flicker ($1/f$) noise characterization is important because certain types of circuits in telecommunications systems are particularly sensitive to this low-frequency noise (see Section 1.4). Low $1/f$ noise is critical for good system performance, particularly when it is injected into one of the inputs of a mixer, in which the $1/f$ noise can distort the spectrum [6]. Bipolar transistors and FETs exhibit very different levels of flicker noise. Therefore, it is important to characterize these devices when considering their use in circuits sensitive to this type of noise.

2.2.5 Large-Signal Measurements

Determining how active devices behave at different power levels is also an important consideration when designing telecommunications systems. Many RF amplifiers are designed to operate in the weakly nonlinear region, where power added efficiency (PAE) peaks. The PAE is defined as the ratio of the additional power provided by the amplifier to the DC power [7], as described in Equation (2.6):

$$\mathrm{PAE} = \frac{P_{RF\mathrm{out}} - P_{RF\mathrm{in}}}{P_{\mathrm{DC}}} \tag{2.6}$$

Large-signal measurements provide output power, gain, and efficiency information at a given input power level. Figure 2.12 shows the results of large signal mea-

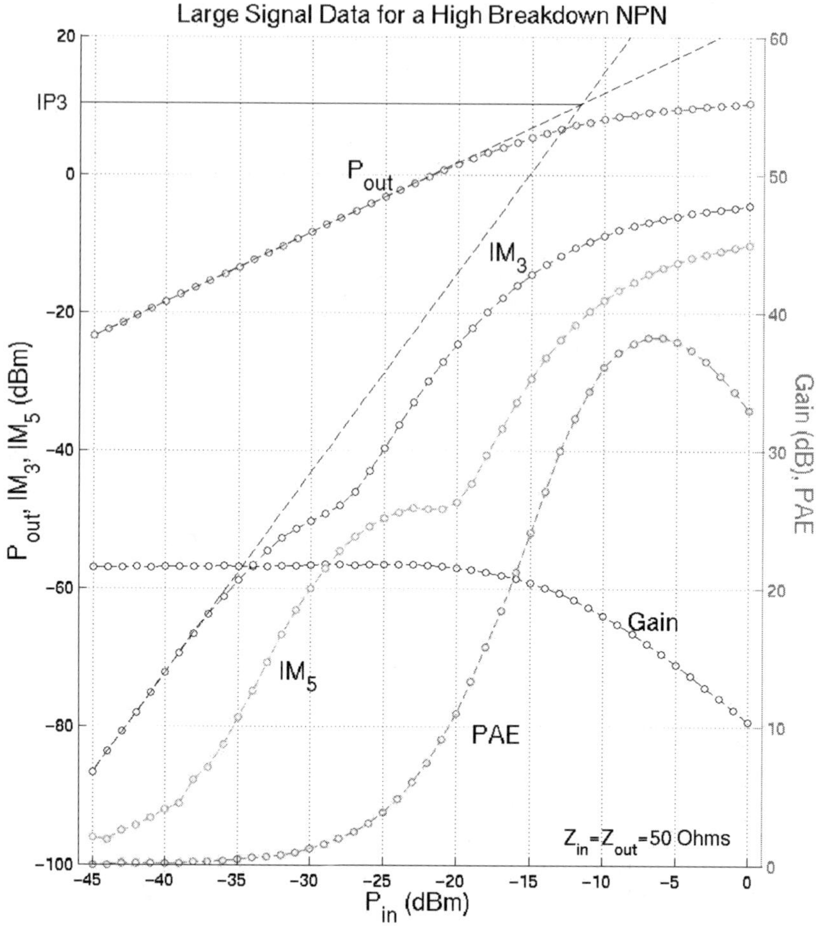

Figure 2.12 Measured large-signal and intermodulation distortion data for an IBM SiGe HBT. The source and load terminations were set to 50 Ω, and the determination of IP3 is illustrated

surements made on a SiGe HBT using an ATN Microwave LP2™ Load-Pull system. For these measurements, both the input impedance and output impedance were set to 50 Ω. The figure illustrates the power levels at which the device enters compression and the PAE in that region. The 1-dB gain compression point, usually given in terms of output power, is an important quantity when considering the dynamic range of the transistor.

In addition to power-level considerations, impedance matching throughout the system is an essential aspect of RF circuit design. Thus, it is also important to explore how the input and output impedance presented to each device in the system affects that device's performance in the circuit. Furthermore, a designer may want to determine the necessary impedance-matching conditions to achieve a specific desired performance from the device. The ATN Microwave LP Load-Pull system is used to make these measurements. Once either the desired input or output impedance is determined, contours can be generated for the other termination to illustrate trade-offs that must be made between, for example, maximum power-added efficiency, and maximum output power. Figure 2.13A shows the output power contours versus load termination for a SiGe transistor whose input impedance was conjugately matched for maximum gain. Figure 2.13B shows the PAE contours for the same device. Note that the optimum load termination for maximum output power and that for maximum PAE are near each other on the Smith chart, suggesting that a designer will be able to use this device in the circuit without a significant trade-off between output power and PAE.

2.2.6 Distortion

Many components of telecommunication systems will receive numerous signals closely spaced in frequency at their inputs. The nonlinearities inherent in all active devices lead to certain undesirable effects, such as intermodulation and harmonic distortion that lead to the transfer of power to other frequencies near the frequency of interest. For a device with two signals at its input, one at a frequency f_1 and the

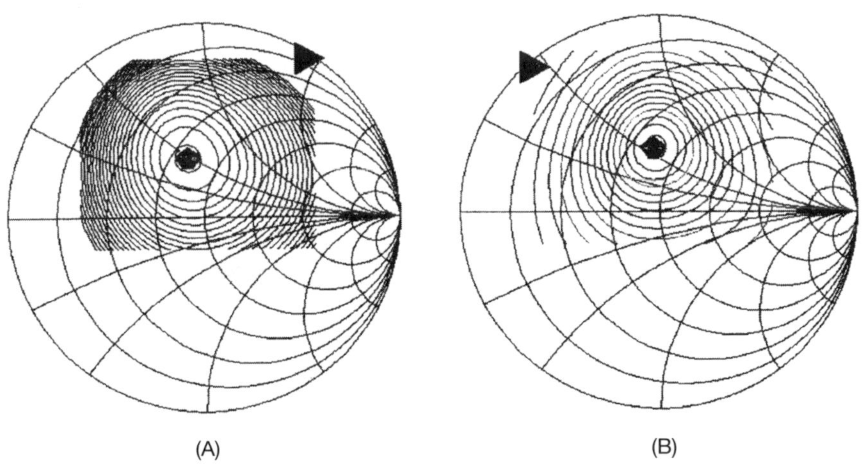

Figure 2.13 Output power and PAE contours vs. load termination for a SiGe transistor. Note that the optimum load impedance for both are very close together, and there is a minmum trade-off when matching for maxium output power and maximum power-added efficiency. (A) Measured large-signal and intermodulation distortion data for an IBM SiGe HBT. The source and load terminations were set to 50 Ω, and the determination of IP3 is illustrated. (B) Output power (dBm) vs. load termination for a SiGe Transistor. Contour start: 14.5 dBm; contour step: 0.25 dBm; frequency = 900 MHz; source impedance set to complex conjugate match.

other at a frequency f_2, it is traditionally the third-order (at frequencies $2f_1–f_2$ and $2f_2–f_1$) and fifth-order ($3f_1–2f_2$ and $3f_2–2f_1$) intermodulation products that are of most concern because they are near the two frequencies of interest (f_1 and f_2), and therefore will be the most difficult to filter out of the system. Therefore, RF and telecommunications applications, such as power amplifiers, require devices that exhibit highly linear operating characteristics. Two-tone measurements must be performed on the device offerings in the SiGe BiCMOS and RF CMOS technologies to fully analyze the linearity of these devices. The ATN LP2 Load-Pull system at IBM is used to make these measurements. The third-order and fifth-order intermodulation products are commonly measured and the third-order intercept point (IP3, or sometimes P_{IP}), an important figure of merit for describing linearity, is obtained. The intermodulation products are shown in Fig. 2.14 for a SiGe HBT; the extrapolation to IP3 is also illustrated. The third-order intercept point is then used to determine the spurious free dynamic range (DR_f), defined as the difference between the output power at the fundamental and the output power at the third-order intermodulation product when the output power at the third-order intermodulation frequency is equal to the minimum detectable output signal, $P_{o,\mathrm{mds}}$ [8]. The spurious DR_f is given by Equation (2.7):

$$DR_f = \tfrac{2}{3}(\mathrm{IP3} - P_{o,\mathrm{mds}}) \tag{2.7}$$

and illustrated in Fig. 2.14. Thus, the third-order intercept point provides the power amp designer with a metric for determining a distortion-free operating range for the device.

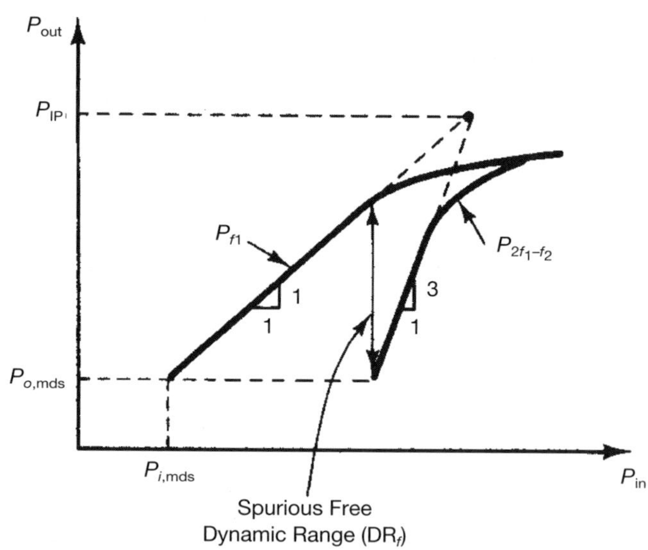

Figure 2.14 Spurious DR_f illustrated for a transistor exhibiting third-order intermodulation products [8].

2.2.7 Test Sites

The key to generating accurate, scalable, full-featured models, as described in the next section, lies in the availability of test-site structures from which to make measurements. This means that the modeler must "cover their bases" early in the technology development process by ensuring that any conceivable structure that they might need has been incorporated into the test site. For example, test-site characterization macros needed to construct an NFET model would include:

1. A length and width array macro for DC extraction/optimization of the Berkeley short-channel insulated-gate FET model (BSIM) parameter set

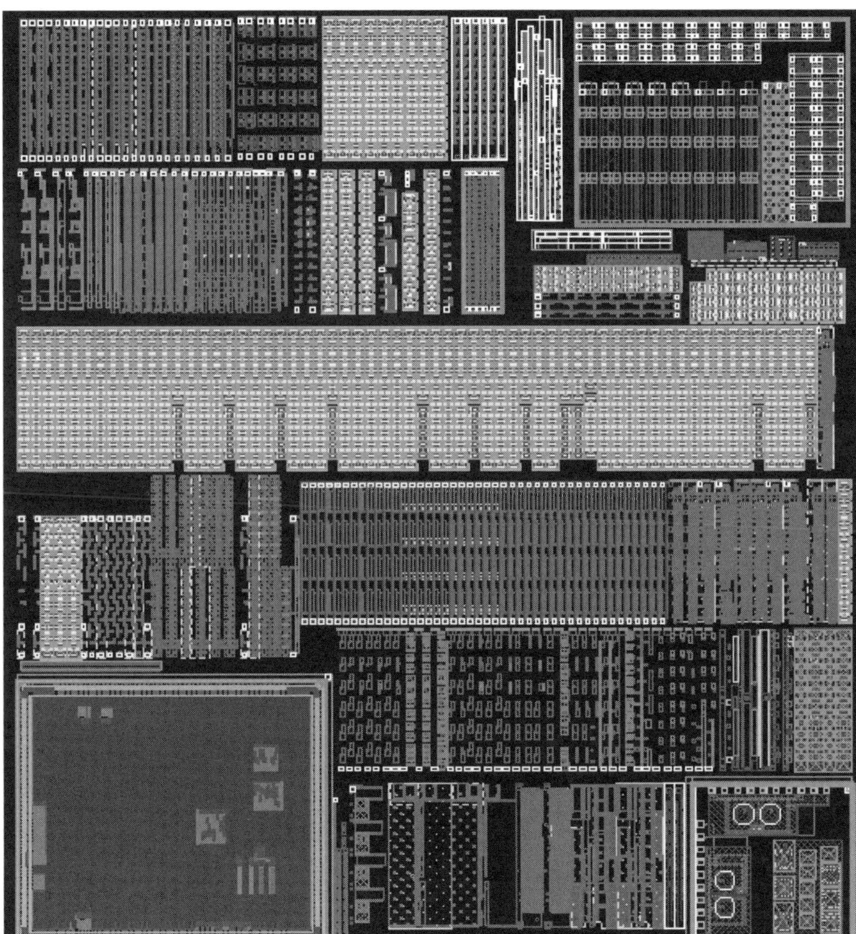

Figure 2.15 0.25-μm test site. This test site includes all process technology, in-line test macros, and modeling macros. The technology has not only the standard 0.25-μm digital CMOS elements but also the RF FET layouts and passives.

2. A capacitor array to extract gate oxide, overlap, and source/drain (S/D) diode capacitance and leakage
3. A set length and width array macros to measure V_{th} and mobility mismatch
4. A set of RF S-parameter structures of varying length, width, and number of finger configurations
5. Open/short/thru "de-embedding" structures to go with the S-parameter macros just mentioned
6. DC and S-parameter gate-resistance extraction macro
7. Macros to gauge N-well proximity and other process-specific effects

The vast number of structures required to cover the characterization needs of all the devices in a given technology can be enormous. A typical SiGe modeling test site can be as large as 20 mm × 20 mm. Figure 2.15 shows a top-level view of the "Granite" test site, which is the primary modeling test site for the 0.25-μm FETs.

2.2.8 Summary

We have described many of the key elements in characterization in the development of compact-device models. Discussions have included in-line wafer measurements; DC and AC measurements, noise characterization, large-signal measurements, and distortion. We have also briefly discussed the importance of test sites, and IBM's DMACS software system.

2.2.9 References

1. G. Freeman, J. Kierstead, and W. Schweiger, "Electrical Parameter Data Analysis and Object-Oriented Techniques in Semiconductor Process Development," in *Proceedings of 1996 BCTM,* pp. 81–88, 1996.
2. M. Peterson, "Device Measurement and Characterization System: Creating Unity in a Diverse Characterization World," *MicroNews,* vol. 7, no. 2, pp. 29–34, May 2001.
3. Paul J. Van Wijnen, *On the Characterization and Optimization of High-Speed Silicon Bipolar Transistors,* Cascade Microtech, 1995.
4. M. S. Gupta, "Power Gain in Feedback Amplifiers, a Classic Revisited," *IEEE Trans. Microwave Theory and Techniques,* vol. 40, no. 5, pp. 864–879, May 1992.
5. NP5 System Operating Manual, ATN Microwave, Inc., 1990.
6. D. Greenberg, S. Sweeney, C. LaMothe, K. Jenkins, D. Friedman, B. Martin, G. Freeman, D. Ahlgren, S. Subbanna, A. Joseph "Noise Performance and Considerations for Integrated RF/Analog/Mixed-Signal Design in a High-Performance SiGe BiCMOS Technology," IEDM paper, 2001, pp. 22.1.1–22.1.4, December 2001.
7. S. A. Maas, *Nonlinear Microwave Circuits,* IEEE Press, New York, p. 370, 1997.
8. G. Gonzalez, *Microwave Transistor Amplifiers, Analysis and Design,* Prentice-Hall, Englewood Cliffs, New Jersey, pp.174–180, 1984.

2.3 COMPACT-MODEL DEVELOPMENT: ACTIVE DEVICES

A compact model is a mathematical model that predicts the electrical characteristics of a device as a function of the conditions and constraints applied to it. In the case of the MOSFET device, a compact model predicts the output current (Ids) and its derivatives (gm, gds, gmb) as a function of temperature, voltage bias, channel length, and device width. A compact model may be composed of a single element, such as an ideal resistor, or a complex network of interdependent sources, resistors, capacitors, and diodes used to model a BJT. To simulate a model or circuit, containing more than one element, a simulation program with IC emphasis (SPICE) solver can be used, such as AVANT! HSPICE® or Cadance Design Systems (CDS) Spectre®. Implementing the compact model may require the extraction of numerous parameters, as is the case for a junction diode, or it may require the elaborate extraction and optimization needed to extract the dozens of parameters that go into a BSIM-based MOSFET model.

The primary goal of the IBM compact model development team is to provide physics-based scalable models that are fully integrated into the overall design-kit environment. The models must be capable of predicting the full complement of device characteristics and behavior as a function of bias and temperature, and must represent the statistical process window of the technology being modeled. Another important consideration is to make use of industry-standard or common elements in building the model to allow for more efficient translation of the models to different simulators while maintaining consistency in simulation results.

The emphasis on models that are physics based dictates that the development effort employs direct parameter extraction (as discussed in Section 2.2), rather than empirical or numerical optimization, whenever possible. It is also important to make use of process information obtained from the technology development team, such as vertical profile dimensions or doping concentrations. A more physical model is better able to predict results for conditions beyond those used during the initial model-parameter extraction, such as different device geometries. The demands of the RF/analog design environment have also led to the use of more complex subcircuit topologies for both active and passive devices to better predict high-frequency behavior. However, it is important to minimize the number of elements required in order to maximize simulation efficiency.

All of the key building blocks for the development of the scalable, statistical, and physics-based models are described throughout this section. These are highlighted in Fig. 2.16. Note that these inputs are required regardless of the type of device that is being modeled.

2.3.1 Statistical Modeling

The compact device models developed by IBM for the SiGe BiCMOS technologies have the ability to support standard Monte Carlo (statistical), process-corner, and wafer-specific simulations. The basic structure of the model library makes use of a

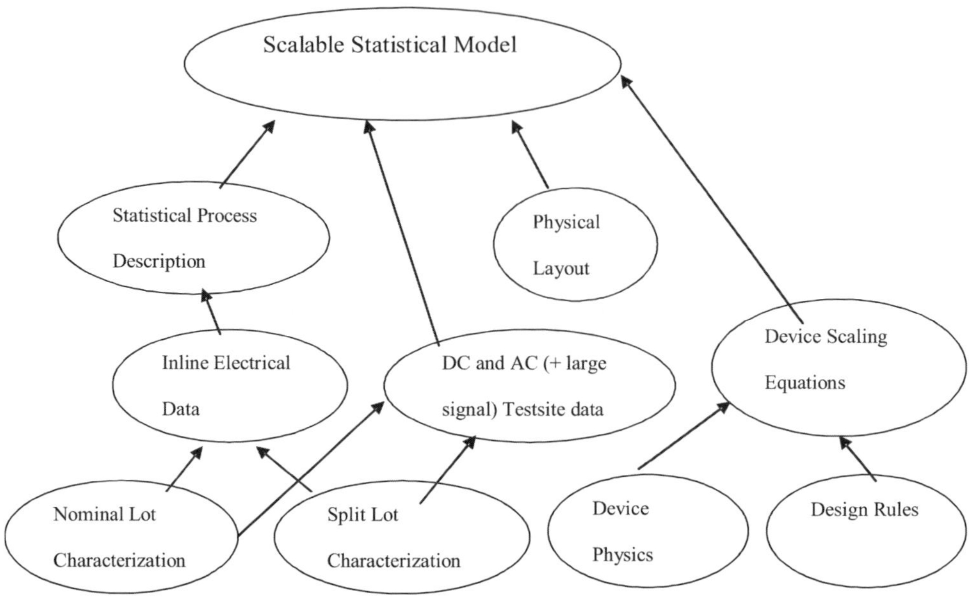

Figure 2.16 IBM Compact-Model Methodology. A view of the key sources of information used in the development of the physics-based compact models.

"skew file," which defines all of the statistical distributions, process-corner parameters, and other model parameters that are shared across multiple devices. The definition of these distributions is dependent on the cooperation between the technology development and compact modeling teams to determine the most dominant process parameter variations and the proper correlation of process and device model parameters. These correlations account for effects across multiple devices that may share a common process step as well as multiple parameters within a single device that exhibit a strong physical correlation.

The primary input for specifying the nominal and tolerance specifications of the process parameters comes from the in-line wafer (kerf) parametric data. These data provide the necessary statistical sample and establishes a direct connection between the skew file parameters and the measurements that are used to place product wafers in the manufacturing line. Additional wafer characterization is used to supplement the in-line data and provide the basis for correlation of the process parameters and key device metrics, such as NPN f_T and f_{MAX}.

The conventional method for process corner simulation has involved using a pair of model parameter sets to represent the process extremes, often referred to as *fast/slow* or *high/low* corners. The assumption in this method requires that, for a fast process corner, the device parameters are skewed so as to maximize active device currents and minimize other capacitances and resistances. While this may be valid for analyzing the process variation of the characteristic time constant of an analog circuit, the drawback of this method for corner-file generation is that this definition of the model parameter combinations will not always yield an extreme in the circuit performance for all types of analog applications.

To enable process-corner simulation, the IBM skew-file approach (patent pending) includes multiple corner parameters corresponding to each of the device types, such as resistors, capacitors, and BJTs. This structure supports simulation of different combinations beyond a single fast and slow corner pair and enhances a designer's ability to assess the sensitivity of the circuit performance. This sensitivity analysis can be done by repeating a simulation with each of the individual corner parameters set to +1 and –1 (corresponding to a ± 1-sigma variation) and comparing the results against the nominal simulation. With only one corner parameter set to be nonzero at a time, the total number of simulations will be twice the number of corner parameters. These single-parameter simulations are done to determine the appropriate sign, positive or negative, for each corner parameter necessary to maximize (or minimize) the overall circuit performance. In this methodology, equal weights are given to the variations of all of the device types, so all of the corner parameters are set to the same magnitude. This magnitude is determined by first finding the 3-sigma variation limits of the circuit performance using a statistical simulation and then setting the magnitude of the corner parameters to match these limits. By using an initial Monte Carlo simulation to calibrate the results from the corner analyses in this way, the designer has an efficient means to account for the effects of the process variation and include the necessary design margin. As a result, this approach provides the benefits of both conventional corner and Monte Carlo simulations and requires only a few additional simulation iterations.

In addition to the process statistics in the skew file, the individual model files also include distributions to represent device mismatch effects. As depicted in Fig. 2.17, the skew-file statistics represent the global process variation across all wafers, while the mismatch represents the local variation observed on a typical wafer. Specific test-site structures are used to measure the mismatch and are designed using good layout practices such as same orientation of near-adjacent de-

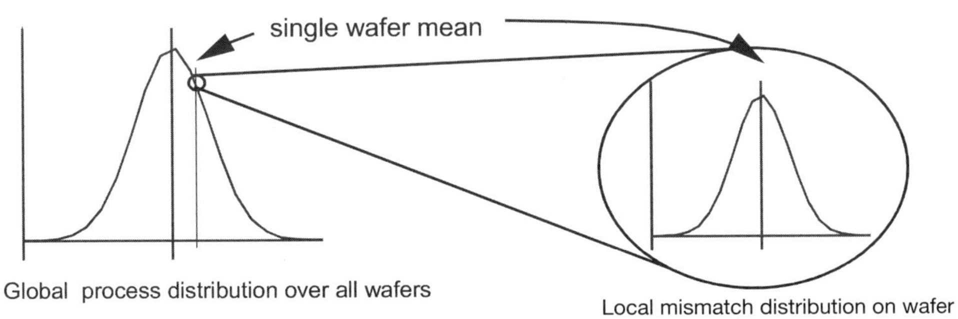

Figure 2.17 Process variation and local mismatch distributions.

vices with symmetric wiring. As with all other aspects of the models, every attempt is made to define these mismatch effects and account for geometric and bias dependence in a manner that is consistent with the physical nature of the devices.

One key aspect of the extraction methodology necessary to maintain the statistical integrity of the final models is the assessment of the hardware used to extract the model parameters. It is important for the modeler to be able to establish any offset that may exist between the defined nominal process values and the measurements of the test-site wafers. Following the device characterization and completion of the model extraction, skew-file parameters are then recentered to represent the nominal process and device specifications. This concept of "recentering" also enables the models to support simulation analyses using skew-file parameter adjustments that are based on a set of process parameters and single bias-point measurements that are taken from the in-line wafer parametric data.

The whole development process is illustrated in Fig. 2.18. Note that model-parameter recentering and the inclusion of the process distributions to enable full statistical simulation follows directly after parameter extraction, as shown in the lower left corner of the figure. Although this diagram refers to the use of the BSIM model in support of MOSFET devices, the overall development flow is applicable to all SiGe BiCMOS technology devices.

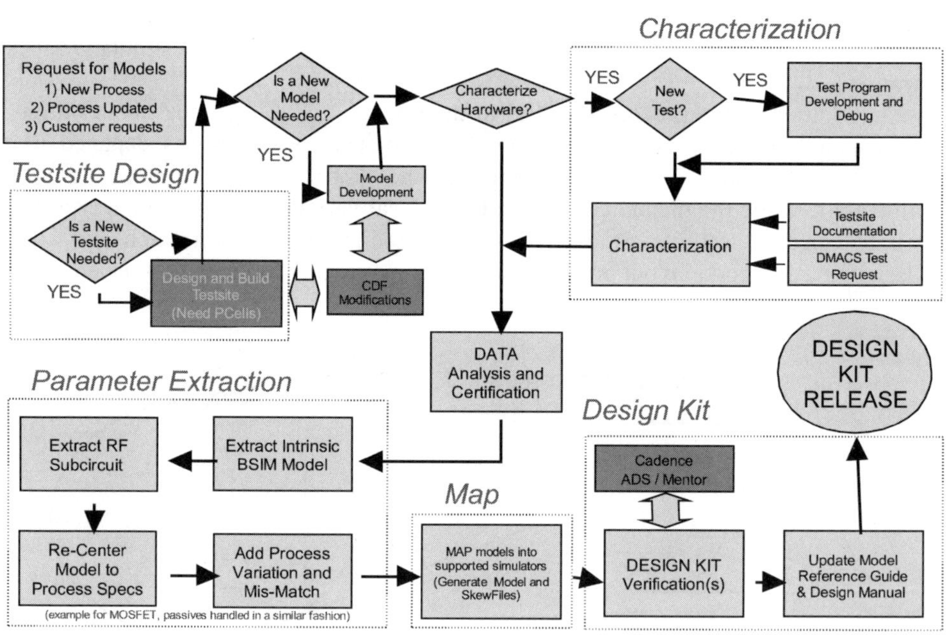

Figure 2.18 Compact-model development. This figure illustrates the basic flow of the steps in the compact-model development process.

2.3.2 HBT

Until the mid-1990s, the semiconductor industry relied almost exclusively on the spice Gummel poon (SGP) model for BJT circuit design. The SGP model included effects important in analog design not found in the earlier Ebers–Moll-type models such as low-current nonidealities and high injection effects, and replaced the underlying physical model with equations based on the more complete integral charge-control relation (ICCR) [1]. But recently, with the revival of BJT and HBT technology for high-speed communications and RF applications, the SGP model was found to be increasingly inadequate and needed to be revised to include more accurate modeling of the physical effects found in high-speed devices operating at high current densities. These effects include better early effect modeling (output conductance), quasi-saturation, avalanche multiplication, thermal self-heating, and accurate transit-time modeling.

This needed revision of the SGP for modern bipolar transistors was addressed by a group of bipolar modelers from across the world who started a committee to establish a public-domain improved bipolar compact model. This activity was very successful and resulted in the development of the VBIC model formally presented in 1995 [2]. Eventually, the Compact Model Council [3] selected the VBIC model as a standard model and helped establish its use in industry. The VBIC model was physically based on the same ICCR that underlies the SGP model, but also included several additional model elements built around the core model. Additional effects modeled include a parasitic PNP bipolar transistor, self-heating, bias dependent early voltages, temperature scaling, a Kull-based model for quasi-saturation, and additional parasitic capacitances found in aggressively scaled modern devices. Also, in contrast to the SGP model that used separate equations to model the transistor in each operating regime, the VBIC model was constructed with continuous smooth functions over all bias ranges for enhanced numerical stability. But in an effort to keep a partial backward compatibility to SGP model, the extra physical modeling structure increased the internal model node count from 3 to 7, and approximately doubled the required number of parameters to 70. The primary recognized inadequacy of the VBIC model revolves around the poor implementation of the Kull model for device operation in strong quasi saturation [4].

IBM's SiGe technology design kits currently integrate both SGP and VBIC models for the SiGe HBT, but the rapidly growing suite of SiGe HBT technologies with an extremely wide range of device performance targets, has caused additional questions on the validity of the physical assumptions used to derive the standard VBIC model. For example, in IBM SiGe technologies, the model must correctly predict the strong quasi saturation and avalanche breakdown of the SiGe 5PA high-voltage (6.4-V) HBT, as well as model non-quasi-static transport and AC current crowding of the recently announced 350-GHz nine-generation SiGe HBT (as described in the Introduction). This is a difficult task for even the most complex models. The VBIC model has performed well, but has known deficiencies in several areas including poor modeling of strong quasi saturation, output conductance, and avalanche breakdown. Additionally, Si-based compact models do not contain the basis for a physically accurate model of the SiGe HBT base charge. Inclusion of the Ge grading layer, modifies the base bandgap, and therefore the intrinsic base (η_{ib}) carrier concentration, altering the critical charge storage components in the charge control relations.

For these reasons, in 2002 two additional HBT models—high current transistor model (HiCUM) and most exquisite transistor model (MEXTRAM)—were under evaluation by IBM and the Compact Modeling Council as potential successors to the current VBIC standard. The HiCUM developed at Ruhr-University in Bochum, Germany, and first implemented in 1981, was developed initially for design of high-speed emitter-coupled logic (ECL) circuits that operate at high current densities [5]. Based on the ICCR, the model was extended to include SiGe HBT structures with the general ICCR (GICCR) that now provides the physical basis for the model [6]. A most important impact of the GICCR is the implementation of weighting factors that account for the change in mobility and intrinsic carrier concentrations affecting the charge storage in the neutral regions affected by the high Ge content. Other important effects present in the HiCUM model include accurate modeling of the quasi-saturation region and extensive physical description of bias-dependent transit times and non-quasi-static behavior.

The MEXTRAM model was developed and implemented at Philips [7], and is also based on the ICCR of the SGP model. Differentiating the MEXTRAM model from VBIC is the modeling of the collector epilayer. MEXTRAM extends the modified Kull model by adding the effects of velocity saturation in the collector at high current densities, correctly predicting quasi-saturation and the onset of Kirk effect. This collector model is also implemented in a smoother mathematical description that is beneficial to the calculation of higher-order derivatives, important for harmonic distortion analysis. MEXTRAM, like HiCUM, has also implemented additional parameters to take into account bandgap grading in SiGe devices, as well as an extra parameter to model the changes in I_B due to neutral-base recombination.

Table 2.2 gives a brief summary of the physical effects included in each of the BJT models and a comparison of the number internal nodes and model parameters required. The existing IBM SiGe design kits have implemented the VBIC model as the primary element within the NPN subcircuit for technologies that typically include both high-f_T and high-breakdown types of devices. The weak avalanche effect in the VBIC model is based on the assumption that the peak E-field occurs at the base–collector interface, which is valid for the highly doped collector of the high-f_T device. For the lower doped collector of the high-breakdown NPN, the peak E-field occurs at the collector–substrate interface. To account for this difference and to overcome this limitation in the VBIC model, the IBM model topology was modified to move the physical location of the weak avalanche current generator inherent to the VBIC model for high-breakdown devices. The standard VBIC avalanche current is still used for the high-f_T devices. While there are distinct differences in the performance of these devices, a single model file has been used with an input parameter that is passed in from the circuit netlist to specify which device type is being modeled. It is another way that the overall modeling methodology tries to reflect the physical realities of the devices, as it allows for

Table 2.2 Comparison of Modeled Physical Effects and Requirements for Current HBT Compact Models

	SGP	VBIC	HICUM	MEXTRAM
HBT/SiGe modeling	—	✓	✓✓	✓✓
Quasi-saturation	—	✓	✓✓	✓✓
f_T modeling	—	✓	✓✓✓	✓✓
Self-heating	—	✓	✓	✓
Substrate modeling	—	✓	✓	—
Base-emitter breakdown	—	✓	✓	—
Parasitic pnp	—	✓	—	-✓
Number of internal nodes	3	7	4	5
Number of parameters	35	80+	90+	67

Source: Data are taken from Berkner [8].

commonality in many of the key calculations and model parameters that are derived from the basic process flow.

Typical model-to-hardware correlation plots help to illustrate the success achieved in using the VBIC model to represent the various device characteristics of the SiGe HBT NPNs. The DC forward Gummel and output curves for the high-f_T NPN in the 0.25-μm SiGe process are shown in Fig. 2.19, and for the high-breakdown NPN in the same process in Fig. 2.20. These figures show the effects of self-heating, weak avalanche current, and the difference in the quasi-saturation behavior of these devices. Note that the model parameter extraction process begins with the fitting of these DC characteristics. Once initial parameter values associated with the basic DC characteristics have been determined, extraction and model optimization continues using S-parameter, noise figure, and large-signal measurements (as described in Section 2.2). Figure 2.21 shows the V_{CB} dependence of f_T vs. I_C, for both high-f_T and high-breakdown devices, as extrapolated from S-parameter data at a fixed frequency. Minimum noise figure and RF power gain examples for the high-f_T NPN are included in Fig. 2.22. Finally, the initial results of the model vs. large-signal data correlation can be seen in Fig. 2.23 for both device types.

Another key aspect in the development of the SiGe HBT models is the inclusion of extensive geometric scaling equations. While the more advanced BSIM-based MOSFET models have multiple parameters to represent short channel, or narrow-channel, effects, the built-in scaling of VBIC is not adequate to support the range of layout geometries offered by the SiGe design-kit parameterized cells (PCELLs). Use of a single "area" factor to scale all of the current-density, capacitance, and resistance parameters does not give the flexibility necessary to accurately differentiate the dimensional changes in the various regions within the device. Separate calculations are included as a part of the subcircuit model to determine the proper effective dimensions, such as intrinsic and extrinsic junction areas or perimeters, to generate the final set of VBIC parameters given a specific emitter size and device layout configuration.

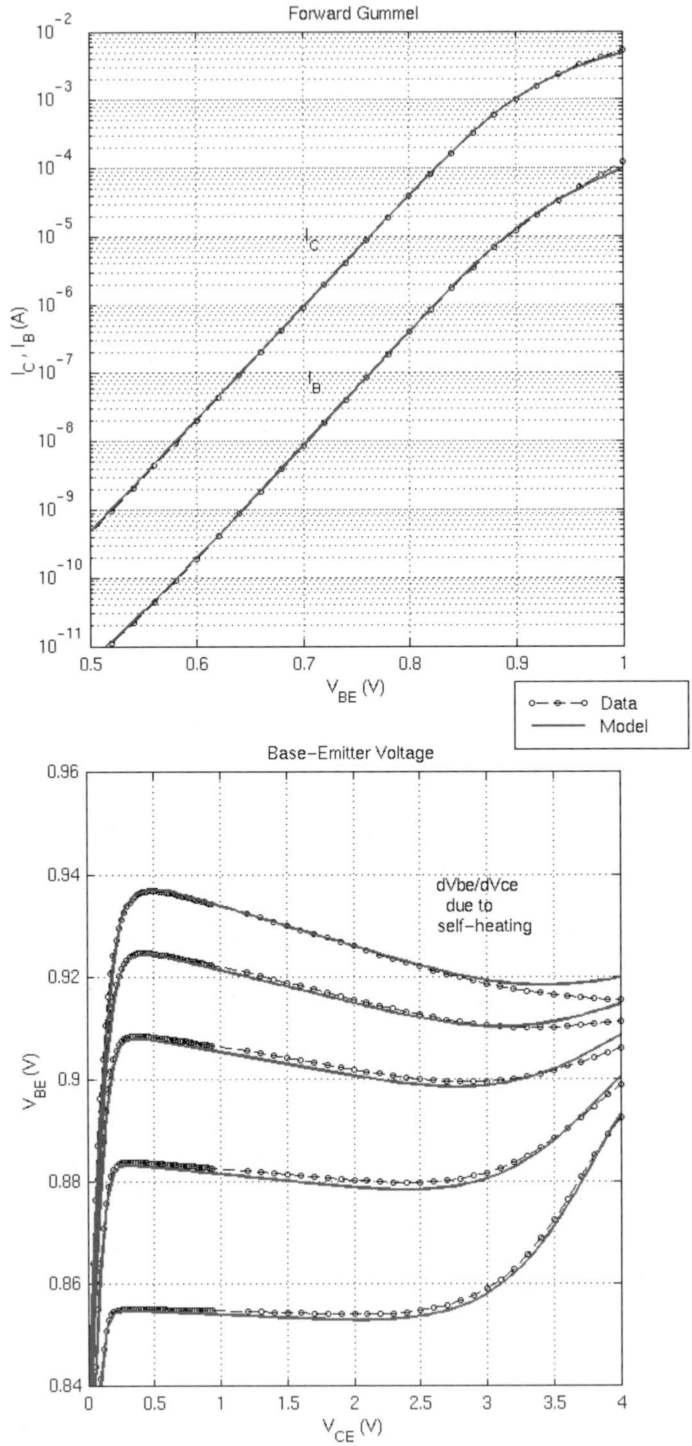

Figure 2.19 DC characteristics of high-f_T NPN in 0.25-μm SiGe process.

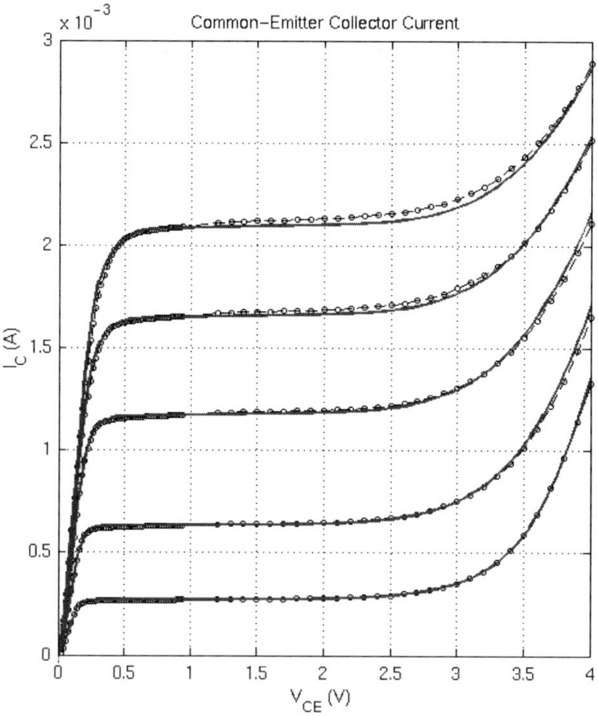

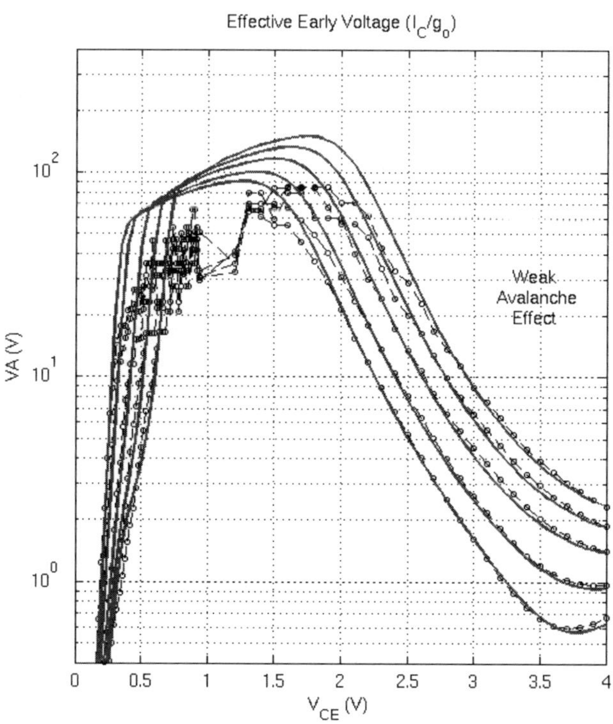

Figure 2.19 *Continued*

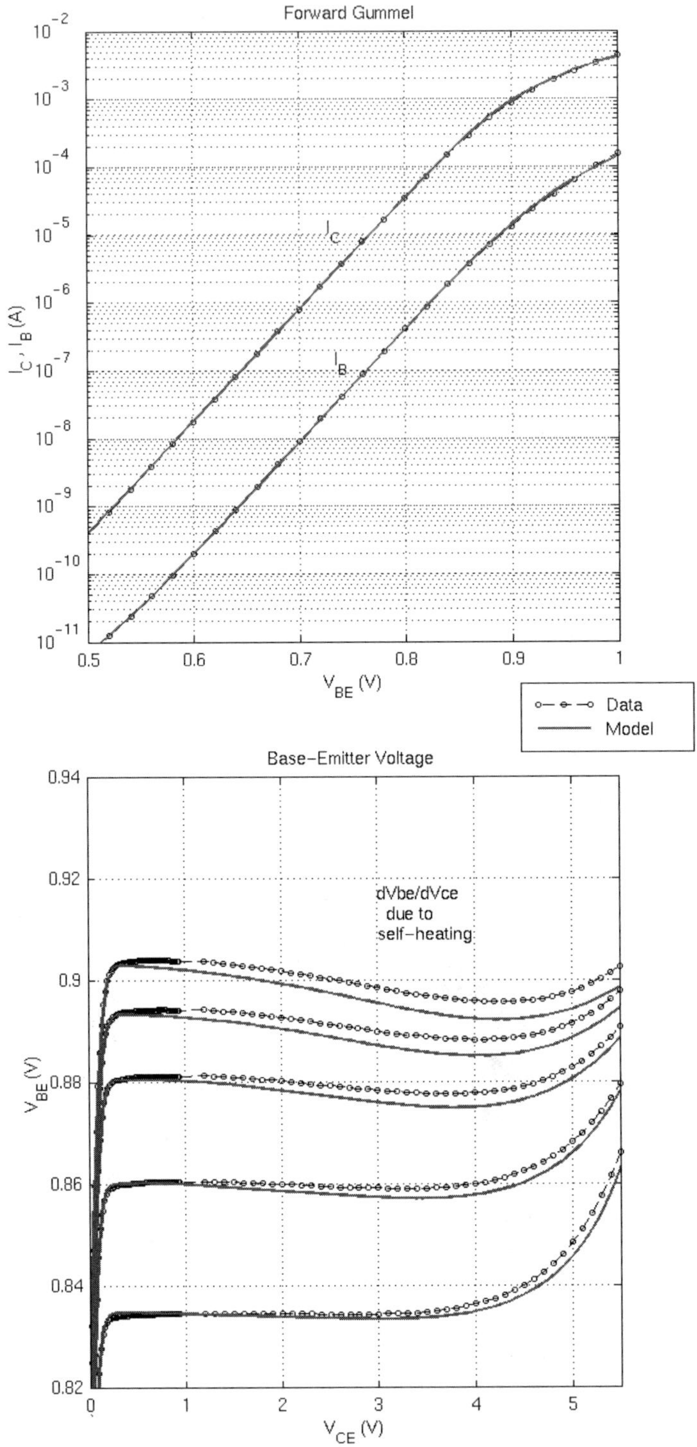

Figure 2.20 DC characteristics of high-breakdown NPN in 0.25-μm SiGe process.

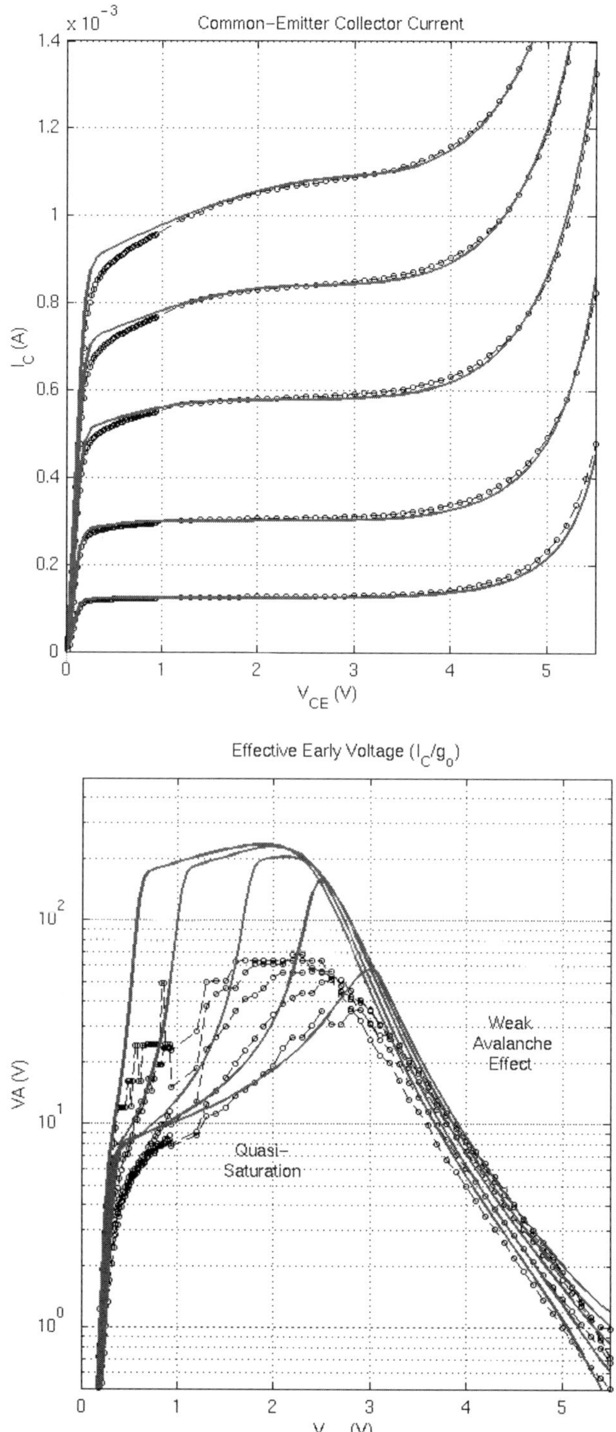

Figure 2.20 *Continued*

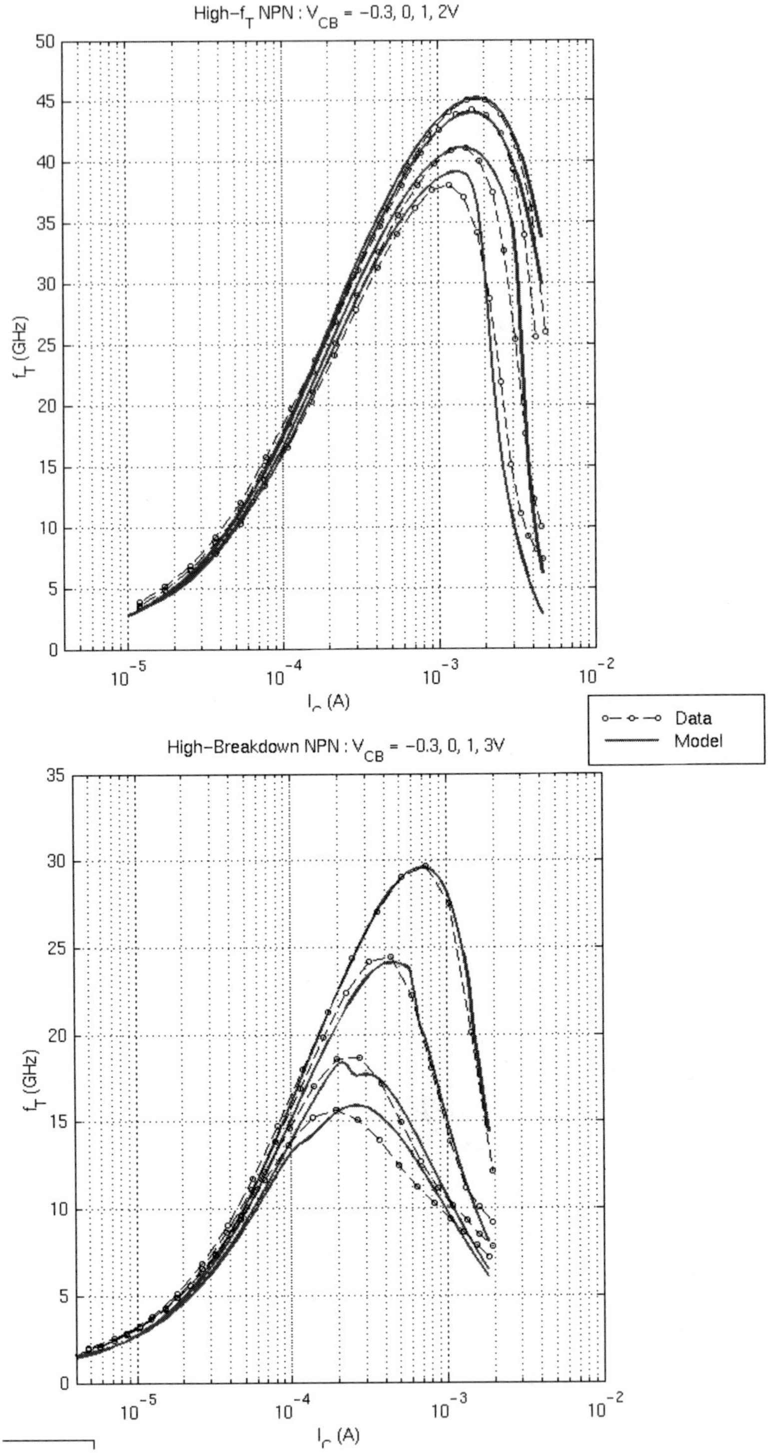

Figure 2.21 f_T characterisitics vs. V_{CB} for high-f_T and high-breakdown NPNs.

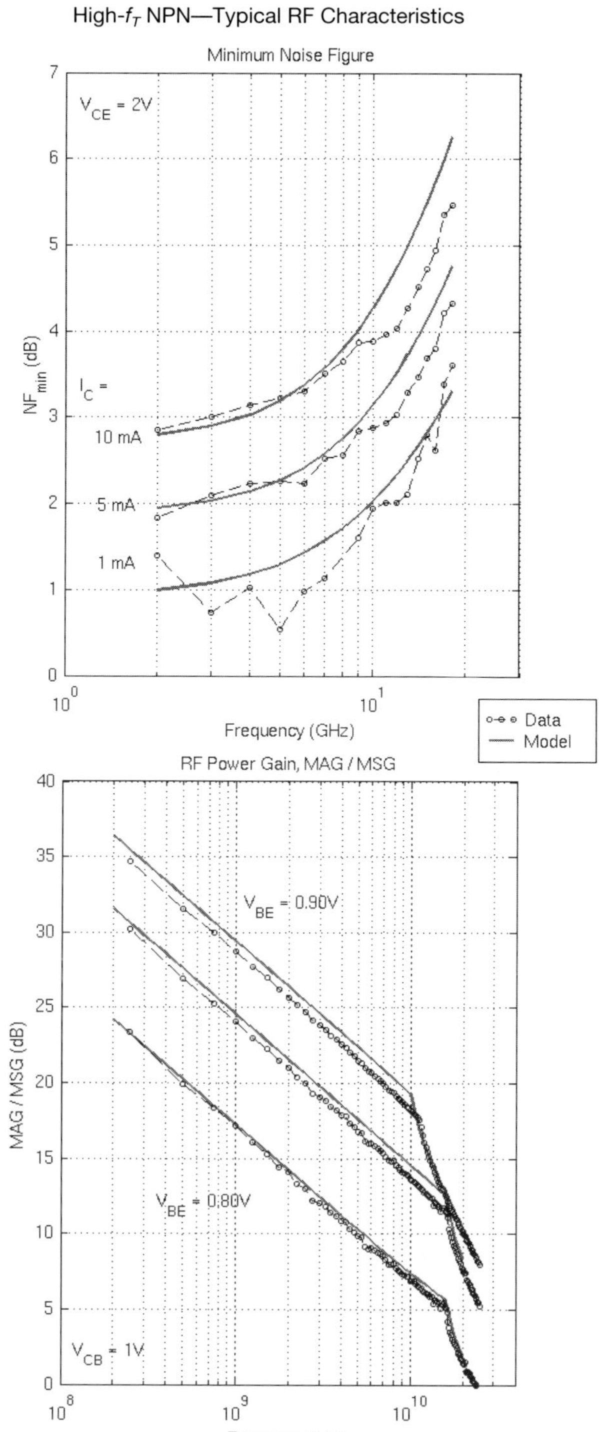

Figure 2.22 Minimum noise figure and RF power gain for high-f_T NPN.

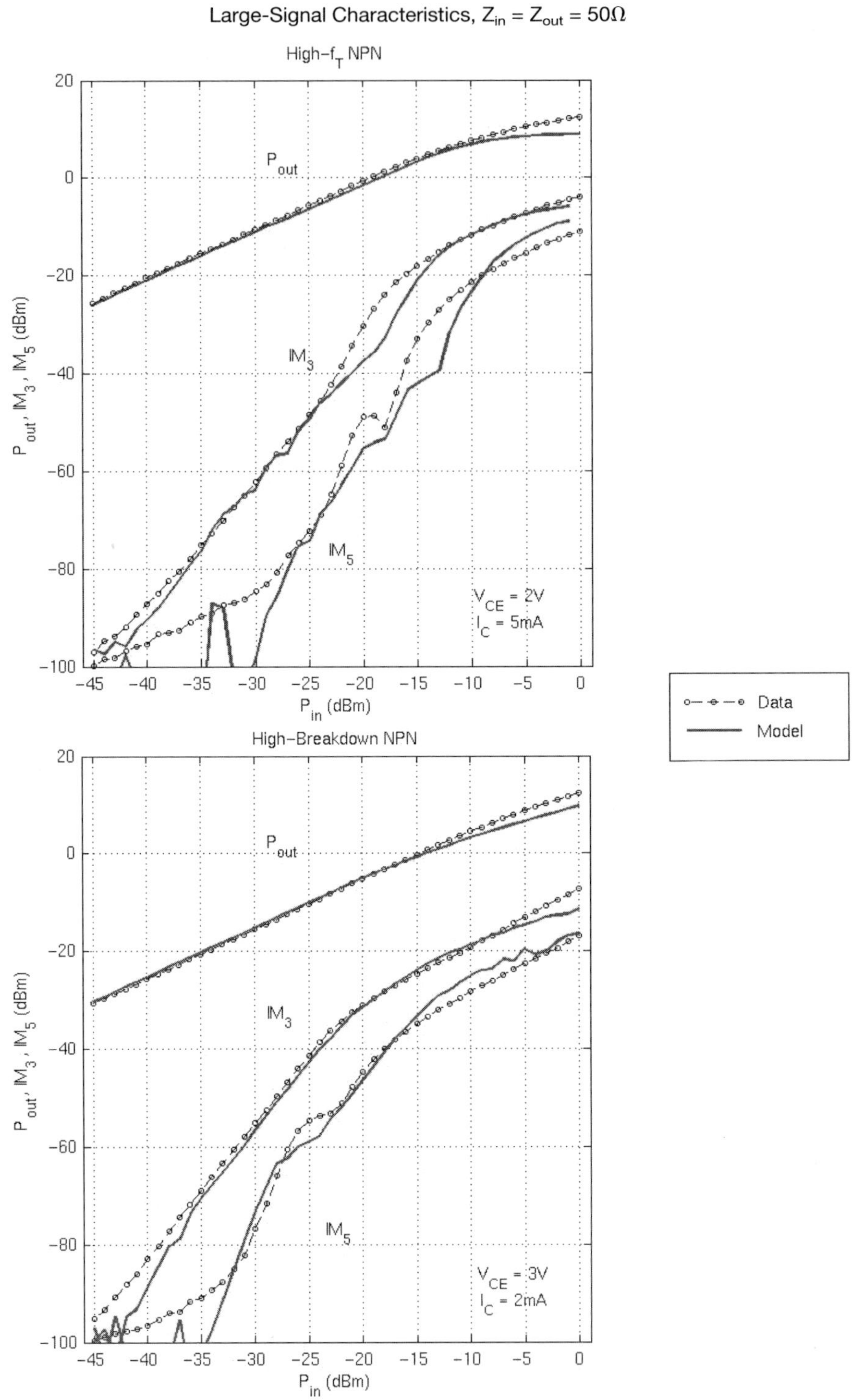

Figure 2.23 Large-signal characteristics for high-f_T and high-breakdown NPNs. (IM3 and IM5 = third-order and fifth-order intermodulation, respectively.)

Modeling V_{BE} and beta mismatch of the SiGe NPNs is important for providing designers with information necessary to assess performance in many typical small-signal analog applications such as high-speed A/D converters, bandgap voltage references, and differential circuits. The scaling equations as implemented also include statistical distributions to represent low current V_{BE} mismatch as a decreasing function of increasing emitter area. Another factor is defined to account for the increase in V_{BE} mismatch that is observed as the current bias increases. A third distribution is used to represent the beta mismatch, which is also modeled as a function of emitter area.

2.3.3 MOSFET

The technical development of the MOSFET compact model has closely followed the increased demands placed on it by circuit designers. As MOSFET-based designs have gone from purely digital, to analog, to analog-RF, MOSFET models have become increasingly more complex.

Traditionally, the first MOSFET model most electrical engineers encounter is the so-called digital model. In the digital model, the primary focus is on accurately predicting the *ON* current (Ids @$V_{gs} = V_{ds} = V_{dd}$) and the switching speed of the device. Issues such as scalability and Ids accuracy near the threshold are not addressed, since focus is on minimum channel-length (i.e., fast) devices that are either on or off. This model (usually MOS level 1, 2, or 3) serves as a good introduction point because it is "physically based." By physically based, we mean that the equations that the model is composed of have their basis in semiconductor-device physics. This is in contrast to an empirical model where the equations may be composed of splined polynomials whose coefficients have been optimized to provide the best fit between the measured data and the simulation results.

Because CMOS technologies have progressed into the submicron region, the limitations of the existing digital models became more and more apparent. The following list highlights many of the physical effects that the existing models are unable to predict [9]:

- Short and narrow channel effects on threshold voltage
- Nonuniform doping
- Mobility reduction due to vertical field
- Bulk charge effects
- Velocity saturation
- DIBL
- Channel-length modulation (CLM)
- Substrate current-induced body effect (SCBE)
- Subthreshold conduction
- Source/drain parasitic resistance

The inability of the digital model to deal with these effects results in a model that does not scale with channel length and/or width. In fact, the operating region and voltage bias dependencies of these effects meant that the drain current is not accurately modeled even for a fixed dimension device. Even worse, the derivatives of the drain current (gm, gds, and gmbs) become discontinuous due to the "splining" approach that has been applied to the digital-model equations. In addition to the inherent inaccuracies, many analog circuit simulations do not converge to a result when these digital models are used. This represents a disaster for the analog designer whose designs depend upon an accurate representation of MOSFET drain current, gain (gm), and output impedance (1/gds).

In the late 1980s and early 1990s, these issues drove several attempts to create an analog model. Of these, the BSIM series of models out of the University of California at Berkeley are the most widely accepted and used. In developing BSIM, a "start from scratch" approach was used that placed the emphasis on three areas of importance:

1. *Device Physics* The robustness of a compact-device model can usually be traced to how much of its fundamental basis is tied to the physics of the device.
2. *Scalability* Width and length scalability are incorporated into nearly all of the equations that compose the structure of the BSIM model. This enables a single model, and thus a single-model parameter set, to be used to predict the performance of a device over a wide range of geometries.
3. *Robustness* The equations used to represent the various effects were combined in such a way as to create one continuous expression for I_{ds}. This eliminated the discontinuous derivative problems found in many of the earlier MOSFET models mentioned earlier. From a circuit simulation standpoint, the robustness of the BSIM model can also be tied to its formulation as a charge-conserving model. In a charge-conserving model, the nodal equations are written in terms of charge instead of capacitance. These equations are considered balanced when all of the charges sum to 0. Capacitance-based models are not charge conserving, because capacitance is an incremental quantity that only accurately predicts the change in charge versus voltage for infinitesimally small changes in voltage [10].

BSIM itself has become an evolution in model development. The initial BSIM model was an offshoot of the compact short-channel IGFET model (CSIM) developed at AT&T during the early 1980s. The focus here was to use "mathematical conditioning" to implement the model equations in a way that was robust and efficient from a simulator standpoint. In the initial BSIM model, the base-model parameters were set up to represent a large device (wide/long), and small geometry effects were treated as corrections to the base model. In BSIM2 [11] an attempt was made to make the model easier to scale as a function of device size by removing the "small geometry corrections" from the intrinsic model and creating a set of extrinsic-model parameters for each key effect that were summed together during the model call to create a composite model parameter. This set of parameters included a base parameter, a width parameter, and a length parameter for each effect being modeled. For example, the flatband voltage was modeled as ($V_{FB} = XV_{FB} + LV_{FB}/L + WV_{FB}/W$). The problem with this approach is that it treated the variation of the model parameters as a linear variation with W and L. This was a poor assumption and the scalability of the BSIM2 model was limited as a result.

The MOSFET model used by IBM in its SiGe based design kits (in 2002) is the third-generation BSIM3 model from U.C. Berkeley (BSIM3v3.2). Although its name implies that it is a continuation of the earlier BSIM and BSIM2 models, it was really a total rewrite from scratch. The goal here was to introduce more accurate physics-based equations into the model in a way that was still mathematically robust from a simulator standpoint.

In general, the best compact models are those whose representative equations have their basis in the physics behind the devices they are representing. These models tend to scale better and to be more accurate at biases and geometries outside the bounds being measured. This was the approach used in developing the BSIM3 model. Where possible, the equations used to represent the effects just listed were based on the solution of Poisson's 2D equation for the distribution of charge across the channel of a MOSFET. As a result, the fundamental equations for threshold voltage, mobility, and velocity saturation tended to be very physically based.

The semiempirical nature of the BSIM model dictates that a combined extraction and optimization approach should be used to extract the BSIM parameter values. First, the physical "process" parameters, such as oxide thickness, base threshold voltage (long/wide), delta L, delta W, and series resistance, are extracted from either single- or multidevice measurements. Next, an optimization approach is used to come up with a set of BSIM parameters that minimize the model-to-hardware error across the device size, bias range, and temperature space being fit. To date, two different optimization approaches have been used. The more "traditional" approach is the so-called local optimization method. In local optimization, parameters are fitted, a few at a time, by varying their value to minimize the model to hardware difference over a very specific range of device size and operating region where the effects of these parameter are dominant. This "local" optimization approach is preferred over the standard "global" optimization approach because it tends to avoid the local minima phenomena that can occur when optimizing a very large number of parameters over a very large space. The second optimization method used more recently at IBM is a genetic algorithm (GA) -based approach [12]. The GA avoids using the "stair climbing" optimization approach that causes other global optimization approaches to become trapped in local minima. The GA is able to optimize all of the BSIM parameters at once by means of a "fitness function" that weights its target criteria accordingly. The GA approach is very central-processing-unit (CPU) intensive, but it minimizes the human effort needed to extract the models. The key to success lies in the definition of the fitness function.

Length and width scalability have been built into the threshold voltage equation, effective mobility equation, and most of the other fundamental equations behind the BSIM model. However, if the physics behind a device does not exactly match the physics behind the scaling equations, then the models may not scale across the entire geometry range being offered in a given technology. To get around this problem a technique known as *binning* has been incorporated into the BSIM model. As its name implies, binning allows the width/length geometrical space to be broken up into several regions or bins. Figure 2.24 shows a case where the BSIM model for a device is broken up into nine bins, three in the width dimension times three in the length dimension. This approach enables the modeler to extract nine separate BSIM parameter sets to cover the entire *W*/*L* space. Since each of the nine BSIM parameter sets only has to cover its *W*/*L* region, the scalability requirements place on the intrinsic physics-based scaling are relaxed. The downside of using this approach is the increased development overhead associated with creating and maintaining binned models. Through 2002, the SiGe development team has not released any BSIM-based MOSFET models that have used binning. This does remain a possibility in the future, as decreasing channel lengths continue to challenge the intrinsic scalability of the BSIM model.

When a designer places a MOSFET device of given *W*/*L* in their schematic, they have a level of confidence that the Ids or gm or gds value predicted by the model is valid for the device they will be laying out. If the device they are using has an RF PCELL (described below), the designer has confidence that the device parasitics are also represented accurately within the model. If a designer is using a standard "digital" MOSFET PCELL, they must rely on an estimation of device parasitics that have been incorporated into the design kit. To predict the source/drain (S/D) parasitic capacitance, an estimate of the area and perimeter of both the source and drain is passed to the S/D diode model contained within the core BSIM model subcircuit. To estimate the size of the S/D diffusions, we make the following assumptions: (1) contact sizes and spacing are at a minimum, thus enabling the smallest possible diffusion; (2) in the event of an even number of gate "fingers," there will be one less drain diffusion than source diffusion; (3) applied photo bias is nominal [13]. Within the SiGe BiCMOS 6HP design kit, the customer has the choice of two different parasitic estimation approaches. In approach one, the S/D area and perimeter estimates

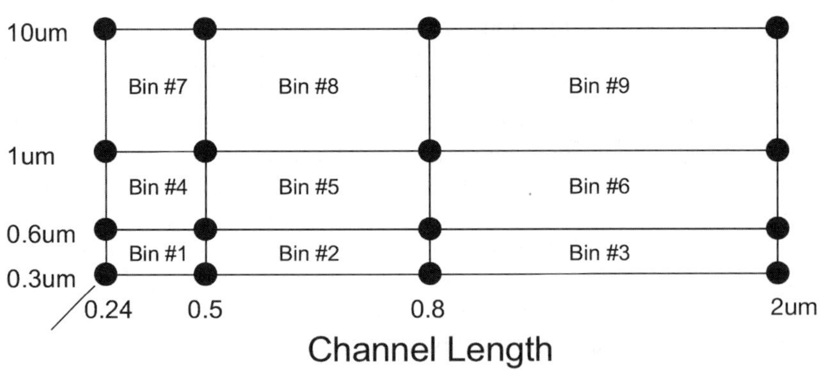

Figure 2.24 Binning the MOSFET model.

are calculated in the cadence Component Description Form (CDF) associated with each schematic level device instance. The values of drain area (AD), drain perimeter (PD), source area (AS), and source perimeter (PS) are then passed into the device model. In the second approach, the values of AD, PD, AS, and PS are calculated in the device model itself. Each approach has its own advantages and disadvantages. Calculating the area and perimeter values in the CDF simplifies the model code and reduces overall simulation time, since the values are calculated once prior to netlisting. Calculating the area and perimeter in the model enables statistical process variation to be applied to the area and perimeter during Monte Carlo simulation, and allows the user to sweep the channel length (or width) as a design variable during simulation and still have the estimated parasitics included.

In addition to increased focus on analog model accuracy, the BSIM generation of compact-device models has also placed an increased focus on modeling noise. Prior to BSIM, the noise contributions of a MOSFET were modeled as the standard kt/q thermal-noise contribution. In BSIM, flicker noise ($1/f$ noise), channel thermal noise, induced gate noise, and thermal noise associated with parasitic gate and diffusion resistances are all modeled.

In concept, the jump from an analog model to an RF-quality model is not really as big as the jump from a digital model to the analog model. From a designer's perspective, the RF model really involves extending the analog model accuracy from the low-frequency region into the RF region. At low frequencies, the capacitive load of the gate on the preceding circuit element is important. At high frequencies, the complex load composed of the gate capacitance, the channel resistance, and diffusion parasitics all couple together to form a complex load on the former stage. From an output impedance standpoint, a similar analogy can be made. At low frequency, the output resistance is dominated by the channel resistance. At high frequency, the S/D parasitics and the bulk resistance contribute largely to the complex output impedance. This presents a unique problem to the MOSFET modeler. At low frequency, the performance of the MOSFET can be adequately represented by modeling the intrinsic device (Ids), the overlap capacitance, and the S/D parasitic diode. This can be done because these effects are independent of the device layout (as long as the number of fingers and multiplicity are know). At high frequency, the layout cannot be ignored. The approaches used to wire to the gate and contact the substrate greatly affect the high-frequency performance of the device. This creates a problem for the MOSFET modeler. If they do not know the exact layout of the MOSFET, how can they model its high-frequency performance? Our answer is to provide two MOSFET PCELL layouts to the designer. The first is the standard "digital" layout where the designer is free to wire to the gate and to the substrate. The model provided with this device will be verified at lower frequencies. The second PCELL will be similar to the one shown in Fig. 2.25. In this PCELL, the gate wiring and substrate contact scheme are defined and controlled. The model provided with this device scales as a function of channel width, length, and number of fingers at both low frequency and high frequency. By using this layout and its associated RF model, the designer can be sure that the RF model they are using represents the device they have designed. The purpose of this "RF PCELL" is not to provide designers with

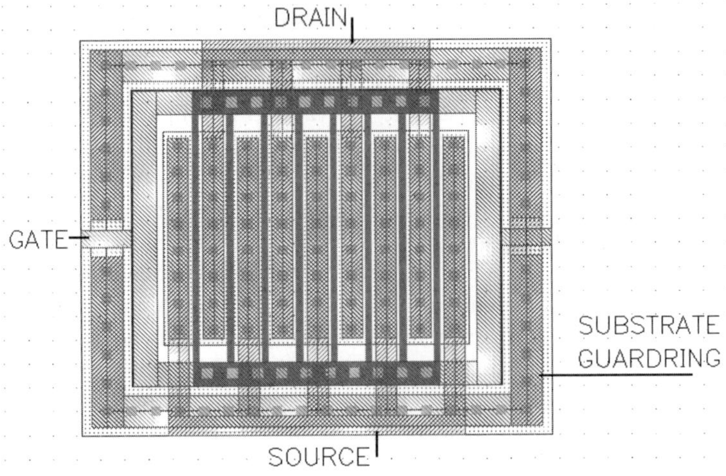

Figure 2.25 Schematic Overview of an RF MOSFET Model.

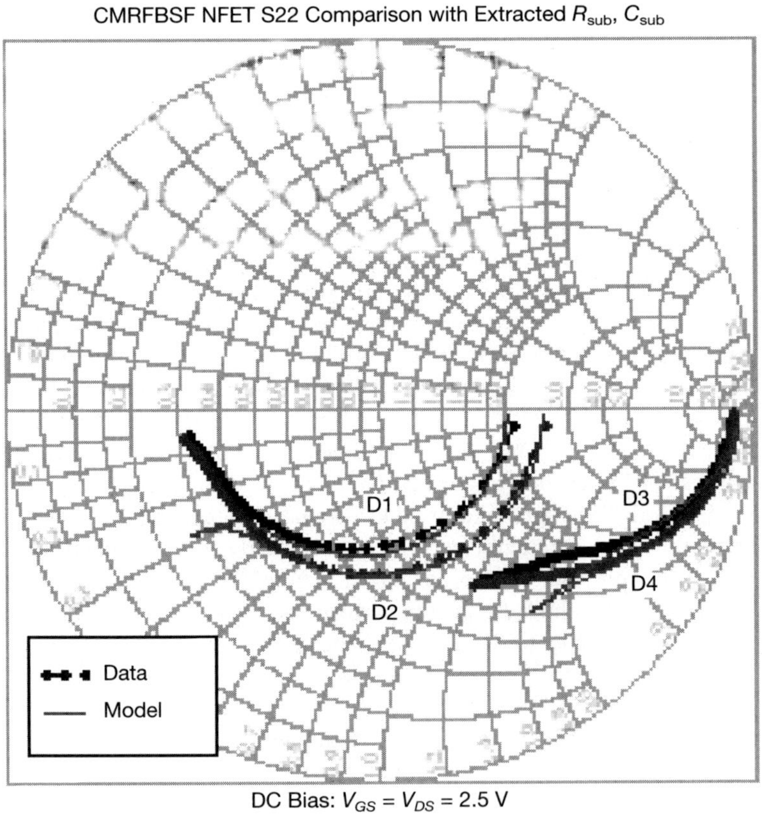

Figure 2.26 Example of an RF MOSFET PCELL Based Device Layout. D1: $W = 300$ μm, $L = 0.24$ μm, Single Contact. D2: $W = 300$ μm, $L = 0.24$ μm, Ring Contact. D3: $W = 80$ μm, $L = 0.5$ μm, Single Contact. D4: $W = 80$ μm, $L = 0.5$ μm, Ring Contact.

the highest performance layout. It is used to provide them with a controlled device configuration that has been well characterized and modeled.

The plot shown in Fig. 2.26 gives an indication of how critical the substrate circuit is to the accuracy of the model at high frequency. This figure shows S22 S-parameters as a function of frequency for two different layout configurations. Devices D1 and D2 are both 300-μm-wide NFETs with a 0.24-μm channel length. Device D1 has a single substrate contact, while the substrate contact for D2 is composed of a ring that encircles the entire device. Devices D3 and D4 are both 300-μm-wide NFETs with a 0.5-μm channel length. Device D3 has a single substrate contact, while the substrate contact for D4 is composed of a ring that encircles the entire device. Figure 2.27 demonstrates two key points. First, the differences in the hardware curves for D1 and D2 indicate a significant variation in S22 as a function of frequency for a minimum channel-length device. This demonstrates that output impedance at the drain of the device can vary significantly depending upon the proximity of the substrate contact. Second, the model-to-hardware comparison of all four devices shows that the effects can indeed be modeled using a model structure similar to the one shown in Fig. 2.27. The problem is that the model parameters needed to model the substrate resistance effects are layout dependent, as previously described. Therefore, a unique set of substrate network parameters must be used for the unique layout shown in Fig. 2.25.

2.3.4 Summary

In this section, we have discussed the topic of compact-device models of HBTs and MOSFETs. Statistical modeling has been presented as a key aspect of both active de-

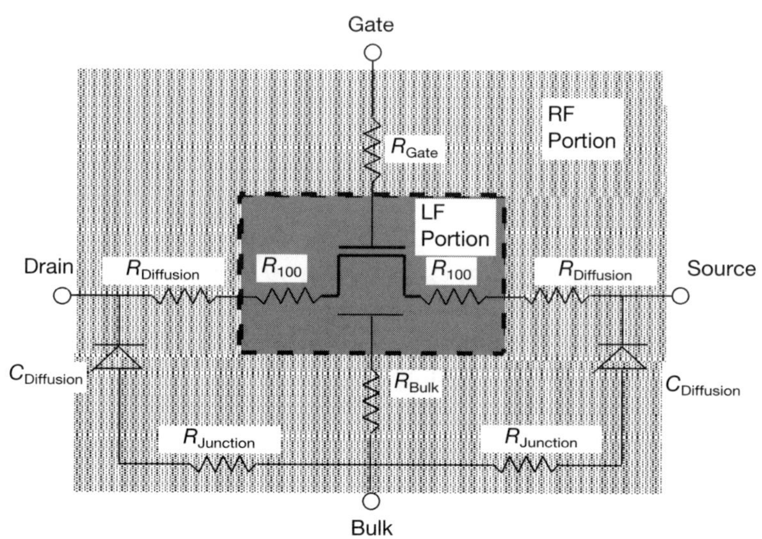

Figure 2.27 Comparison of single vs. ring substrate contact using S22.

vice families. The concept of the RF FET PCELL-based model has also been discussed in detail. The next Section (2.4) expands this discussion to advanced passive devices.

2.3.5 References

1. H. K. Gummel, "A Charge Control Relation for Bipolar Transistors," *Bell Syst. Tech. J.,* vol. 49, pp. 115-120, 1970.
2. C. C. McAndrews, J. A. Seitchik, D. F. Bowers, M. Dunn, M. Foisy, I. Getreu, M. McSwain, S. Moinian, J. Parker, D. J. Roulston, M. Schröter, P. van Wijnen, L. F. Wagner, "VBIC95, the Vertical Bipolar Inter-Company Model," *IEEE J. Solid-State Circuits,* vol. 31, pp.1476–1483, 1996.
3. http://www.eigroup.org/cmc/.
4. G. M. Kull, L. W. Nagel, S. Lee, P. Lloyd, E. J. Pendergast, H. Dirks, "A Unified Circuit Model for Bipolar Transistors Including Quasi-Saturation Effects," *IEEE Trans. Electron Devices,* vol. 32, no. 6, pp. 1103–1113, 1985.
5. http://www.iee.et.tu-dresden.de/iee/eb/comp_mod.html.
6. M. Schroter and H.-M Rein, "A Generalized Integral Charge-Control Relation and Its Application to Compact Models for Silicon-Based HBT's," *IEEE Trans. Electron Devices Lett,* vol. 40, pp. 2036–2046, 1993.
7. http://www.semiconductors.philips.com/models/.
8. J. Berkner, "Bipolar Parameter Extraction," in *2001 Bipolar biCMOS Circuits and Technology Meeting Short Course Notes,* IEEE, 2001.
9. W. Liu, X. Jin, J. Chen, M. C. Jeng, Z. Liu, Y. Cheng, K. Chen, M. Chan, K. Hui, J. Huang, R. Tu, P. K. Ko, C. Hu, *BSIM3v3.2.2 MOSFET Model User's Manual,* p. 2–1, University of California, Berkeley, 1999
10. K. Kundert, *The Designer's Guide to SPICE & SPECTRE,* pp 167–176, Kluwer Academic Publishers, 1995.
11. D. Foty, *MOSFET Modeling with SPICE,* pp. 213–215, Englewood Cliffs, New Jersey, Prentice-Hall, 1997.
12. J. Watts et al., "Extraction of Compact Model Parameters for ULSI MOSFETs Using A Genetic Algorithm," in *Tech. Proceedings of Second Intl. Conf. on Modeling and Simulation of Microsystems, 1999,* pp. 176–179, 1999.
13. S. Parker et al., "Parasitic Estimation in Submicron Multifinger MOSFETs," *MicroNews,* vol. 7, no. 3, pp. 32–36, 2001.

2.4 COMPACT-MODEL DEVELOPMENT: ADVANCED PASSIVES

In the previous section, we discussed the topic of compact modeling of active devices, specifically, HBTs and FETs. In this section, we continue the discussion to IBM's implementation of compact models for advanced passive devices, including inductors, capacitors, and varactors.

2.4.1 Inductors

Inductors are essential devices in many RF circuits (VCOs, impedance-matching networks, etc.). Their on-chip implementation decreases packaging parasitics, reduces the number of pins required, enables more compact circuit boards, and paves the way for SOC solutions. Any design flow for RF applications would be incomplete without adequately optimized and modeled on-chip inductors [1].

An inductor's performance is best judged by its inductance and Q (quality factor) [2] across a frequency spectrum of interest. Q is inversely proportional to power dissipated by the inductor. Therefore, power losses associated with the inductor should be minimized for increased performance. The two main causes of power loss in the inductor are (1) resistance of the metal (or metals) with which the inductor is built, and (2) loss within the substrate underneath the inductor due to capacitive and/or magnetic coupling into the substrate. Losses associated with the inductor's series-resistive elements increase with frequency due to two primary effects; skin-effect loss and magnetic field-induced proximity-effect loss.

There are a variety of BEOL options with which inductors can be implemented. While for low Q, low-frequency applications inductors can be built using regular metal interconnects, for most RF applications very-low resistance one or more thick Cu or Al metals (to minimize losses due to series resistance) must be used. Furthermore, the inductor metalization should be separated from the substrate with as thick a dielectric stack as possible, preferably with a low dielectric constant (to minimize capacitive coupling, and hence losses within the substrate underneath the inductor). A Faraday shield [3] between the inductor and the substrate should also be an option for designers, as it may increase the peak Q of an inductor, (notably) at the expense of a reduced self-resonant frequency and added layout complexity.

Designers seek accurate compact models for all device types, and the inductor is no exception. The inductor model should be inclusive to all the various configurations possible for the device. The model should support various metalization options, whether the inductor is built with wiring interconnects or dedicated thick Cu metalization, various dielectric stack heights, possible Faraday shielding (ground plane) options, and a range of values for the parameters defining the inductor planar geometry. For an octagonal spiral, which is the preferred shape for on-chip inductors, the planar geometry parameters are outer dimension (d), turn width (w), turn spacing (s), and number of turns (n). Figure 2.28 shows a visual image of an on-chip inductor that outlines the planar geometry parameters (x, w, s, n).

The low-frequency inductance of an on-chip spiral can be calculated using various methods, as outlined in [4]. The preferred method is known as current sheet approximation [4]. The inductance, L, is found by using Equation (2.8):

$$L = \frac{\mu_o n^2 d_{\text{avg}} c_1}{2} (\ln(c_2/\rho) + c_3\rho + c_4\rho^2) \tag{2.8}$$

where,

$\mu_0 = 4\pi 10^{-7}$ permeability of free space;

$d_{\text{in}} = (x - 2nw - 2*(n-1)*s)$ the inner diameter of the spiral;

$d_{\text{out}} = x$ the outer diameter;

$d_{\text{avg}} = (d_{\text{in}} + d_{\text{out}})/2$ the averaged diameter;

$\rho = (d_{\text{out}} - d_{\text{in}})/(d_{\text{out}} + d_{\text{in}})$ the fill ratio,

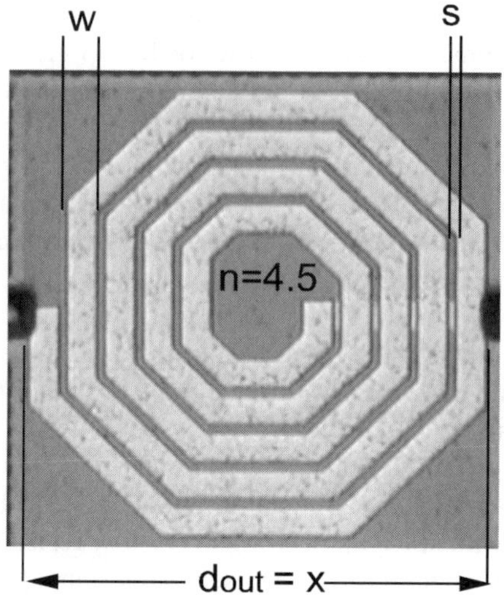

Figure 2.28 Inductor visual image outlining planar geometry parameters.

and $c1$, $c2$, $c3$, and $c4$ are layout-dependent coefficients (for an octagonal spiral, they are chosen as $c1 = 1.27$, $c2 = 2.29$, $c3 = 0$, and $c4 = 0.19$).

The inductance calculated this way does not include the vertical thickness of the spiral metalization (the self-inductance of a metal diminishes with increased thickness). In an accurate inductor model, the metal thickness and its effect on the inductance should be addressed by use of scaling constants.

The fact that the resistance and inductance of an on-chip inductor are frequency dependent should also be addressed in the model. Skin-effect losses are caused by higher current levels on the surface of the metal as frequency increases. Since the cross-sectional area through which the current flows decreases with increasing frequency (it flows in an increasingly narrow annular ring), the conductor's effective resistance and inductance undergo a change with frequency. The fact that the current flows toward the surface of the metal causes the resistance of the conductor to increase as a function of the square root of the frequency. This surface current flow also excludes the magnetic field from the inside of the conductor, causing the self-inductance to reduce slightly with increasing frequency, until it is completely determined by the magnetic field external to the conductor.

The enhanced magnetic field that exists in the central and outer portions of the spiral tends to cause nonuniform current flow in the turns. Inner turns tend to have current flow only on the innermost edge of each turn, while the outer turns tend to carry current on their outermost edges. Figure 2.29 shows electromagnetic simulations of two inductors ($n = 1.5$ and $n = 2.5$). The current flow is shown at a frequency of 2 GHz. The simulations are done using Sonnet™, a planar electromagnetic (EM) simulator that accurately handles vertical current flow. Such simulators are

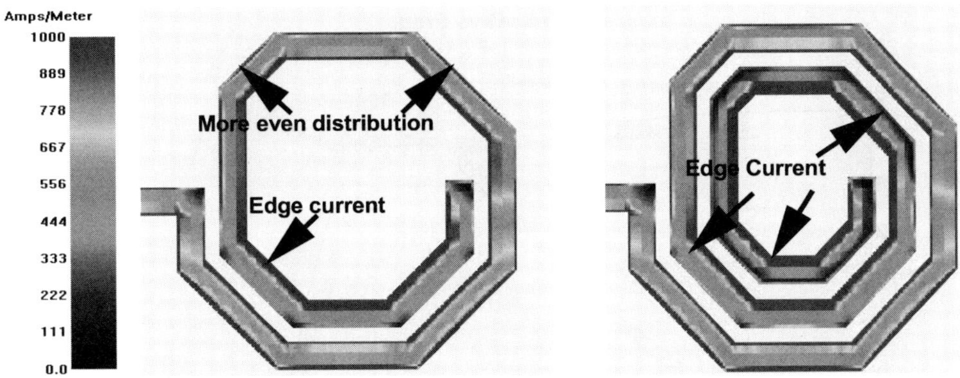

Figure 2.29 Electromagnetic simulation of two inductors at 2 GHz showing nonuniform current distribution.

known as 2.5D simulators (compared to 3D, where current flow is allowed in all directions).

This effect is frequency dependent, as the induced voltage in the turns (eddy current) increases with increasing frequency. This nonuniform current flow is typically called "proximity effect" and tends to cause the effective spiral resistance to rise faster, with increasing frequency, than can be attributed to skin effect alone. In addition, the net inductance will decrease due to an effective reduction in the radius of the spiral caused by the current crowding to the innermost edge of the inner turns. The proximity effect can be the dominant loss mechanism at frequencies of interest for multiturn spirals. The modeling of the frequency dependence of inductance and resistance can be quite challenging. The standard approach is calculating the resistance and inductance at a predetermined frequency point. This frequency point can be chosen from the frequency spectrum of operation, for which the inductor is expected to dissipate the most power. While this approach produces accurate results for single-frequency simulations, for broadband simulations it is rather inaccurate. A better approach is implementing the frequency dependence via a network of fixed resistors and inductors. Such networks then would have the desired frequency dependence in its effective resistance and inductance. Circuit implementation of skin and frequency effects (CISP) produces accurate results for all frequencies of interest and is superior to the standard approach that is only accurate at a single frequency point. Figure 2.30 shows measured and simulated inductor quality factors and inductance versus frequency. Simulations are done with a standard model, where the model is accurate at around 5 GHz, and also the CISP model, where there is broadband accuracy.

A scalable inductor model provides a mapping from all the input model parameters (i.e., type of metalization; dielectric stack properties, type of Faraday shield, if any; planar geometry parameters) to a subcircuit for which the circuit element val-

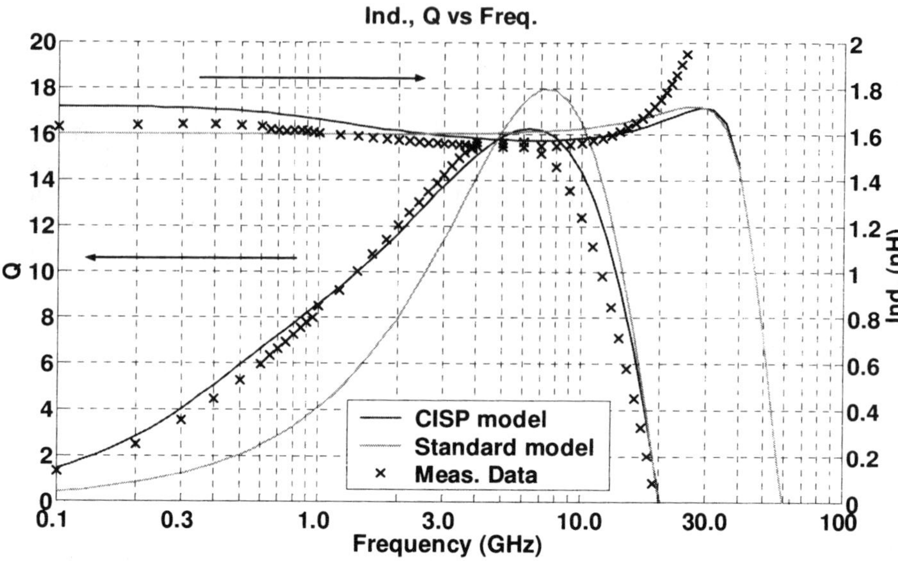

Figure 2.30 Inductance and inductor Q vs. frequency. Data and two models are shown.

ues are functions of the input model parameters. The subcircuit should employ CISP to capture the frequency dependence of the resistance and inductance, should include all the various capacitances associated with the device, and should account for substrate-related losses. Such a subcircuit would then be an accurate representation of the device for the frequency spectrum of interest. Figure 2.31 contains a schematic of a general inductor model subcircuit. In the 2002 IBM model, the block for series inductance and resistance elements was replaced by a CISP network that greatly improves the broadband accuracy.

The inductor models meet the requirements outlined earlier. Along with a design automation tool flow, described in Section 4.1, the designer receives a scalable inductor model that spans a very wide range of inductances (from hundreds of picohenries to hundreds of nanohenries) with quality factors over 30 (available with standard IBM dual-metal BiCMOS and RF CMOS offerings). Users can see on-the-fly the low-frequency inductance and peak Q frequency values as they alter the PCELL parameters (the input model parameters mentioned earlier). The user receives full documentation on the model showing how the device is predicted to behave with statistical process variations, changes in temperature, and how the model correlates with measurements for a wide variety of device configurations. The designer is also given guidance on how to achieve the desired inductance, peak Q, peak-Q frequency, and self-resonant frequency, which is critical for efficient use of the model. The IBM inductor is widely scalable. It employs CISP to capture skin and proximity effects, and therefore is accurate for all frequencies of interest. It also supports various ground-plane, back-end metalization, and di-

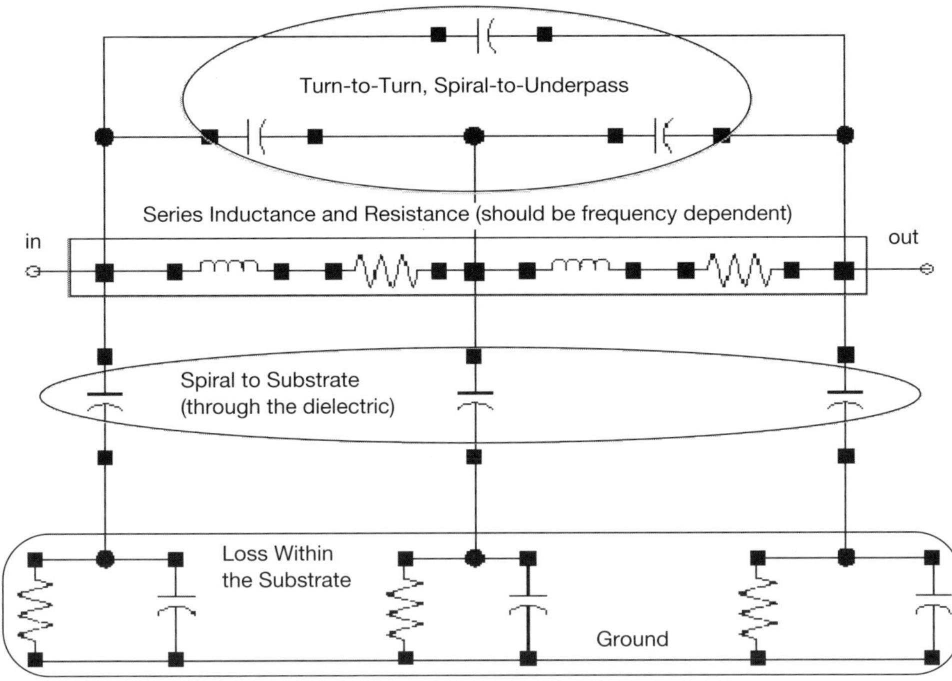

Figure 2.31 Inductor-model subcircuit.

electric options. In its development and verification, designer/user feedback, EM simulations (e.g., Sonnet™) and on-wafer S-parameter measurements are utilized. For geometries that are not yet covered by our scalable broadband model, such as symmetric cross-coupled interleaved structures [5], we provide S-parameter blocks and/or geometry-optimized subcircuits that can be used to model the specific structure of interest.

2.4.2 Capacitors

The SiGe BiCMOS technologies support different types of capacitors, including MOS, poly-poly, MIM, and high capacitance density dual-capacitance devices. The variety of the device offering allows designers to optimize their circuit performance and layout by selecting a capacitor based on the requirements for density, maximum voltage use, voltage-bias sensitivity, or higher Q. Each of these capacitor types requires a unique subcircuit topology to represent the physical nature of the structures. In addition to a main capacitor element, each model includes R, L, and C elements, as necessary, to account for the fringe effects, parasitic capacitance under the bottom plate of the capacitor, series resistance of the top and bottom plates and, in the case of the MIM, the inductance associated with the metal plates. Element values are calculated using a physics-based approach and provide for a high degree of model accuracy. The SiGe BiCMOS capacitor models allow the utmost in designer flexibility. For example, in technologies that offer several BEOL options, the MIM model selects the correct model element values based upon the BEOL chosen by the designer. An example of the basic model topology for both MOS and MIM capacitors is shown in Fig. 2.32.

Each element of the subcircuit in the model is coded to support the scalable dimensions of the capacitor, based on the defined width and length of the structure. Capacitance data are used to extract parameters for the voltage and temperature dependence of the main capacitor element. The final *RC* or *RLC* configuration of the subcircuit is verified with S-parameter measurements. As can be seen from the plots in Fig. 2.33, the MIM capacitor has a higher Q, while a MOS capacitor of similar geometry yields a higher capacitance. These results illustrate the trade-off in density and performance that different capacitor types often provide. Model to hardware correlation is ensured through the useful frequency range for the device, generally tens of gigahertz for MIM devices and several gigahertz for MOS and poly-poly capacitors. This provides designers with the largest range of device operation possible.

Capacitor models also provide complete statistical simulation capabilities. By characterizing devices of various geometries, accurate matching equations that scale across the entire range of device sizes are developed. Variations in process pa-

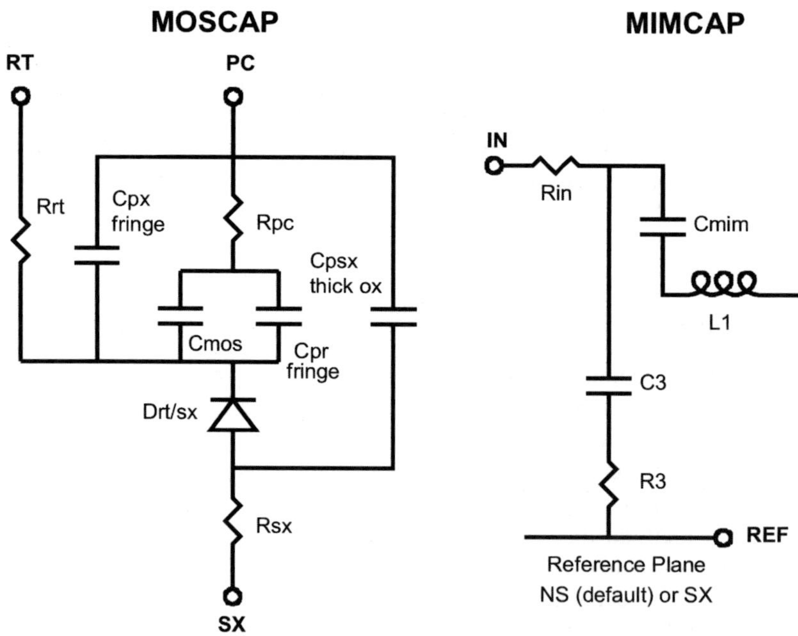

Figure 2.32 Typical MOS and MIM capacitor-model subcircuits.

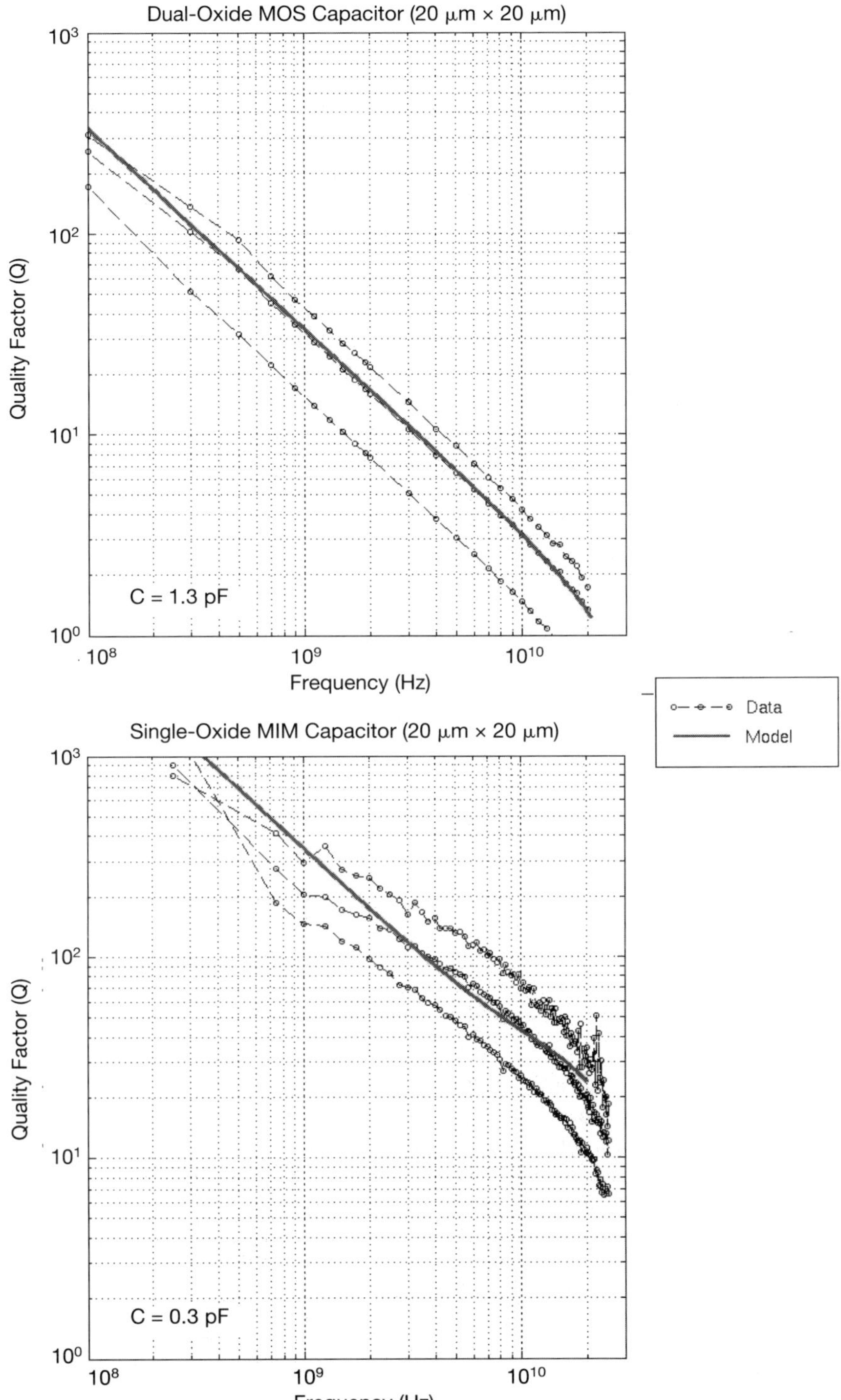

Figure 2.33 Frequency response of MOS and MIM capacitors.

rameters are also modeled and allow for the use of corner or Monte Carlo simulations to predict capacitor performance across wafers. Process parameters that the capacitors share with other devices utilize the same distribution function. For example, the gate poly of a MOSFET is also the same poly that forms the top plate of a MOSCAP device. By utilizing the same distribution for this poly resistance, both devices will have properly correlated simulation results.

2.4.3 Varactors

While competitive technology comparisons often focus on the performance of the active devices, the features of the passive devices, such as varactors, MIMs, and inductors, are essential for offering a complete technology solution. VCO designs are dependent on key varactor characteristics, including the linearity, tuning range, and capacitance density. Current technology offerings include CB junction, hyperabrupt (HA) junction, and MOS accumulation varactors.

The model topology, process parameters, and scaling equations for the CB junction varactor are taken directly from the structure of the NPN model. An example of this commonality is that the calculations for the extrinsic collector resistance in the NPN are also used for the CB varactor cathode resistance. The parasitic capacitance of the cathode to substrate is also modeled in the same manner as the collector–substrate capacitance. Modifications are made to reflect the PCELL layout options for multiple anode and cathode devices, but the physical correlation between the NPN and CB varactor is preserved. The useful range of operation for the CB varactor is out to a few tens of GHz.

The HA varactor is a P^+ source to N-well junction in which there is an additional cathode implant to improve the linearity and tuning range. This is modeled using the standard equations for reverse-bias capacitance of a diode. Layout options allow the designer to specify anode length and width and number of anodes, or anode width and number of anodes and total capacitance. Like the CB varactor, the useful range of operation for the HA varactor is out to a few tens of GHz. The devices with the best Q are those with a number of narrow anodes as opposed to one large-area anode.

An nMOS varactor is a tunable capacitor that uses a thin-oxide NFET in an N-well with the N^+ source and drain shorted together. The variation of the capacitance is controlled by the gate to diffusion potential, which takes the silicon surface under the gate from the depletion into the accumulation region. The typical device model that has been supported for standard CMOS technologies makes use of a high-order polynomial to represent the behavior of the capacitance as a function of the applied voltage. Although it is possible to achieve a reasonable fit of the *CV* curve, this implementation is highly nonphysical in nature and has been prone to causing convergence problems during typical circuit analyses, as the polynomial can be numerically unstable near the boundaries of the voltage range over which the varactor is biased.

The most recent efforts to model this nMOS varactor for the SiGe technologies have focused on the definition of a more physical subcircuit, which is shown in Fig. 2.34. The improved topology represents two capacitors connected in series; the first is a fixed-oxide capacitor and the second represents the voltage-dependent capacitance of the depletion region in the silicon below the gate oxide. While a varactor layout using a large polysilicon area would yield a high capacitance with better tunability, multiple-finger devices of smaller area are often wired together instead of a single large device to obtain a higher Q. This layout configuration dictates that the model must properly account for the fringe capacitance as a function of the total width and length of the polysilicon gate. An additional term is included to represent the capacitance of the closely spaced metal lines used to connect the gates and the S/Ds of the multiple-finger devices. Finally, the subcircuit includes a diode element to model the parasitic capacitance of the N-well-to-substrate junction. This basic building block is replicated three times and connected with resistance elements that represent the metal wiring of the multiple-finger gate and S/D contacts.

The plots in Fig. 2.35 show a typical *CV* curve and the frequency response for a large multiple-finger nMOS varactor. Although the model fit of the capacitance steadily degrades in the depletion region ($V < -1$ V), the use of this device is restricted to a maximum reverse-bias voltage of –0.5 V due to the instability of the inversion layer. As a result, the model's inability to properly predict the voltage dependence in this region is not significant. The models are valid through a few tens of GHz, which is beyond the useful range of operation by several GHz.

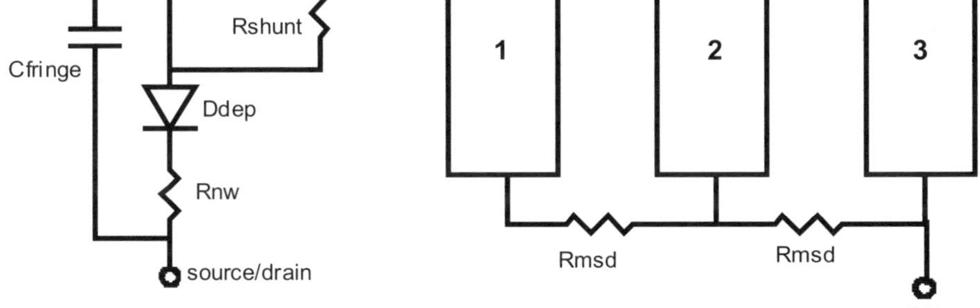

(A) Single segment of MOS varactor model topology

(B) Schematic of full subcircuit with segments connected by metal line resistances

Figure 2.34 nMOS varactor-model subcircuit. (A) Single segment of MOS varactor-model topology, (B) schematic of full subcircuit with segments connected by metal line resistances.

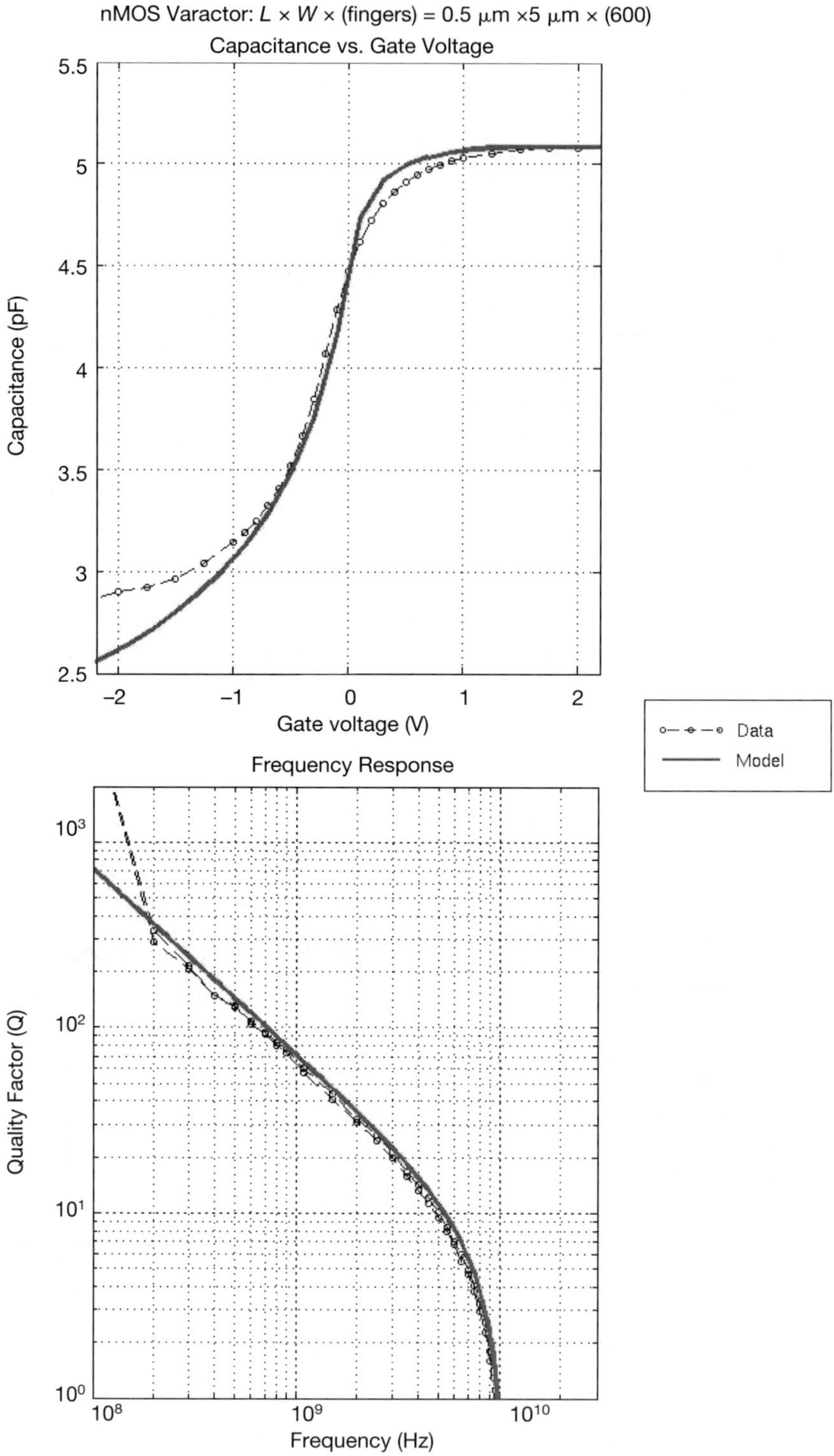

Figure 2.35 Capacitance vs. voltage and frequency response of an nMOS varactor.

2.4.4 Resistors

Key issues for modeling the different types of diffused and polysilicon resistors available in SiGe technologies include accounting for the resistance of the silicided end regions, calculation of the parasitic capacitance, and determining the proper partitioning of the capacitance and resistance across the elements of the subcircuit. Standard DC measurements of resistance across voltage bias and temperature conditions are made on a set of resistors with varying width and length dimensions to extract the parameters necessary to describe the total resistance as a function of geometry, bias, and temperature.

As with other passive devices, the frequency response of the model is verified using S-Parameter measurements. The actual number of *RC* element segments and the division of the total resistance and capacitance across these elements dictates the frequency dependence predicted by the model. The standard model subcircuit contains four resistance elements to represent the body and end regions of the resistor. Parasitic capacitance is split across three capacitors connected from the nodes between the resistors to the body of the resistor, labeled as substrate (SX), in the diagram in Fig. 2.36. The existing model topology has been successful in predicting the resistance rolloff across frequency for both polysilicon and TaN BEOL resistors, as shown in Fig. 2.37. Note that for resistors of similar geometry, the polysilicon resistor begins to roll off just above 1 GHz, while the TaN resistor maintains its low-frequency resistance value up to nearly 10 GHz.

Additional model features include resistance mismatch as a function of geometry, support for parallel or series bars, and the flexibility to netlist resistors by specifying width and length or width and resistance.

2.4.5 Summary

In this section, we have presented the issues in a broad range of passive compact device models, including inductors, capacitors, varactors, and resistors. In Section 3.3, we also discuss IBM's transmission-line model implementation.

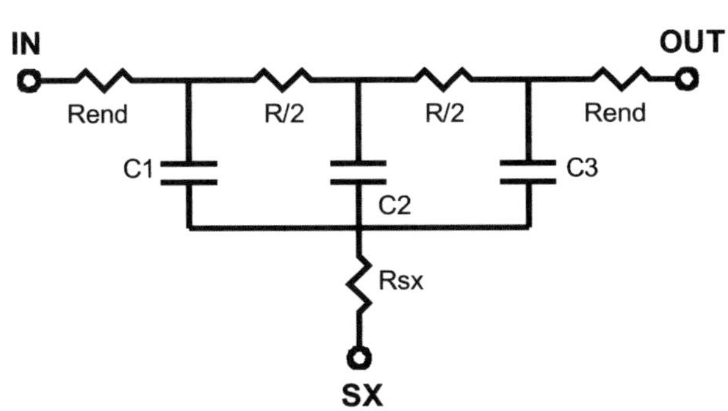

Figure 2.36 Typical resistor model subcircuit.

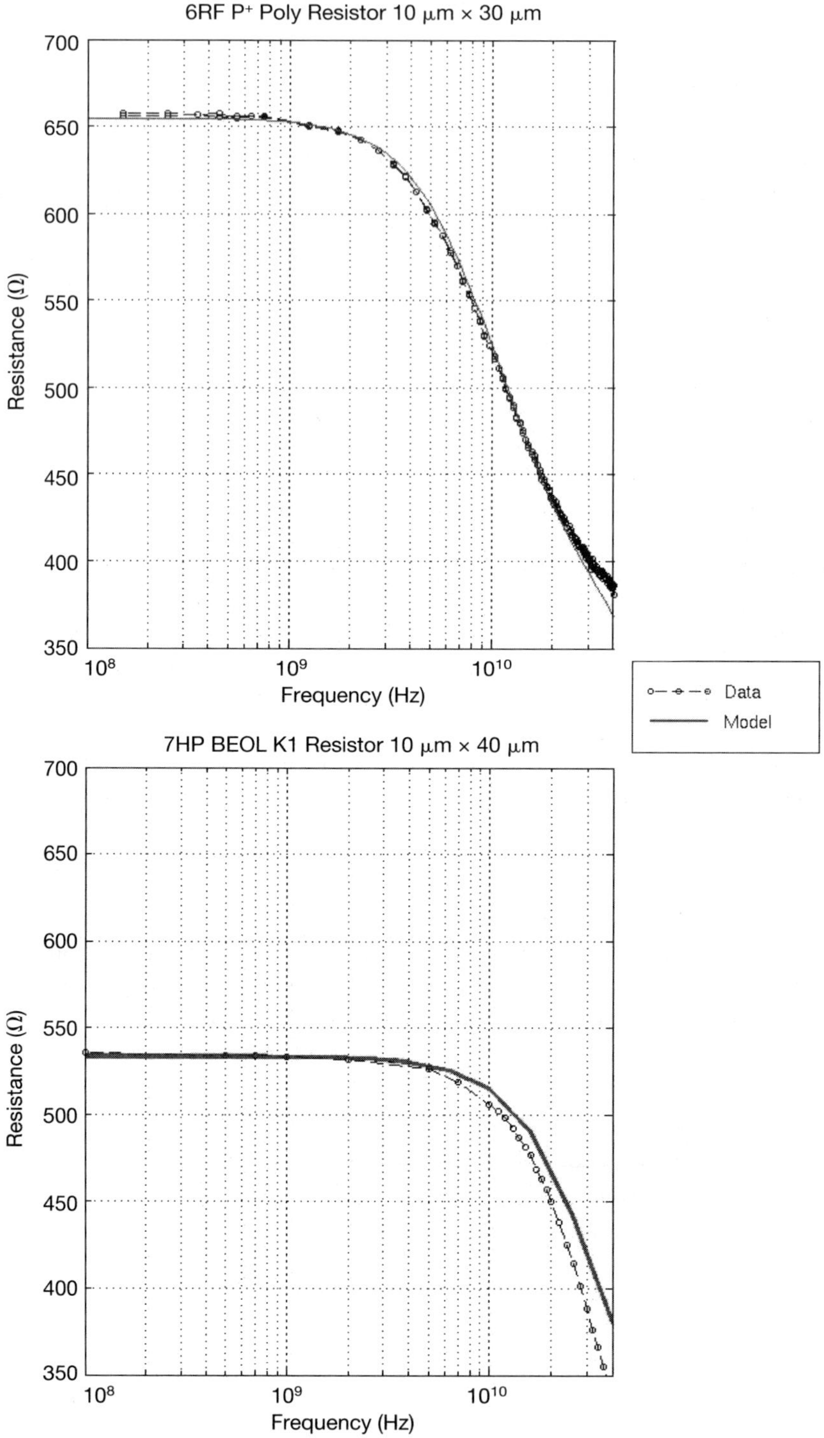

Figure 2.37 Frequency response for a polysilicon and TaN thin-film resistor.

2.4.6 References

1. J. R. Long and M. A. Copeland, "The Modeling, Characterization, and Design of Monolithic Inductors for Silicon RF IC's," *IEEE J. Solid-State Circuits,* vol. 32, pp. 357–369, March 1997.
2. K. O, "Estimation Methods for Quality Factors of Inductors Fabricated in Silicon Integrated Circuit Process Technologies," *IEEE J. Solid-State Circuits,* vol. 33, pp. 1249–1252, August 1998.
3. C. P. Yue and S. S. Wong, "On-Chip Spiral Inductors with Patterned Ground Shields for Si-Based RF IC's," *IEEE J. Solid-State Circuits,* vol. 33, pp. 743–752, May 1998.
4. S. S. Mohan, M. M. Hershenson, S. P. Boyd, and T. H. Lee, "Simple Accurate Expressions for Planar Spiral Inductances," *IEEE J. Solid-State Circuits,* vol. 34, pp. 1419–1424, October 1999.
5. J. R. Long, "Monolithic Transformers for Silicon RF IC Design," *IEEE J. Solid-State Circuits,* vol. 35, pp. 1368–1382, September 2000.

3

DESIGN AUTOMATION AND SIGNAL INTEGRITY

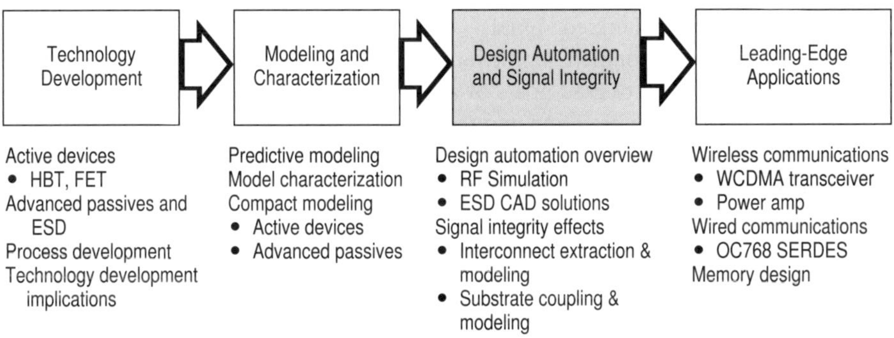

OVERVIEW

In Chapter 2, we presented details of IBM's modeling and characterization methodologies. The next steps in the enablement involve best-in-class design automation solutions, including the CAD tools environment, RF simulation algorithms, ESD CAD solutions, and signal integrity solutions for interconnect and substrate modeling. These offerings form together with the compact models to form the design enablement for the customer. As such they are complex software engineering projects requiring very high quality and efficiency, and in the case of signal integrity the need for effective and efficient modeling usable by the design community.

The chapter presents an overview of the IBM design-automation methodology, the ESD design automation offering, interconnect modeling requirements and solutions, and substrate isolation and modeling solutions.

ISBN 0-471-44653-X

- Section 3.1 is an overview of IBM's design automation methodology.
- Section 3.2 discusses design-automation environment for IBM's world-class ESD offering.
- Section 3.3 introduces the complex topic of interconnect modeling and extraction, including transmission line modeling and substrate interactions with interconnects.
- Section 3.4 discusses issues in substrate modeling and isolation, through coordinated TCAD and test-site activities at IBM.

3.1 DESIGN AUTOMATION OVERVIEW

Circuit designers interface with technology through a collection of design tools linked together in a common design framework. The discussions in this chapter focus on a device-level design-automation system in which all devices are indicated by basic layouts and base–compact models as opposed to higher-order design systems concerned with timing closure, methodology checks, etc. The key components of this device-level toolset are schematic capture, layout, physical verification tools, and thus the design flow outlined in this chapter pertains to custom circuit design: analog, digital, RF, and mixed-signal.

The key CAD environment elements for an RF/mixed-signal custom design are shown in Fig. 3.1. The overall goals of an effective CAD framework are to reduce or eliminate hardware fabrication and redesign iterations through optimal tools and design methodologies, and to adapt the design flows to keep pace with the growing demands of the leading-edge designs created by our customers. Also listed in Fig. 3.1 are the key components provided in IBM's RF/AMS device-level design kits, culminating from the technology development work in Chapter 1 and the modeling and characterization work described in Chapter 2. These components form pieces of a robust world-class design kit methodology, which is referred to throughout this book.

Figures 2 to 4 show in more detail how many of the components in the design kit can be used in the design analog (custom) circuit design flow. The key aspects shown in these figures are referred to later this chapter. For further details of how the software design tools work and are used in the design flow, the reader is referred to the specific EDA tools vendor.

3.1.1 Design Entry and Simulation

For entry of a design into the CAD framework (see Fig. 3.2), a library of hardware characterized and modeled device primitives such as bipolar junction transistors, metal-oxide semiconductor FETs (MOSFETs), resistors, capacitors, inductors, diodes, and transmission lines are assembled. A database contains component information for schematic representation, parameter definitions, and netlisting.

A designer uses industry-standard symbolic representations of these active/pas-

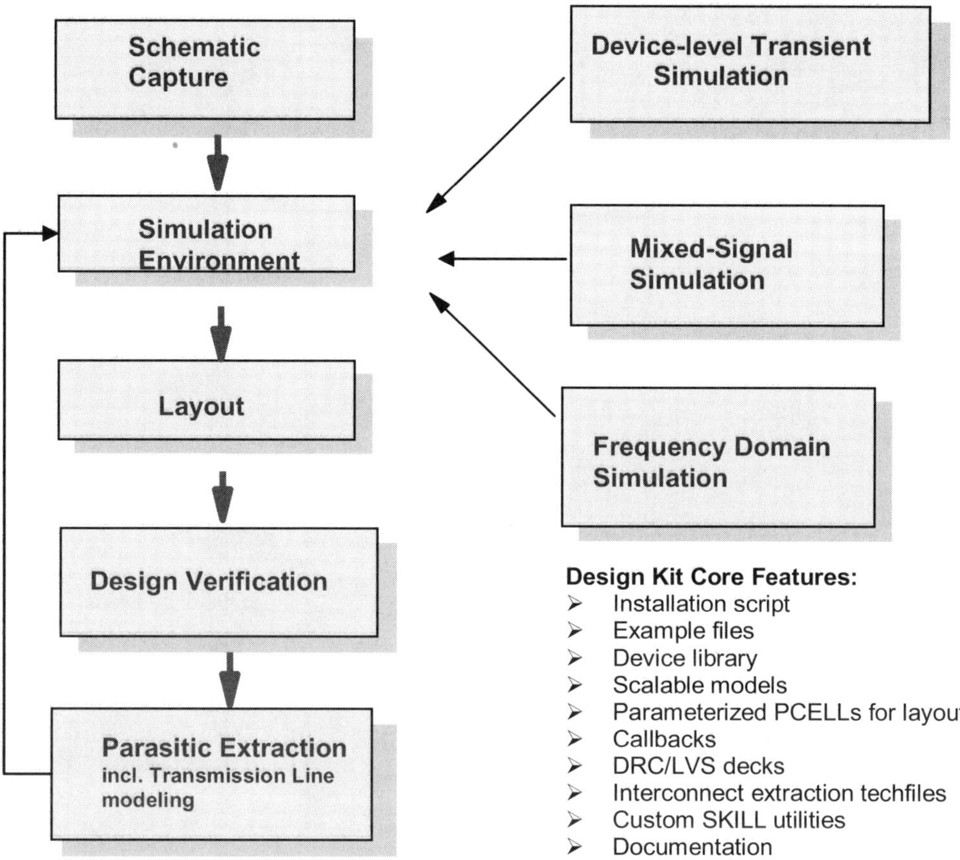

Figure 3.1 High-level view of IBM's RF/mixed-signal design flow and its enablement.

sive components to input the electrical representation (schematic) of the design. The artwork used to define a component can be customized for the application of the element. For example, if a MOSFET device is to be used for a high-voltage application, a parameter is set and some characteristic of the artwork, such as color or shape, is changed to indicate use and modeling.

Each device element is defined by a model name with ports and a set of parameters that express the device dimensions and electrical properties. Parameters can include such characteristics as geometric dimensions, device current rating, resistance/capacitance/inductance, device model parameters and subcircuit construction, reliability information, and device options such as ground planes, trenches, and guard rings. Parameters are defined to ensure proper device selection and electrical verification and to pass device information to the layout engineer.

Each device is defined with a procedure to create a netlist entry of the component

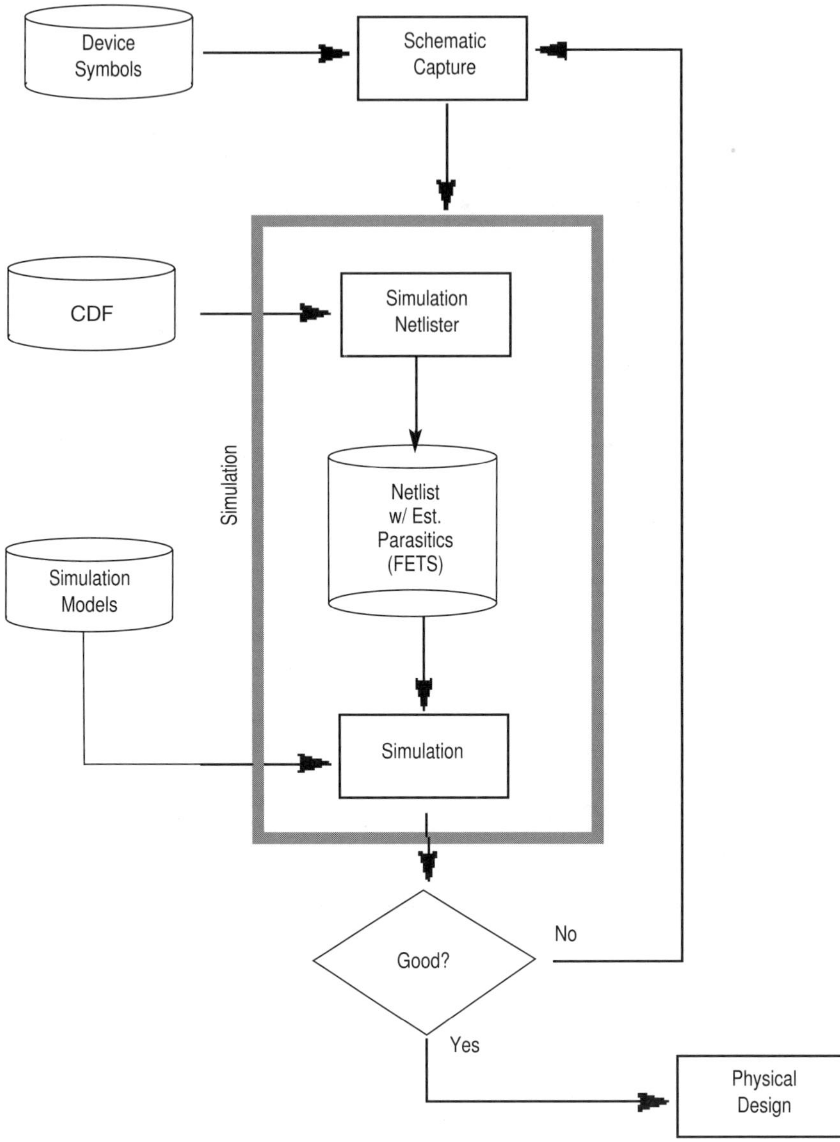

Figure 3.2 Steps in the analog schematic design flow, using the design-kit components.

for a given circuit simulator. Device parameters can be entered as variables to enable a simulation sweep of the parameter to optimize designs. Parameters such as resistance values can be defined as a variable, and a series of simulations automatically scales the resistance. The netlist procedure will write a netlist for each component with the connectivity nodes and parameters for a given circuit. These procedures are customized to incorporate specific enhancements such as device multiplicity.

3.1.2 Parameterized Cells (PCELLs)

Schematic-driven layout provides a head start to physical design (see Fig. 3.3). A schematic-driven layout tool places the components in a predefined aspect ratio. The net connectivity information and device parameters are copied along with the device. In this way, devices are laid out exactly as they are designed in the schemat-

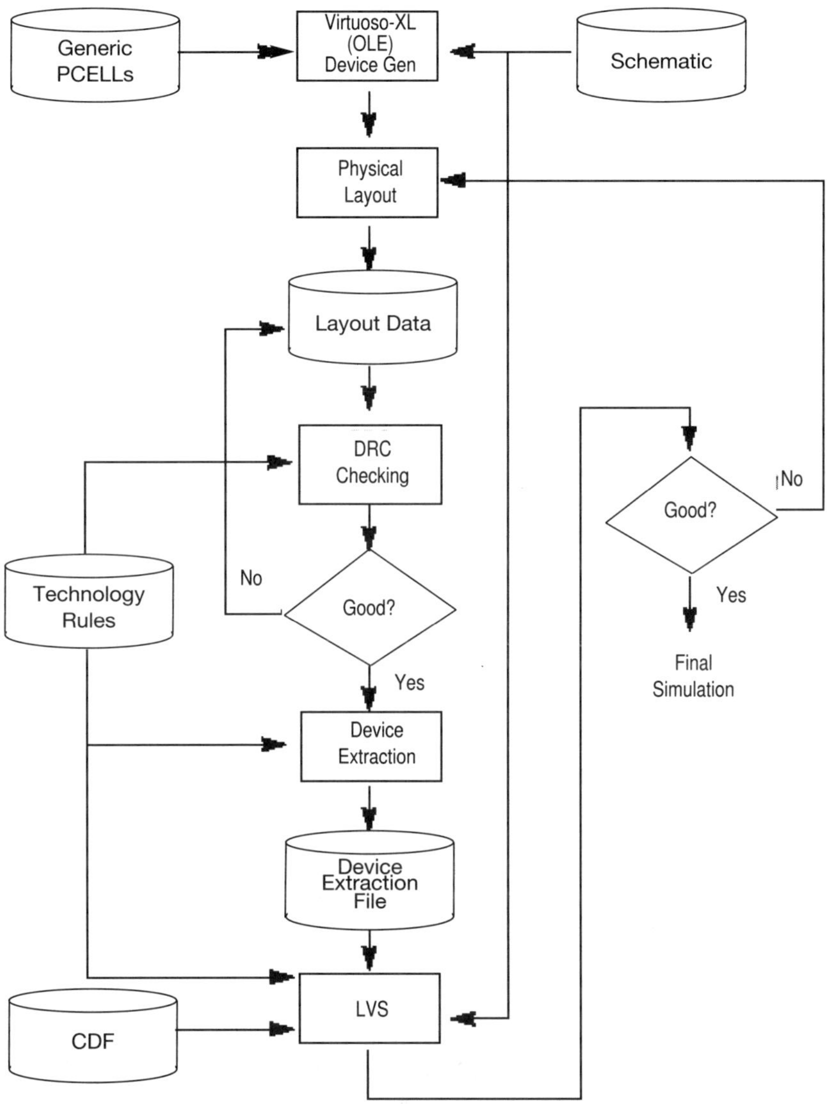

Figure 3.3 Steps in the analog physical design and verification design flow, using the design-kit components.

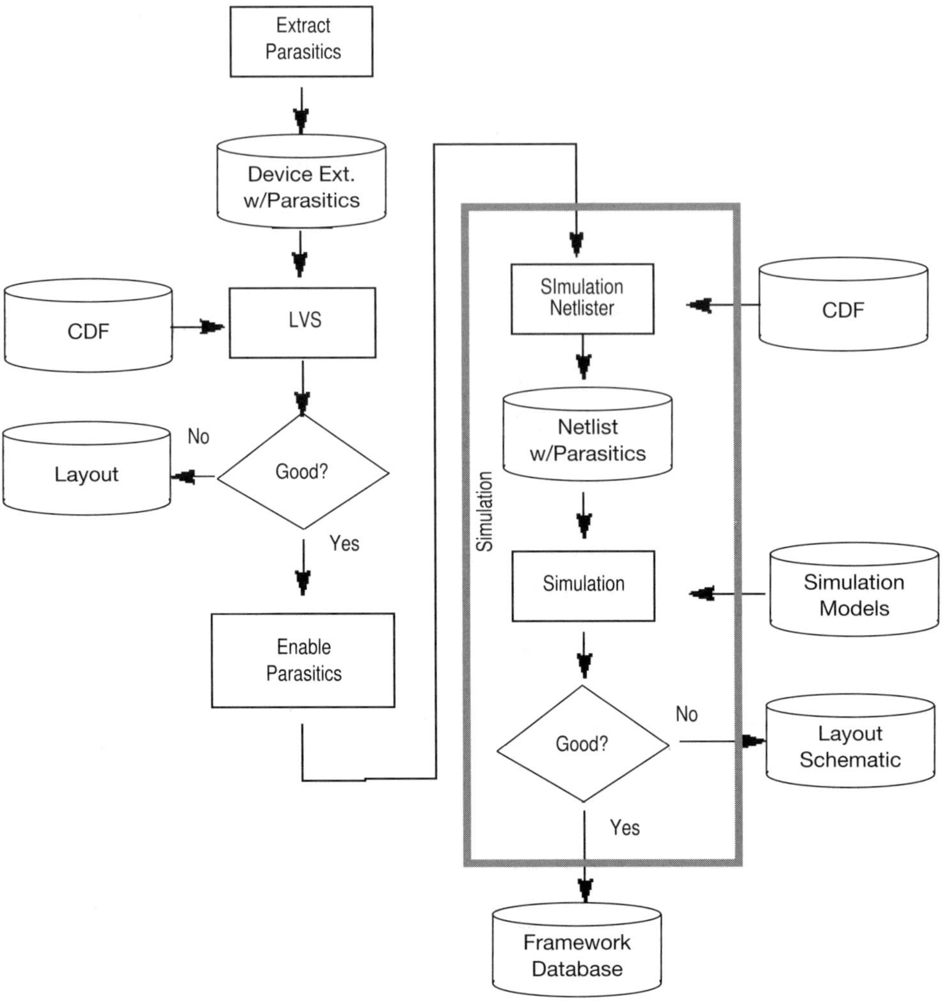

Figure 3.4 Steps in the analog parasitic extraction design flow, using the design-kit components.

ic and with device and net names and parameter selection. Figure 3.4 shows the past LVS flow.

Device primitives are defined as PCELLs. These are programmable component layouts that stretch according to parameter input. Layout options such as guard rings, well connections, ground planes, and multistripe connections are passed to the PCELL as optional parameters. Figure 3.5 shows various inductor PCELLs that have been implemented, mapping to the compact models described in Section 2.4. Figure 3.6 shows how a PCELL can be used to generate a multifingered MOSFET or a multiplicity of MOSFET devices, mapping to the compact models described in Section 2.3.

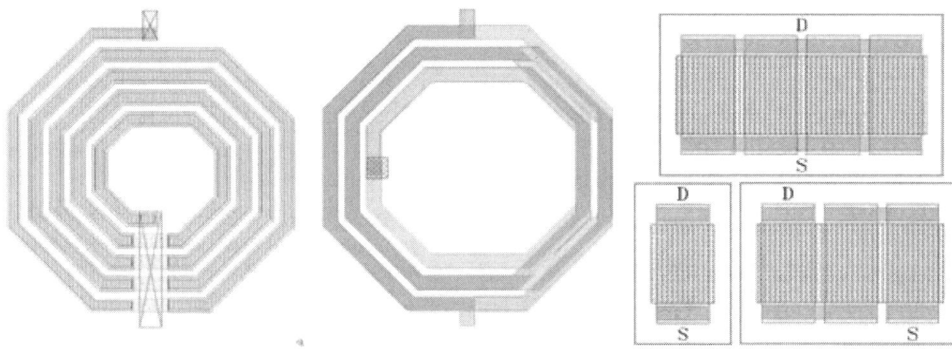

Figure 3.5 Various inductors, implemented as PCELLs, are stacked or in series with various metal widths and number of turns. The resistors can be defined with a series of parallel bars for matching and form de facto considerations.

When the schematic is transferred to the layout, these parameters will build the PCELL component as defined by the circuit designer. The PCELL is designed according to the design rules and, when placed, the component is design-rules-checking (DRC) correct by construction. The design rules can be input to the database for access by all tools within the framework. The PCELLs are created accessing the defined technology rules for spacing and device construction. These

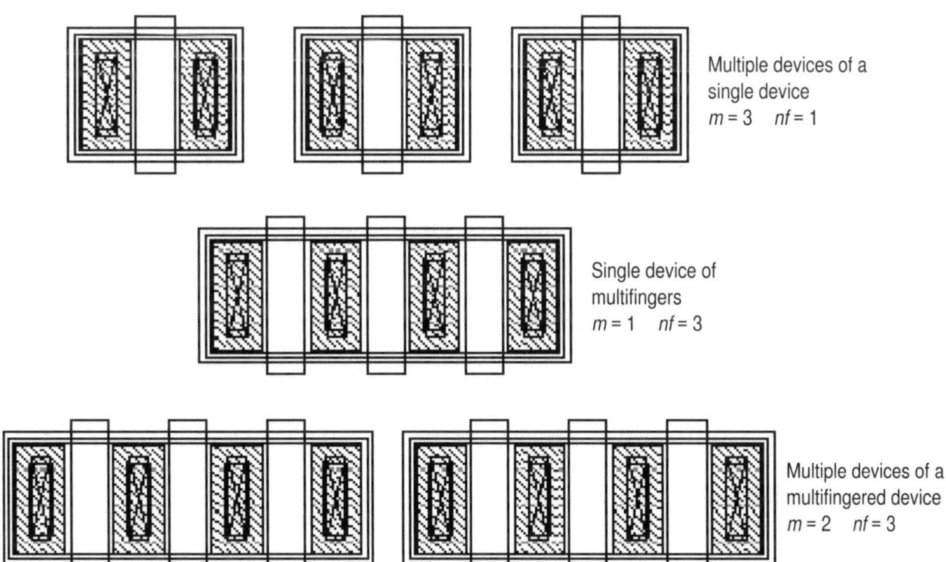

Figure 3.6 Multiplicity vs. multifingered MOSFET PCELL example.

are input to the PCELL as variables, enabling easy migration to technologies with a database update.

Net information is input to a wiring-aid utility that highlights nodes and wires to be connected. Autowiring tools are available but, at high frequencies, custom wiring is recommended to control current flow and parasitics. Layout versus schematic (LVS) ensures that all nets are correctly wired.

Guard rings are designed for circuit isolation and latch-up protection by collecting potential carriers flowing in the substrate [1]. The manual construction of these elements is tedious, and it is difficult to satisfy the process design rules. Additionally, to enclose all circuitry requires intricate polygon design of the guard-ring paths. The flexibility of the framework interpretive language allows design-rule-correct construction of these guard-ring paths to be drawn around sensitive circuitry. Further, the metal may be cut in a designated region to allow wiring between design stages.

3.1.3 Design-Rule Checking (DRC)

Design-rule verification involves checking mask layer interactions to determine that their fabrication meets manufacturing processing and tooling requirements. The design rules are provided in design manuals, written by the technology development engineers. SiGe design rules are often derived from the CMOS-equivalent processes. For example, many SiGe 7HP design rules are the same as the CMOS 7SF 0.18-μm design rules.

Process layers are checked for width, length, area, and separation limits and overlay tolerances with other process steps, like contacts within diffusions, polysilicon, and metals. Robust manufacturability also involves voltage and current considerations, density, and uniformity. Some manufacturing processes use high static charges that may require design protection in the wafer fabrication and final packaging of products; this defines another set of critical integration checks. Proper device construction is verified to ensure all required design levels are present. Layers that may alter the device characteristics are checked for omission from the device, marking improper abutment or device interaction. Design-rule violations are flagged as errors or warnings by the DRC tool. Interpretive procedures are provided for analysis.

Copper wiring brings the requirement of local density to improve soft-metal processing [2]. The manufacturing requirements include minimum density percentages of certain layers and the ability to cut holes to reduce high density and preserve wide-metal integrity. This requires specialized verification procedures to ensure density levels regionally and globally. An analysis of the foundry fill routines must be made available to the designs prior to data submission for customizing fill patterns on existing circuitry to allow critical circuits to be protected [3]. An example is shown in Fig. 7.

3.1.4 Layout Versus Schematic (LVS)

LVS checking proves that the physical data to be fabricated is consistent with the schematic design that has been simulated (see Figs. 3.3 and 3.4). A netlist is created

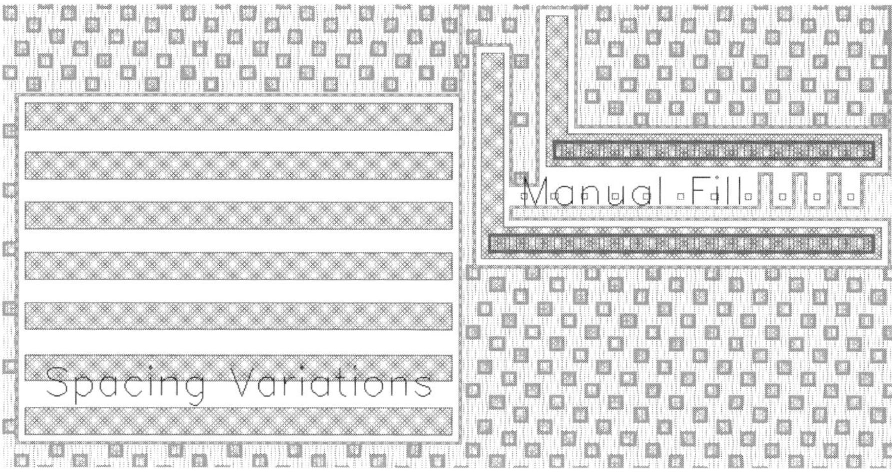

Figure 3.7 Typical layout obtained from the use of an automated pattern-filling process for achieving desired metallization density in nonsensitive circuits.

by identifying devices with parameters and connectivity for each section. These netlists are compared for matches and the differences in parameters or connectivity are identified. Although translators exist for import of third-party tools into the main CDS design framework, it is important to have a common framework and netlisting environment to ensure that the sections accurately represent the data.

The layout must be processed through device recognition and parameter measurement extractions in order to accept variable styles used within the design and find all the shape interactions that make up devices and connectivity. Many devices use nonstandard parameters, such as ground plane, levels of metal, and simulation frequency, and custom compare procedures may be required to process the results and report any differences. Standard compare procedures may be used for simple devices. Custom programs to analyze the results enable the designer to diagnose design errors more efficiently.

Multiplicity is the repetition of like-sized devices with identical node connections. Device symbols define multiplicity as a parameter to the device to indicate the number of devices in parallel. By this convention, *m* number of devices are simulated, yet a single device is shown in the schematic. Device models accept this parameter; however, device wiring parasitics and device mismatch are not applied. Schematic-driven layout creates *m* number of devices and tags each device with a recognition shape. Device extraction recognizes devices in multiplicity separately from nonmultiplicity devices. LVS will compare the multiplicity parameter from the schematic and check the quantity of devices and exact device connection to that in the layout.

As technology and design complexity have increased, it has been necessary to adopt higher-performance *hierarchical* checking tools. *Flat* checking tools typically are limited to designs containing less than tens of thousands of devices. Hierarchical tools enable us to check designs containing millions of devices, and are particularly useful for when there are high levels of integration.

3.1.5 Regional Substrate Methodology

Device extraction for many of IBM's SiGe processes includes complete terminal checking for N-type devices in a P-type substrate for NPN and NFET devices. Implementation of this has required an additional layer to be added to the technology file called SXCUT. This layer (shown in Fig. 3.8) can be used to separate the entire chip substrate into regions and label the substrate node for each region uniquely. The remainder of the chip substrate is given an SXCUT region implicitly, which allows design without SXCUT to be processed as normal.

There are two methods to identify the substrate nodes on the substrate sections as defined by the SXCUT layer. Either method can be used, or they may be used together. However, if the labels are in conflict, an error message will be issued.

1. The substrate nodes can be labeled with text on the layer purpose pair (SXCUT label). This allows the simulation substrate node, used for noise analysis, to be mapped into the LVS-compare and ensure the layout agrees with the circuit under simulation. This method works with GDS2 data or DFII data.
2. The substrate nodes can be labeled with a terminal pin on the layer-purpose pair (subdrawing). This allows the simulation substrate node, used for noise analysis, to be mapped into the LVS, compare, and ensure the layout agrees with the circuit under simulation.

Multiple substrate text labels or terminal pins in the same region are flagged as errors. Substrate contacts to that substrate region will be extracted as a resistor between the circuit node and the regions the SXCUT node name. The device subc in the library SiGe 7HP can be used in schematics to connect circuit nodes to substrate-region nodes through a resistor. If a substrate label is not defined, LVS will check its connections, but not give a terminal match or helpful diagnostics. This feature allows more simulation and substrate analysis flexibility, and guarantees through LVS that the layout matches those assumptions. Coupled with the DT isolation of the substrate through high resistance P-wafers, this methodology will allow specific verification of substrate regions.

Figure 3.8 shows a design sectioned into multiple substrate regions. The regions are identified by a rectangle drawn on the SXCUT drawing and labeled with the SXCUT label (SUB_A, SUB_A2, sub!) and also labeled with a terminal on subdrawing with the pin names (SUB_A, SUB_A2, sub!). The regions may be identified by the LVS/extraction tools through the SXCUT label or the subdrawing pin or both. The LVS/extraction tools check for label/pin consistency within a region.

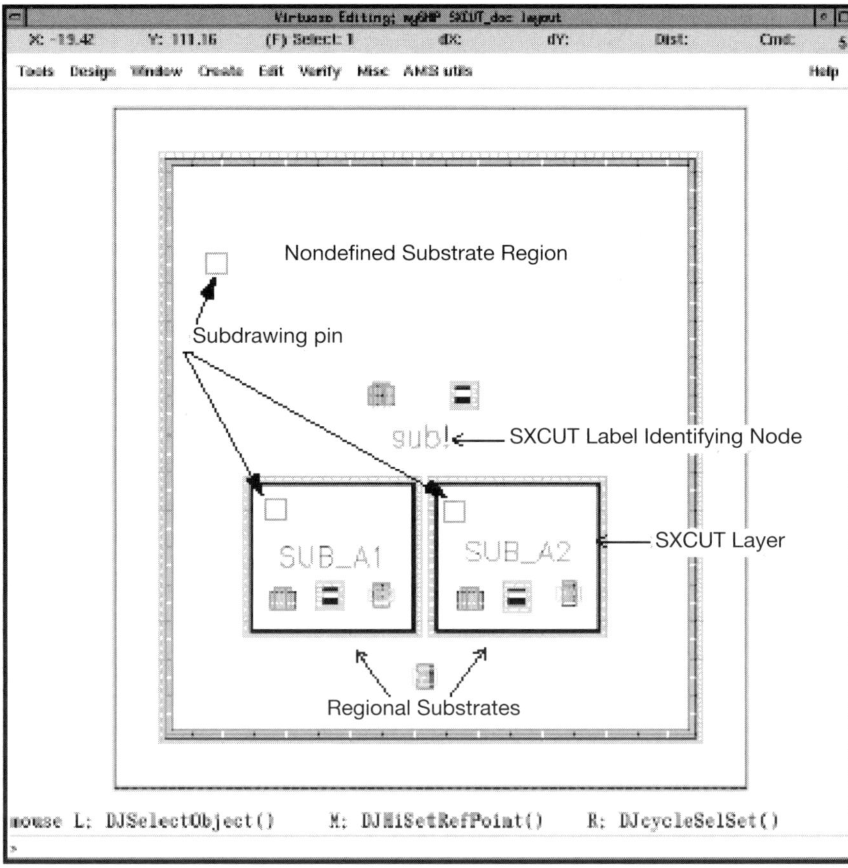

Figure 3.8 Multiple region SXCUT substrate modeling methodology, for interblock isolation.

3.1.6 Simulation Tools Benchmarking

Today, there are numerous commercially available electronic design automation (EDA) tools and flows for accurate and efficient RF IC design. However, two commercially available algorithms dominate the RF simulation industry: periodic steady state (PSS) and harmonic balance (HB). So, given that there is very little choice, which one is better? The answer is cryptic, much depending on the type of circuit being simulated, the frequency of interest, and the device models being used, etc. In this section, we take a look at some of these aspects through simulations on some typical RF circuit blocks, in this case, a simple power amp circuit and a wireline VCO. We present preliminary conclusions about where the strengths of the different algorithms lie, aimed specifically at the designer's concerns and without being technical about the math involved, thus providing a useful starting point for those looking at building best-in-class RF IC simulation flows.

It is important to note that none of the results presented here are intended to bias the reader toward one tool or another. Tools benchmarking is a very complex and time-consuming exercise that requires experience to understand the intended goals. The work presented is purely for algorithm benchmarking, and even then to give a direction rather than a definitive answer.

The design of high-frequency circuitry for digital radio, telephony, optical, and instrumentation applications has increased the demand for simulators that are well suited to predict RF and microwave circuit behavior. The EDA industry has come up with two competing approaches, such as power amplifiers (PAs) [4], LNAs, mixers, and VCOs, for simulating RF ICs. These approaches use the harmonic balance algorithm and the PSS algorithm. Other approaches may exist, but are unknown to the authors, at present.

Harmonic Balance The HB engine uses a native frequency-domain approach. If the input signal is small enough that nonlinear elements in the circuit do not significantly distort the signal output, then the small-signal simulation will give valid results. However, as the input signal becomes increasingly large, then new frequencies will be seen at the output. HB solves for each of these new frequencies. There are several EDA companies that develop and market HB simulation engines. Each of these engines is apparently a derivative of the HB engine with various performance and usability benefits. In this work, the Agilent advanced-design system (ADS) tool was used (2001 version).

Periodic Steady State PSS is effectively an RF simulation extension to a transient simulation engine, with the assumption that a periodic signal exists in the system. Cadence's well-known SpectreRF simulation tool is the tool that implements the PSS algorithm [5]. This tool was used in this benchmarking project (version 446).

The efficiency of each type of simulation (HB versus PSS) depends largely upon the types of elements in the circuit and the complexity of the signal at the input. For example, passive elements, such as resistors, inductors, and capacitors, are linear except in the extremes of their operating ranges, whereas diode elements exhibit nonlinear electrical characteristics at large input signals. Active devices are made up of P-N diode junctions, and are therefore nonlinear as well. Note that both the HB and PSS algorithms first require the calculation of a DC operating point.

Results from Benchmarking Effort of Power Amp Circuit [5] We present some results from our benchmarking effort for a simple power amplifier circuit [5]. Figure 3.9 shows a comparison of S-parameter simulation results from both the PSS engine and the HB engine. As can be seen, the correlation between the HB engine and the PSS engine is very good. Interestingly, the HB engine simulation was completed in 8.4 s, while the PSS engine simulation ran in just 0.34 s. Figure 3.10 shows a power-transfer simulation result of the same power-amplifier circuit. Virtu-

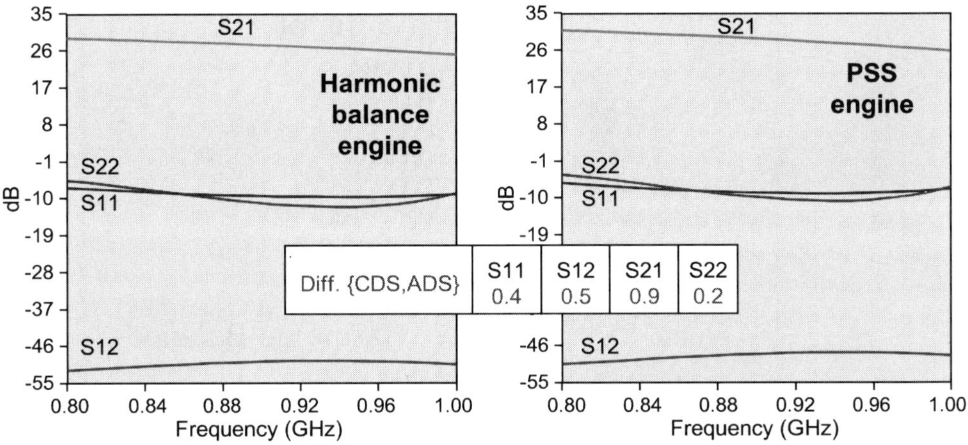

Figure 3.9 S-Parameter simulation results for an SiGe PA circuit [4]. Comparison of the harmonic balance engine vs. the PSS engine.

ally identical results are produced by a PSS simulation and single-tone HB simulation. Figure 3.11 shows a comparison of IM3, IM5, and power using both quasi-PSS (QPSS) and a two-tone HB analyses. The circuit under test is a PA for use in a radio, for example. Both simulation results are in very close agreement with IM3 and IM5 components to about 0 dBm. In our study the PSS simulation fails to converge beyond 0 dBm for full-power two-tone intermodulation product estimation, while HB converges to 20 dBm. This is obviously an important consideration for nonlinear large-signal PA circuit simulation. The PSS results also show a different

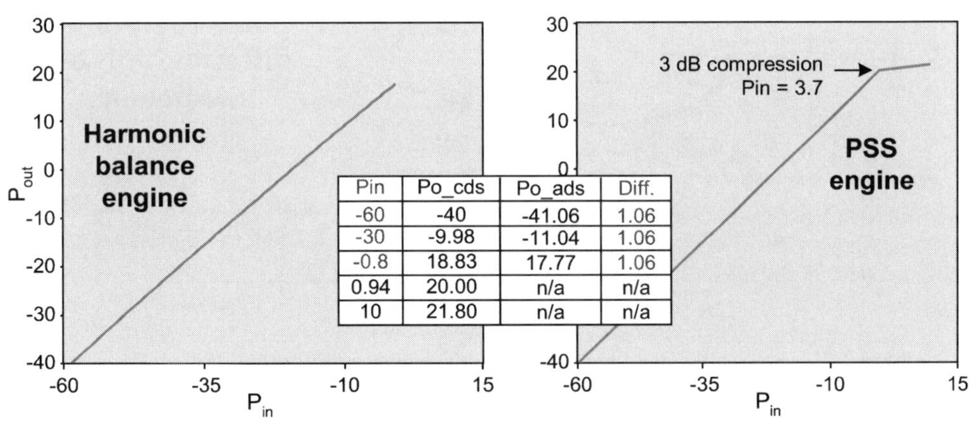

Figure 3.10 P_{out} vs. P_{in} results for PSS and HB (one-tone) simulations for an SiGe PA circuit [4].

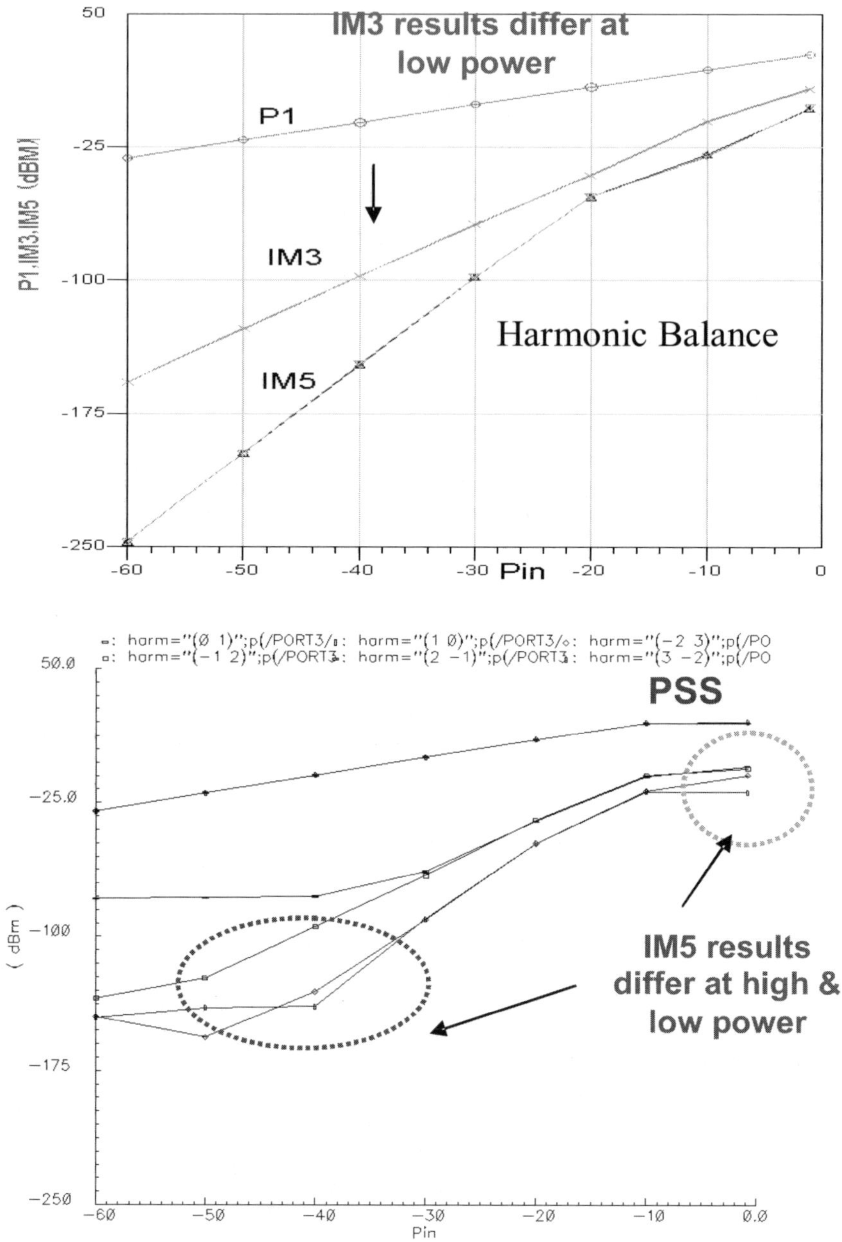

Figure 3.11 Frequency- and time-domain results comparison for IM3, IM5, and power using both SpectreRF's QPSS and a two-tone HB analysis.

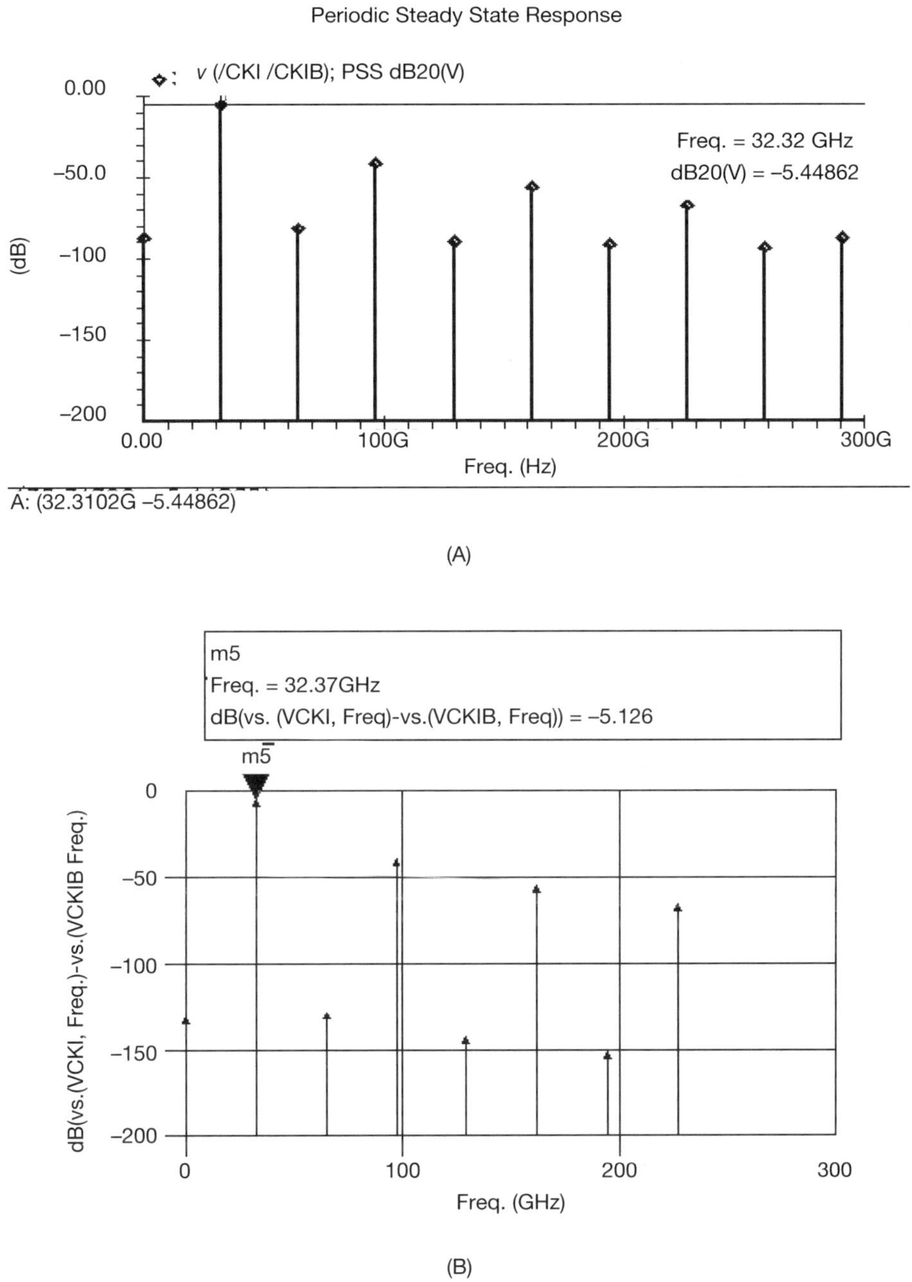

Figure 3.12 (A) Single-tone frequency-domain spectrum using the PSS algorithm; (B) Single-tone frequency-domain spectrum results using the HB engine.

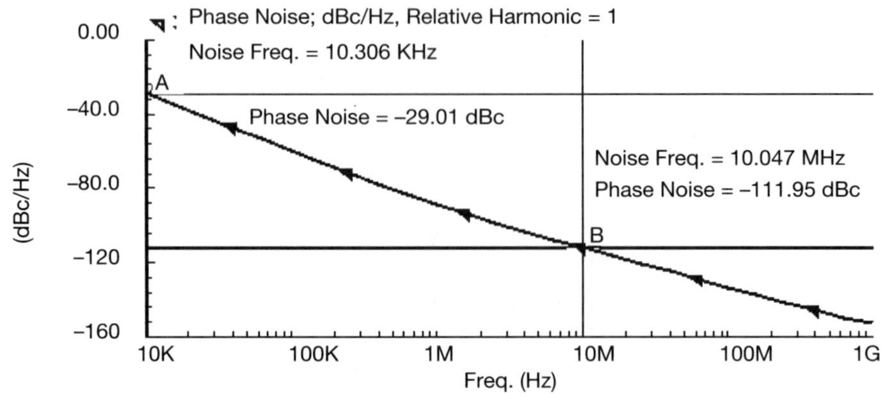

A: (10.3062K –29.0109) delta: (10.037M –82.9402)

B: (10.0473M –111.951) slope: –8.26347u

(A)

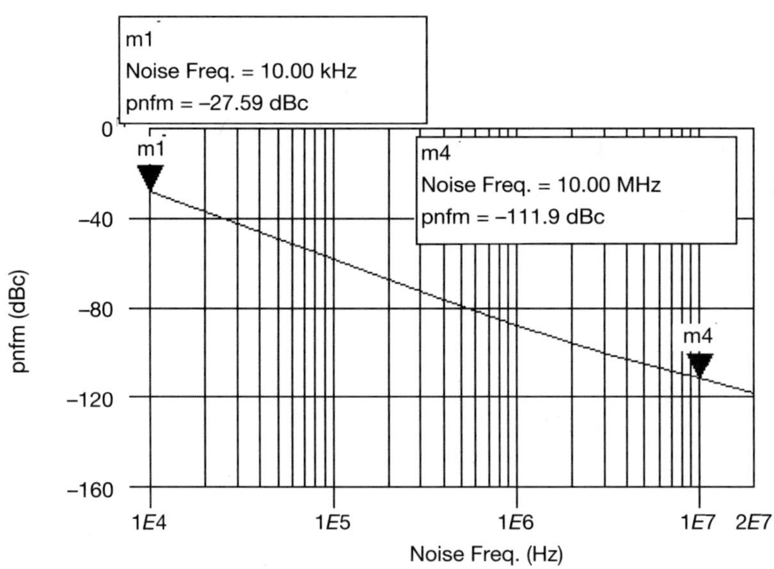

(B)

Figure 3.13 (A) Phase noise results using the PSS algorithm; (B) Phase noise results using the HB engine.

set of results for each of IM3 and IM5, as the upside and downside images (in the frequency spectrum) do not have the same magnitude. We are still unsure why this is happening (in 2002).

Results from Benchmarking Effort of VCO Circuit Here, we present results from the benchmarking of a VCO circuit, designed to run in an OC768 20-GHz phase-lock loop (PLL). Similar to the PA discussed earlier, the simulation was run on the schematic only, so that issues, such as different levels of parasitic extraction accuracy, do not invalidate the benchmarking.

Figure 3.12A and 3.12B show the single-tone frequency-domain spectrum results. The results show a frequency of 32.32 GHz for the PSS engine, and 32.37 GHz for the HB engine. These results are very close, although a 50-MHz difference in performance can affect the overall system performance in certain cases. Figure 3.13A and 3.13B show the phase noise results for the circuit. At 10-kHz, the PSS engine gives a value of –29.01 dBc versus a value of –27.59 dBc using the HB engines. At 10 MHz, the PSS engine gives a value of –111.95 dBc and the HB engine gives a value of –111.9dBc. These results are again close, although there is some difference at 10 kHz.

3.1.7 Summary

We have discussed a fully integrated RF/mixed-signal design-kit methodology, as supported at IBM. Key components of IBM's SiGe design kits have been presented in detail. Over time, IBM continues to provide well-tested and effective design kit solutions to its customers, especially in the areas of high-frequency communications design and highly integrated mixed-signal SOC design.

RF simulation benchmarking results also were presented. While HB simulation is well suited for certain nonlinear RF circuits, such as the large-signal simulation of PA circuits, the PSS engine may have benefits for highly nonlinear circuits and small-signal simulations. The PSS engine was also observed to perform relatively better on VCO circuits. Also, transient simulators can perform very well for other types of circuit simulations, such as larger circuits with some custom (or standard cell) digital content.

3.1.8 References

1. "Guard-Ring Layout," in *BiCMOS7HP Design Manual,* Section 3.28.3, IBM Technology Development, February 2002.
2. "Copper Pattern Density and Layout Requirements," in *CMOS7SF Design Manual,* Section 2.10.2, IBM Technology Development, 2000.
3. S. Strang and Donald Jordan, "Custom Verification Using Diva and DFII," Proceedings of International Cadence User Group Conference, December 2001.
4. P. D. Tseng, L. Zhang, G. B. Gao, M. F. Chang, "A 3-V Monolithic SiGe HBT Power Amplifier for Dual-Mode (CDMA/AMPS) Cellular Handset Applications," *J. Solid-State Circuits,* vol. 35, no. 9, Sept. 2000.

5. K. Kundert, "Introduction to RF Simulation and its Application," *J. Solid State-Circuits,* vol. 34, no. 9, Sept. 1999.

3.2 ESD: BEST-PRACTICE CAD IMPLEMENTATION

Traditionally, ESD has been custom-designed for each product release. In SiGe, IBM has developed an automated ESD design process in the CDS environment using autogenerated hierarchical parameterized cells for custom layout and schematic generation. With this automated design capability, ESD designs can be varied in form factor, structure size, and loading effects to allow for compatibility with the design needs, practices, and functional objectives of each customer. From this hierarchical PCELL ESD kit, it is possible in only minutes to synthesize a complete ESD design for a product chip for any power supply or chip architecture. HBM, machine model (MM), and transmission-line pulse (TLP) results demonstrate the ability to develop automated ESD design systems.

For mixed-signal and RF technology, it is important to have precision models for digital, analog, and RF circuitry. Precision models and ESD evaluation in RF CMOS, GaAs, SiGe, and InP will play a key role in the development of RF applications [1–9]. In SiGe technology, state-of-the-art production transistors have reached cutoff frequency values of 200 GHz [10–14], as described in earlier chapters. At these frequencies, the ability to optimize both the RF design and ESD is critical. Additionally, ESD protection circuits for input nodes must also support quality DC, AC, and RF model capability in order to codesign ESD circuits for these analog and RF circuits. With the growth of the high-speed data rate transmission, optical interconnect, wireless, and wired marketplaces, the breadth of applications and requirements is broad. Each type of application space has a wide range of power-supply conditions, number of independent power domains, and circuit performance objectives. As a result, an ESD design system that has DC and RF characterized models, design flexibility, automation, ESD characterization, and satisfies digital, analog, and RF circuits is required to design and cosynthesize ESD and RF performance. In this section, an automated ESD CAD design system is shown that achieves these objectives in a 120/100-GHz f_T/f_{MAX} SiGe 0.18-μm BiCMOS- and ASIC-compatible CMOS technology [13].

3.2.1 ESD Design CAD System

Traditionally, ESD designs are custom designed using graphical systems. ESD ground rules and structures are typically built into the designs, requiring a mandatory custom layout. Custom design for digital products, such as SRAMs, microprocessors, ASIC development, and foundry technologies, has been the standard process of implementation. This design practice does not allow for the flexibility needed for RF applications. A difficulty in the design of RF ESD solutions is that traditionally, specific designs are fixed in size in order to achieve verifiable ESD results for a technology. A difficulty with analog and RF technology is that a wide range of circuit applications exists where a one-size ESD structure is not suitable due to loading of the circuit. A second issue is that for the cosynthesis of the circuit, the circuit must be designed to properly evaluate the RF performance objectives. RF characterization of a network that is flexible with the device size is important for the evaluation of the trade-offs of RF performance and ESD. A third issue, for RF mixed-signal designs, is the presence of both analog and digital circuits. In these environments, there are some products that primarily use digital CMOS circuits, and some that are purely bipolar implementations. Some applications prefer CMOS-based ESD networks, and others are motivated to use bipolar-based ESD networks.

To address these issues, an ESD CAD strategy is developed to fulfill the objective of design flexibility, RF characterization and models of ESD elements, automation, and choice of ESD network type. Our design system used a hierarchical system of parameterized cells that are built into higher-level ESD networks. Lowest-order PCELLs are RF and DC characterized. ESD verification, DC characterization, schematics, and LVS are completed on the higher-order circuits.

Diode, bipolar, and MOSFET hierarchical cells are used to establish both CMOS MOSFET-based ESD networks and SiGe bipolar-based networks. The parameterized cells, known as PCELLs, are constructed in a CDS design system. The PCELLs are growable elements that fix some variables, and pass some variables to the higher-order circuit through "inheritance." From base PCELLs, ESD circuits are constructed for input pads, V_{DD}-to-V_{SS} power clamps, and V_{SS}-to-V_{SS} power clamps. In these categories, both the CMOS-based and the BiCMOS SiGe-based implementations exist.

The ESD design system is developed to allow for the change of circuit topology as well as structure size in an automated fashion. Layout and circuit schematics are autogenerated, with the user varying the number of elements in the circuit. The circuit-topology automation allows for the customer to autogenerate new ESD circuits and ESD power clamps without additional design work. Interconnects and wiring between the circuit elements are also autogenerated.

3.2.2 ESD Input Circuits

ESD input circuits consist of different input circuits for digital, analog, and RF devices. Since different circuits may be required for the different applications, the input circuits are segregated to allow for distinctions of these different signal types.

For a CMOS digital circuit, a double-diode design PCELL is created. The double-diode design is a PCELL that allows for the inheritance of four parameters: number and the width of the diode fingers for the "up" and the "down" diode element. The double-diode design is a second order hierarchical implementation that consists of two p^+/n-well PCELLs (or a p^+/n-well PCELL with an n-well-to-substrate diode PCELL implementation).

The p^+/n-well PCELL is an ESD-optimized design whose ends are fixed in design style and whose metal, contact, isolation, length, and finger number are growable. Metal bussing is automatically growable, with the width of the diode structures using algorithms associated with number of fingers and design pitch.

For an RF SiGe-based implementation, two p-n varactors are used. The SiGe-based double-diode circuit utilizes the base–collector junction of the varactor PCELL. This PCELL circuit is a hierarchical PCELL that contains the two varactor PCELLs, power rails, and growable interconnects. SiGe varactors produce excellent ESD performance because of the low resistance subcollector.

3.2.3 ESD Rail-to-Rail Ground Solutions

In mixed-signal RF applications, functional circuit blocks are separated to minimize noise concerns. Digital noise affects both the analog and DC circuitry impacting the noise figure (NF). Designers need the ability to estimate the noise and stability of the circuit in the presence of multiple circuits and ESD networks. To eliminate noise, digital circuit blocks are separated from the analog and RF circuit blocks without a common ground or power bus. The introduction of the ESD elements between the grounds can address the ESD concerns, but increases the noise and stability implications. As a result, the cosynthesis of the ESD and noise concerns needs to be flexible to address both issues.

As part of the ESD CAD design system, a hierarchical parameterized cell is designed that forms a bidirectional SiGe varactor string that can vary the number of varactors and the physical width of each varactor. For example, a design may use four varactors in one direction and two in the other direction between the ground rails. The automated ESD design system has the ability to adjust the design size and the number of elements. In digital circuits, the design decision is typically decided based on the digital DC voltage separation required between the grounds; in RF circuits, the design issue is the capacitive coupling at high frequency. As more elements are added, capacitive coupling is reduced. In our ESD design system, the interconnects and wires automatically stretch and scale with the structure size. Algorithms are developed that autogenerate the interconnects based on the number of diodes up versus diodes down. As elements are added, both the graphical layout and physical schematics introduce the elements maintaining the electrical interconnects and pin connection.

3.2.4 Design Flow

To achieve autogeneration of ESD circuits a design flow has been developed (Fig. 3.14). The flow is based on the development of PCELLs for both the schematic and layout cells. The PCELLs are hierarchical; built from device primitives that have been RF characterized and models. Without the need for additional RF characterization, the design-kit development cycle is compressed. Autogeneration also allows for DRC correct layouts and LVS correct circuits.

As an example of the schematic methodology, from the schematic editing screen, the user invokes AMS utils → ESD. From the ESD pull-down, four functions are defined: ESD → Create an ESD element, ESD → Create and Place an ESD element, ESD → Place an existing ESD element, and ESD → Place an ESD schematic.

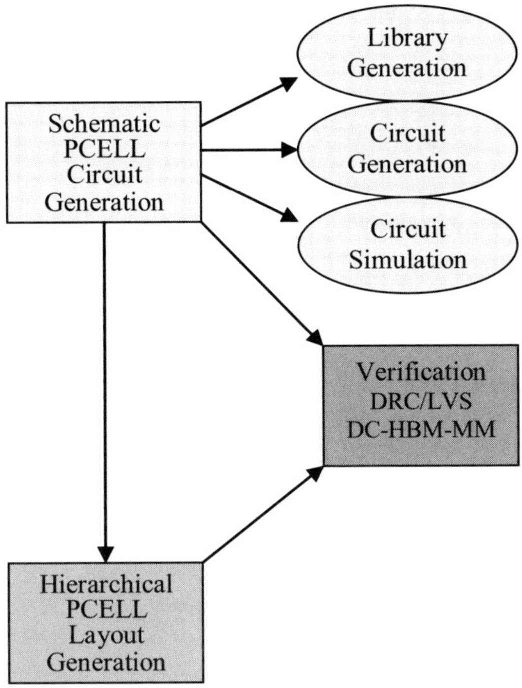

Figure 3.14 IBM's RF ESD design flow, showing the principle of hierarchical PCELL generation in the schematic.

In our ESD CAD design system, the schematic PCELL is generated by the input variables to account for the inherited parameters input values. A problem with schematic autogeneration is the circuit simulation phase. The circuit may be placed as a subcircuit, however, spectre simulation will only allow a single definition of a subcircuit. This prevents the reuse of the schematic PCELL in any other configuration. To retain the ESD circuit variability, a design flow has been built around the schematic PCELL.

There are two methods, one to allow the designer the capability of building an ESD library with the creation of ESD cells. The designer will select the option to "Create an ESD element."

Figure 3.15 shows an example of where the Create-an-ESD-element function initiates creation of an ESD schematic for a parameterized cell of a back-to-back diode string known as "AntiparallelDiodeString." To generate the electrical schematic, the ESD design system requests the "number of diodes up" and the "number of diodes down"; this determines the number of diodes in the string that are used between digital V_{SS} and analog V_{SS} (or RF V_{SS}) for grounds. For power-supply rails, the AntiparallelDiodeString is used between digital V_{DD} and analog V_{DD} (or RF V_{DD}). The design system also requests the number of cathode fingers in

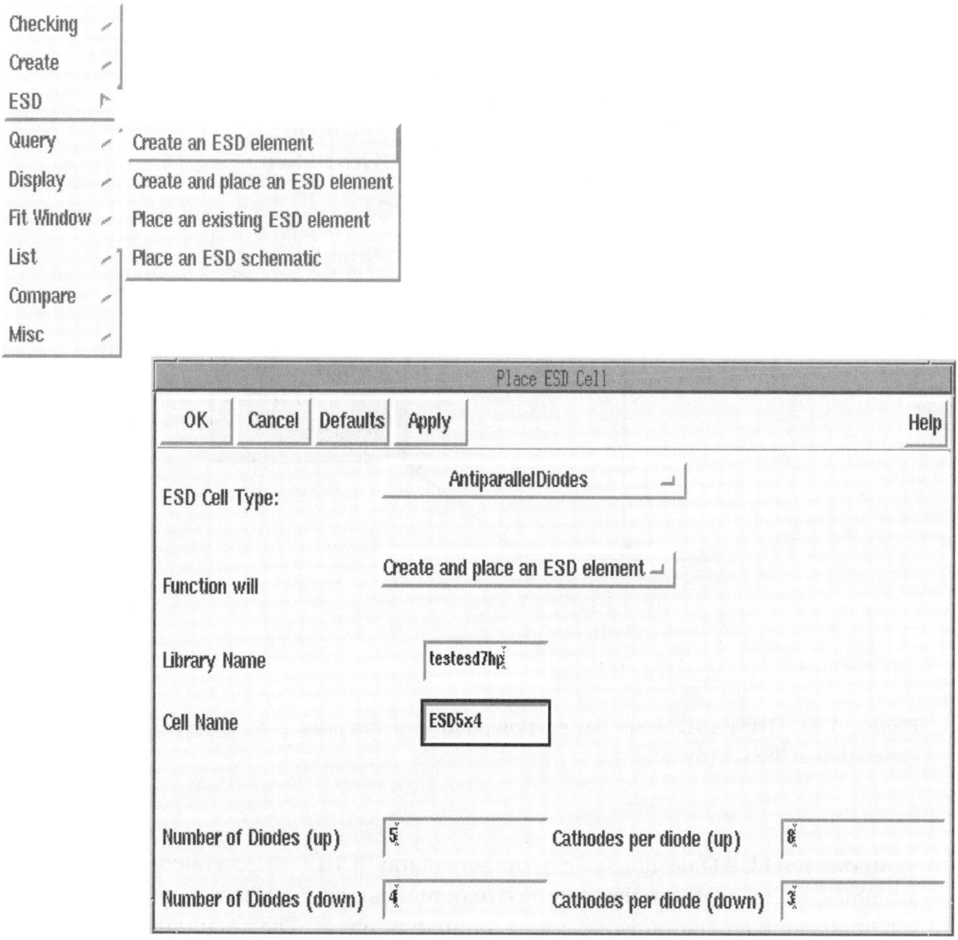

Figure 3.15 Creation of an ESD schematic of an "antiparallel diode string" between grounds from the "schematic method."

the diode structures for the up string and down string. The input parameters are passed into a procedure that will build an ESD cell with the schematic PCELL built according to the input parameters and placed in the designated ESD cell. An instance of the ESD layout PCELL will also be placed in the designated ESD cell. This allows for the automated building of an ESD library creating a schematic, layout, and symbol of the circuit based on the input parameters. This symbol can be placed in the circuit by selecting the "Place an ESD circuit" option.

The second method allows for the autogeneration of the schematic ESD circuit to be placed directly into the design. This procedure, available with the Place-an-ESD-schematic option, will allow the designer to autogenerate the circuit and place it in the schematic (Fig. 3.16). Since these cells are hierarchical, the primitive de-

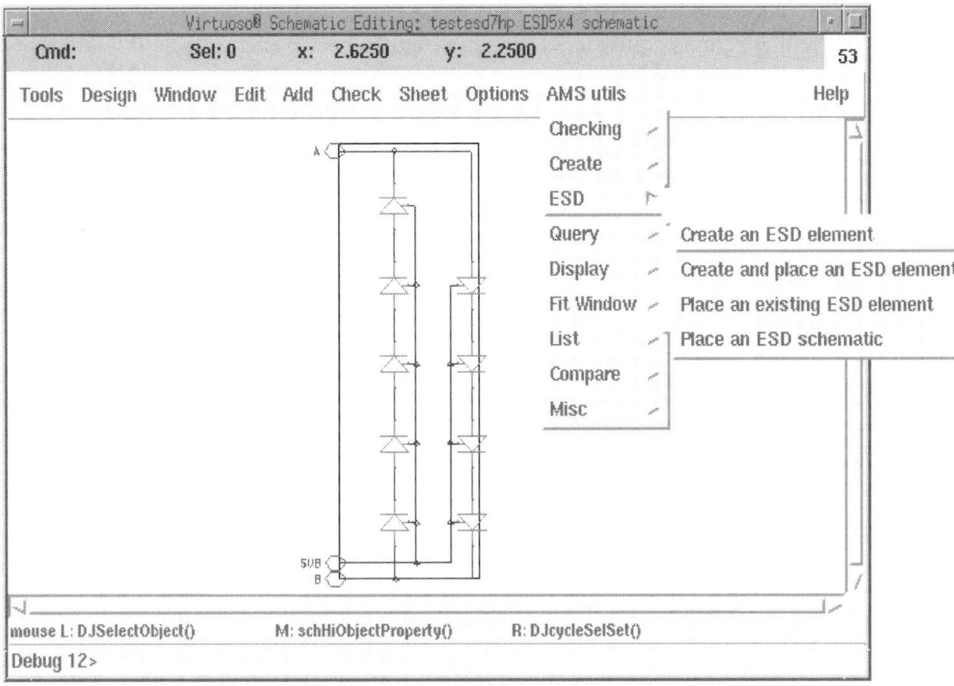

Figure 3.16 Place ESD cell interface, highlighting the inherited parameters and naming conventions.

vices and autowiring are placed by creating an instance of the schematic PCELL, then flattening the element. The instance must be flattened to avoid redefinition of Spectre® subcircuits.

The problem arises during the layout phase of the design. In the schematic due to the flattening, the hierarchy has been removed and only primitive elements remain. During Virtuoso-XL the primitives will be placed and the hierarchy will be lost. To maintain the hierarch, an instance box is placed in the schematic, retaining the input parameters and device names and characteristics as properties, and the elements are recognized and the primitives are replaced with the hierarchical PCELL.

To produce multiple implementations using different inherited parameter variable input, different embodiments of the same circuit type can be created in our methodology. In this process, the schematic is renamed to be able to produce multiple implementations in a common chip or design; the renaming process allows for the design system to distinguish multiple cell views present in a common design.

When the inherited parameters are defined, the circuit schematic is generated according to the selected variables. Substrate, ground, and pin connections are established for the system, to identify the connectivity of the circuit.

The design system can also autogenerate the layout from the electrical schematic, which will appear as equivalent to the previously discussed graphical implementation.

The physical layout of the ESD circuits is implemented with PCELLs using existing primitives in the reference library. The circuit topology is formed within the PCELL, including wiring such that all parasitics can be accounted for in preproduction test-site construction. Figure 3.17 shows an example of the design containing two growable p^+/n-well PCELLs and three stretch lines. The top stretch line has an algorithm associated with the pitch and finger number to move the V_{DD} wire vertically. The lower stretch-line algorithm moves the V_{SS} bus downward as cathode fingers are added to the lower PCELL element. The vertical stretch line allows the input, V_{DD}, and V_{SS} metal to grow with the length of the diode element.

ESD measurements of the diode string network is shown in Table 3.1. The table shows an example of a structure with five diodes in one direction and a return in the opposite direction as a function of the number of cathode fingers.

3.2.5 ESD Power Clamps

A difficulty in supporting a wide range of applications is the variety of power-rail voltage conditions and architectures. Application types vary from power amplifiers, VCOs, mixers, hard-disk-drive circuits, and test equipment. Some chips have negative voltage on the ground connections. As a result, an ESD power-clamp strategy must be suitable for CMOS digital blocks, analog blocks, and RF circuits with a wide variety of voltage conditions as well as negative bias on the substrate. To ad-

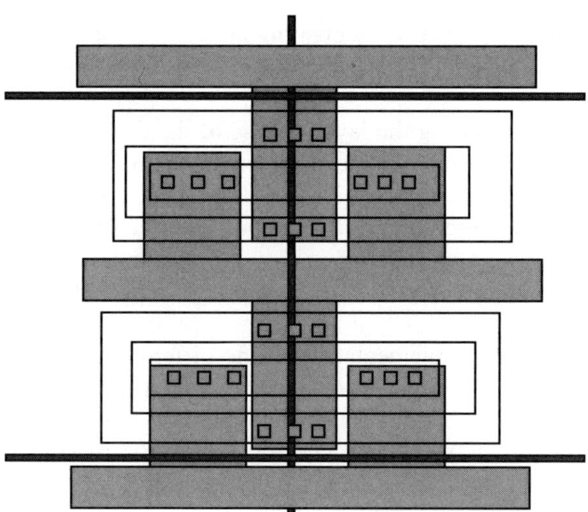

Figure 3.17 Automated parameterized cell hierarchical RF double diode circuit, highlighting the stretch lines and growable segmentation.

Table 3.1 Design System Asymmetric Diode String Parameterized Cell HBM and MM ESD Results

SiGeC Asymmetric Diode String 5:1 Number of Cathodes	HBM Failure Voltage (V)	MM Failure Voltage (V)
2	2300	240
4	3500	390
6	4800	510
8	6000	720
10	7200	750

dress this, our ESD design system has both SiGe bipolar-based ESD power clamps, and CMOS-based ESD power clamps. These ESD power clamps are designed out of parameterized cells and are growable, with flexible voltage and trigger conditions.

3.2.6 CMOS-Based Power Clamps

For satisfying the CMOS digital circuitry, an *RC*-triggered MOSFET-based power clamp is constructed out of parameterized cells. This automated hierarchical *RC*-triggered clamp consists of an NFET, PFET, and MIM capacitor PCELLs. For different size digital blocks and design form factors, the size of the ESD power clamp can be physically varied (Fig. 3.18).

The design is constructed such that the inverter drive network is fixed; the *RC*-triggered network and the output clamp element, on the other hand, are a sub-PCELL of the circuit (Fig. 3.19). The *RC* trigger is a second-order parameterized cell where the resistor is fixed and the capacitor is variable in size. The capacitor element grows to the left. In this fashion, the *RC* can be tuned to a customer's chip design for optimization. The *RC* trigger is then integrated with the fixed inverter network forming the third-order PCELL triggering network. The output clamp segment is automated to change in physical size and grows to the right. The customer

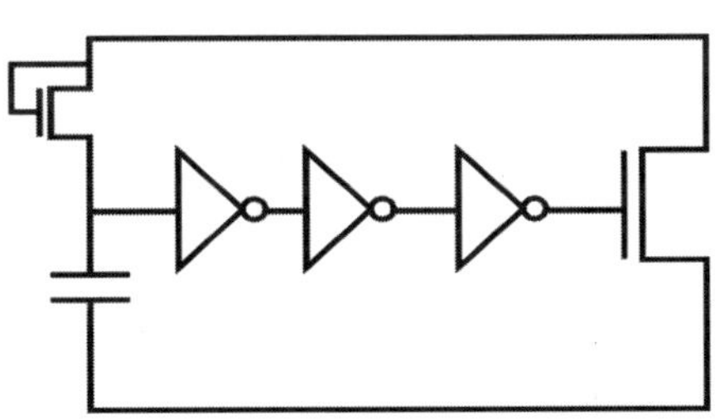

Figure 3.18 Hierarchical parameterized design of the *RC*-triggered circuit. Clamp element is growable for different ESD-size power clamps.

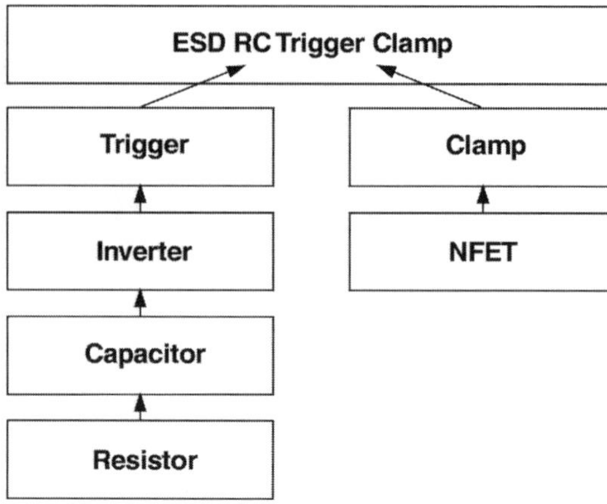

Figure 3.19 *RC*-Triggered ESD power-clamp hierarchical structure.

has two inherited parameters that are passed up to the highest-order circuit; the first is the capacitor size, which provides *RC* tuning, and the second is the size of the output clamp, which provides the ESD robustness of the circuit.

3.2.7 BiCMOS ESD Power Clamps

For the BiCMOS analog and RF functional blocks, automated hierarchical ESD power clamps are designed to allow for different voltage trigger conditions and the size of the power clamp. A first ESD power-clamp circuit has a fixed trigger voltage based on the BV_{CEO} of the trigger transistor, and the output device is a low f_T device with a high BV_{CEO} SiGe HBT npn device [9]. This network is suitable for BiCMOS SiGe chips, bipolar-only implementations, and both zero-potential and negative biased substrates. In our design kit, both high- and low-voltage triggers can be used. The circuit diagram of the BiCMOS SiGe ESD power clamp is shown in Fig. 3.20.

A hierarchical PCELL of the ESD power-clamp circuit is generated allowing for different sizes. In this hierarchical PCELL, the design consists of a resistor ballast, two transistors, and bias-resistor PCELLs. In this design, the trigger transistor and the bias resistor are fixed, while the output clamp transistor is a repetition group consisting of the ballast resistor and the output transistor stage. Hence, the ballast resistor and the output transistor are a subcell inside the full circuit hierarchical design (Fig. 3.21).

Figure 3.22 displays graphically the hierarchical PCELL compilation needed in order to establish the design.

From the schematic approach, the Darlington ESD power-clamp network can be generated from the schematic-cell view. This circuit can be represented by the full schematic in the semiconductor schematic or a facsimile symbol function. Figure

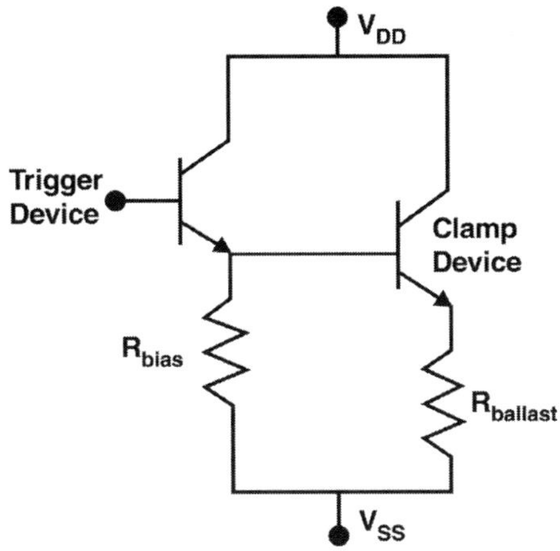

Figure 3.20 ESD Darlington power-clamp network.

3.23 shows a "symbol" that represents the circuit. The box around the symbol of the hierarchical PCELL schematic contains all the inherited parameters and circuit information.

ESD experimental studies are easily generated in test sites by "creating instances" of the ESD power clamps varying the inherited parameters. Table 3.2 shows an example of the power clamp as the "number of clamps" are increased at the output with a fixed-trigger element. ESD results improved with the structure size. Excellent HBM, MM, and TLP testing results are achieved in this ESD power-clamp instance [15].

To address the different power-supply conditions, a level-shifting parameterized subdesign PCELL is created. The second ESD power clamp consists of a level

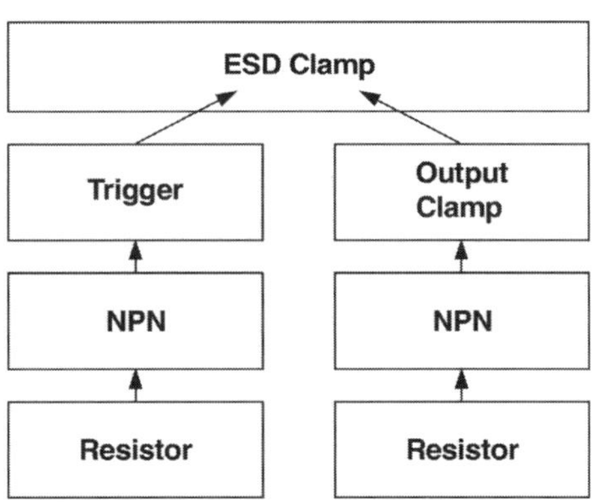

Figure 3.21 Hierarchical structure of parameterized cell ESD power-clamp circuit.

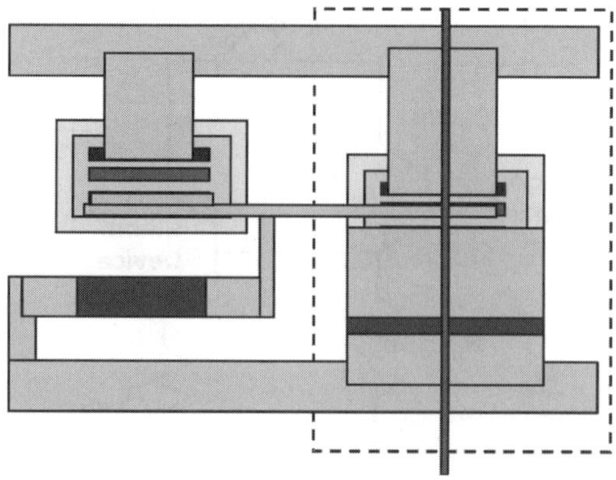

Figure 3.22 Hierarchical ESD power-clamp design utilizing a growable ESD power-clamp output device using repetition groups and stretch-line algorithms utilizing the "graphical methodology."

shifting PCELL, a trigger SiGe transistor, a bias resistor PCELL, and the repetition group of the clamp output and resistor ballast elements (Figs. 3.24 and 3.25). This design has two automation variables: first the trigger condition allows the growing of a string of series SiGe varactors to increase the trigger condition; second, the output clamp and ballast resistor can be increased based on the design area and desired ESD protection level. All interconnects are growable to allow for the diode string

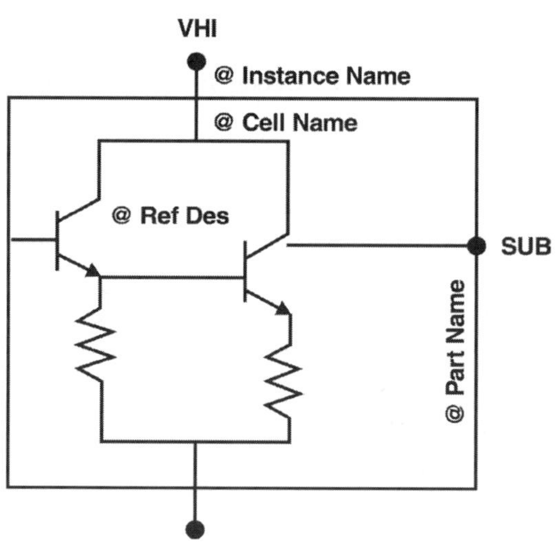

Figure 3.23 An example of a hierarchical parameterized schematic cell "symbol" representing the circuit. The symbol contains all inherited parameter and design information.

Table 3.2 Hierarchical Parameterized Cell ESD Power-Clamp ESD Results with a 120-GHz/100-GHz f_T/f_{MAX} SiGe HBT with Carbon Incorporation

Power Clamp Emitter Length (μm)	HBM Failure Voltage (V)	MM Failure Voltage (V)	TLP Failure Current (A)
50	2500	240	0.7
100	3100	390	1.25
150	4700	480	1.7
200	5000	600	1.8
250	5900	630	2.1

and clamp size to be increased in size automatically. From the schematic methodology, the personalization of the ESD power clamp can be initiated first as a circuit, and then as graphical layout. Figure 3.26 shows a schematic view with two diodes in series with the trigger and three parallel clamp elements.

3.2.8 Summary

The evolution of IBM's SiGe ESD design process has followed a path from a single RF-characterized fixed element, to a family of fixed design-size ESD networks containing RF-characterized PCELLs, to a new methodology of variable design-size ESD networks containing RF-characterized PCELLs. The new concept has full RF characterization and models, as well as HBM, MM, and TLP characterization; this

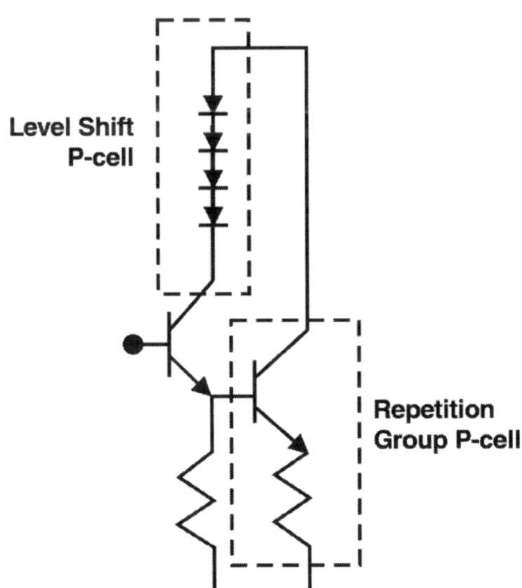

Figure 3.24 ESD power clamp with growable level shift trigger and power-clamp output.

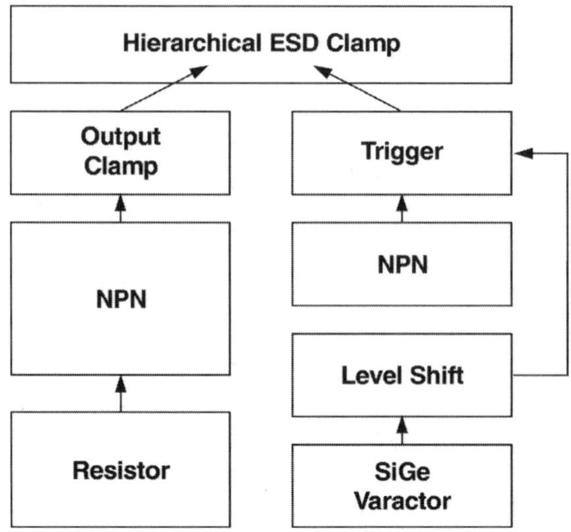

Figure 3.25 ESD power-clamp hierarchy, highlighting the growable regions of the parameterized ESD network.

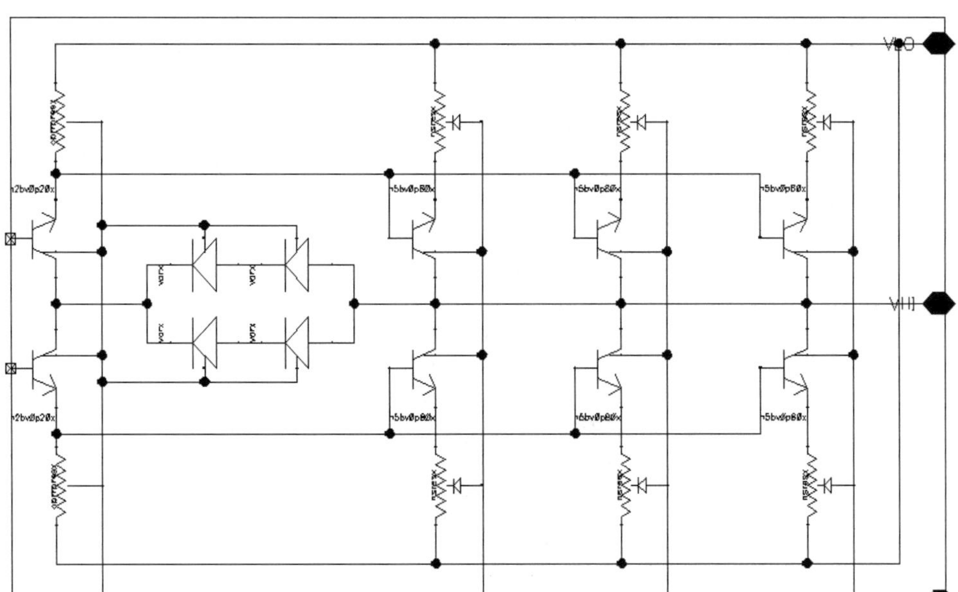

Figure 3.26 Hierarchical parameterized-cell variable-trigger SiGe Darlington power-clamp autogenerated schematic.

allows for automated design change for cosynthesis of ESD networks, and circuit function for digital, analog, and RF circuits. The system allows for autogeneration of the schematic and the layout; it also allows mapping from one to the other. The system also allows for placement of multiple implementations of an ESD network with different circuit topologies in a given chip implementation. The system has produced a significant improvement in ESD design, test site development, and customer release. With its evolution, we anticipate an increased opening of the number of inherited parameters, new ESD-optimized PCELLs, and new ESD circuits. This ESD design concept and architecture improves the state of the art for ESD design, and ESD RF optimization for both RF CMOS and RF BiCMOS technologies.

3.2.9 References

1. U. Konig, "SiGe and GaAs as Competitive Technologies for RF-Applications," in *Bipolar Circuits and Technology Meeting,* Minneapolis, MN, pp. 87–92, 1998.
2. C. Richier, P. Salome, G. Mabboux, I. Zaza, A. Juge, P. Mortini, "Investigation on Different ESD Protection Strategies Devoted to 3.3 V RF Applications (2 GHz) in a 0.18 μm CMOS Process," in *EOS/ESD Symposium Proceedings,* pp. 251–260, September 2000.
3. S. Voldman, "The State of the Art of Electrostatic Discharge Protection: Physics, Technology, Circuits, Designs, Simulation and Scaling," Invited Paper, in *Bipolar/BiCMOS Circuits and Technology Meeting Symposium,* pp. 19–31, September 27–29, 1998.
4. S. Voldman, P. Juliano, R. Johnson, N. Schmidt, A. Joseph, S. Furkay, E. Rosenbaum, J. Dunn, D. L. Harame, and B. Meyerson, "Electrostatic Discharge and High Current Pulse Characterization of Epitaxial Base Silicon Germanium Heterojunction Bipolar Transistors," International Reliability Physics Symposium, pp. 313–316, March 2000.
5. S. Voldman, N. Schmidt, R. Johnson., L. Lanzerotti, A. Joseph, C. Brennan, J. Dunn, D. Harame, P. Juliano, E. Rosenbaum, and B. Meyerson, "Electrostatic Discharge Characterization of Epitaxial Base Silicon Germanium Heterojunction Bipolar Transistors," in *EOS/ESD Symposium,* pp. 239–251, September 2000.
6. S. Voldman, P. Juliano, N. Schmidt, A. Botula, R. Johnson, L. Lanzerotti, N. Feilchenfeld, J. Joseph, J. Malinowski, E. Eld, V. Gross, C. Brennan, J. Dunn, D. Harame, D. Herman, B. Meyerson, "ESD Robustness of a BiCMOS SiGe Technology," in *Bipolar/BiCMOS Circuits and Technology Meeting Symposium,* pp. 214–217, September 2000.
7. S. Voldman, L. D. Lanzerotti, and R. Johnson, "Emitter Base Junction ESD Reliability of an Epitaxial Base Silicon Germanium Heterojunction Transistor," in *International Physical and Failure Analysis of Integrated Circuits,* pp. 79–84, July 2001.
8. S. Voldman, L. D. Lanzerotti, and R. A. Johnson, "Influence of Process and Device Design on ESD Sensitivity of a Silicon Germanium Heterojunction Bipolar Transistor," in *EOS/ESD Symposium Proceedings,* p. 364, 2001.
9. S. Voldman, A. Botula, D. Hui, and P. Juliano, "Silicon Germanium Heterojunction Bipolar Transistor ESD Power Clamps and the Johnson Limit," in *EOS/ESD Symposium,* pp.326–336, September 13, 2001.
10. A. Gruhle, "Prospects for 200 GHz on Silicon with Silicon Germanium Heterojunction Bipolar Transistors," Invited Paper, in *Bipolar Circuit Technology Meeting,* 2001.

11. L. C. M. van den Oever, L. K. Nanver, J. W. Slottoom, "Design of 200 GHz SiGe HBTs," Bipolar Circuit Technology Meeting, pp. 78–81, 2001.

12. J. Kirchgessner, S. Bigelow, F. K. Chai, R. Cross, P. Dahl, A. Duvallet, B. Gardner, M. Griswold, D. Hammock, J. Heddleson, S. Hildreth, A. Irudayam, C. Lesher, T. Meixner, P. Meng, M. Menner, J. McGinley, D. Monk, D. Morgan, H. Rueda, "A 0.18 μm SiGe:C RF BiCMOS Technology for Wireless and Gigabit Optical Communication Applications," Bipolar Circuit Technology Meeting, pp. 151–154, 2001.

13. A. Joseph, D. Coolbaugh, M. Zierak, R. Wuthrich, P. Geiss, Z. He, X. Liu, B. Orner, J. Johnson, D. Ahlgren, B. Jagannathan, L. Lanzerotti, V. Ramachandran, J. Malinowski, H. Chen, J. Chu, P. Gray, R. Johnson, J. Dunn, S. Subbanna, K. Schonenberg, D. Harame, R. Groves, K. Watson, D. Jadus, M. Meghelli, A. Rylyakov, "A 0.18 μm BiCMOS Technology Featuring a 120/100 GHz (f_T/f_{MAX}) HBT and ASIC Compatible CMOS Using Copper Interconnects," Bipolar Circuit Technology Meeting, pp. 143–146, 2001.

14. B. Ronan, S. Voldman, L. Lanzerotti, J. Rascoe, D. Sheridan, and K. Rajendran, "High Current Transmission Line Pulse (TLP) and ESD Characterization of a Silicon Germanium Heterojunction Bipolar Transistor with Carbon Incorporation," International Reliability Physics Symposium (IRPS), Dallas, Texas, pp. 175–183, April 2002.

15. S. Voldman, "Variable Trigger-Voltage ESD Power Clamps for Mixed Voltage Applications Using a 120 GHz/100 GHz (f_T/f_{MAX}) Silicon Germanium Heterojunction Bipolar Transistor with Carbon Incorporation," in *Proceedings of EOS/ESD Symposium,* 2A.1, October 2002.

3.3 INTERCONNECT EXTRACTION AND MODELING

Communications IC designers face many complex design challenges in meeting today's communications standards such as Gigabit Ethernet (IEEE 802.3ae) and wideband code division multiplexing access (CDMA) (third-generation cellular telephone protocol, 3G). These challenges are encountered across the full range of the design methodology from model accuracy and simulation convergence to accurate parasitic extraction and resimulation. It is widely understood that accurate estimation of interconnect effects is required to minimize design iteration [1], in RF, digital, and mixed-signal applications to avoid signal integrity problems for critical nets, through either resistance–inductance–capacitance (RLC) extraction or transmission-line modeling. In this section, we present the full spectrum of solutions, from interconnect extraction to transmission-line modeling, to modeling of signal nets that experience loss to the substrate.

As shown in Fig. 3.37, the metal stack for SiGe BiCMOS processes is rapidly becoming complex. Many times, the effects of these metal interconnects have been treated as an afterthought at the last step of the design phase, with little consideration early in the design flow on the possible impacts of the metal impedance. There are four key changes that have led to the critical need for designers to consider the effects of the interconnects:

- A more complex and varied metal/dielectric stack with denser pitches and thick dielectric/metal add-on modules (analog metal)

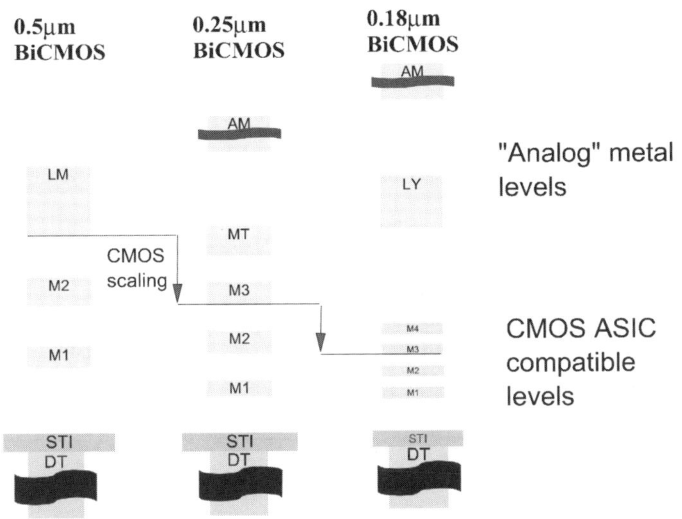

Figure 3.27 Metal stack complexity rises as SiGe BiCMOS technologies scale.

- Higher levels of integration leading to larger ICs and longer chipwide interconnects
- Higher signal frequencies, driven by shrinking technologies
- Lower V_{dd} levels, driven by shrinking technologies, leading to lower SNRs

Add to this list the higher complexities of standards—such as, 3G and SONET—and the intense time-to-market pressures, it is apparent that the effective modeling of the interconnect effects is critical to design success.

Telecommunication designs typically contain many hundreds to thousands of devices with a broad frequency range, and designers need to be careful to use the appropriate tool on the relevant nets. There is an effective accuracy versus speed trade-off that can be performed (as shown in Fig. 3.28) with the following properties:

- 3D field simulation is very accurate but very slow.
- 2D field solution is faster than 3D, but no end effects.
- *RLC* extraction is fast/fairly accurate, until nets get large and frequencies rise. Extraction tools are always needed, as they are integrated into the design environment and are often simple to use.
- Transmission-line models offer good accuracy and performance. Traditionally, transmission lines are employed to control high-frequency distributed impedance effects that become important as the line length approaches a fraction of the wavelength. Increasingly, transmission-line structures are also employed to isolate critical interconnect from substrate either through refer-

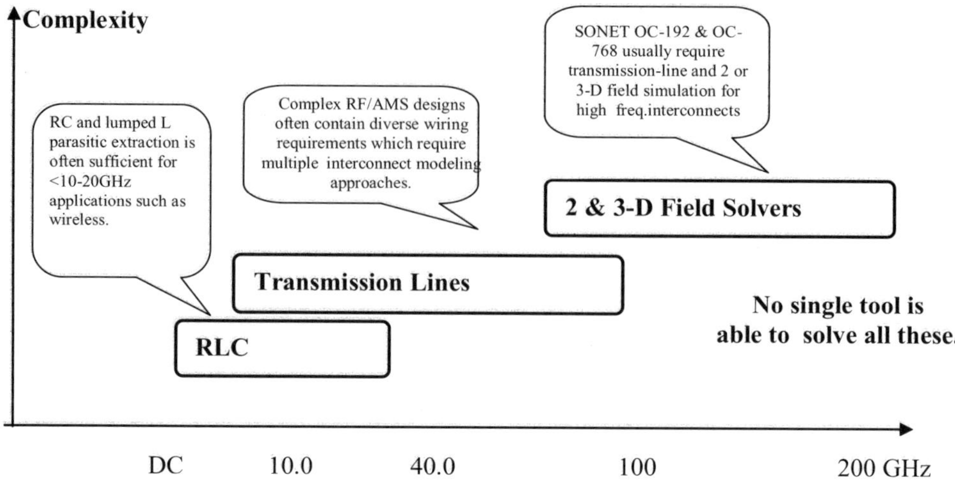

Figure 3.28 Requirement for different interconnect modeling solutions across the RF/microwave frequency range.

ence-plane shielding for single-wire lines or field containment in coupled coplanar structures. Such isolation provides for lower power loss and coupling in the substrate.

Table 3.3 lists some of the trade-offs and applications for the different approaches to modeling interconnect.

When verifying the accuracy of any interconnect modeling/extraction approach, attention needs to be paid to the outliers in the statistical report (see Fig. 3.29 as an example), i.e., why inaccuracies have occurred above the acceptable accuracy range, which is typically 10% for capacitance calculations. For example, five 25% errors in wide-metal testing may only show up as statistically insignificant in a test suite of 1500 structures. However, if those five structures are from a certain size and formation of interconnects, then there may be a bug in the model/tool that needs to be fixed. In contrast, outliers may also be attributed to unusual structures that are seldom encountered in real design layouts.

3.3.1 Transmission-Line Modeling

A general review of wiring design methodologies, measurement techniques, and modeling aspects can be found in [1,2]. In order to analyze these effects, the electrical characteristics of interconnects are estimated using numerical modeling or measurements. Numerical simulations can be performed with different techniques. Among them are well-known Finite Element Method (FEM) [3,4] and finite difference time domain (FDTD) [5] methods, as well as the method of moments (MOM)

Table 3.3 The Various Strengths, Weaknesses, and Applications of the Various Approaches to Interconnect Modeling

Tools	Functionality	Strengths	Weaknesses	Applications
Parasitic *RLC* extraction	*RC* and *L* wiring estimation, in extraction phase of design flow	Useful for large numbers of nets, integrated in design flow	Inaccurate for higher frequencies; tools sometimes inaccurate for wide metal	<10–20 GHz Typically *L* needed above *RC*, for accuracy
Transmission lines	Distributed frequency and time-domain interconnect effects	Arbitrary accuracy and efficiency	Typically poor physical design integration	Wireless and wired > 1 GHz
2D field solution	Modeling of 2D and simple 3D structures	Accurate and efficient for 2D effects	Inaccurate for 3D and fringe effects	Packages, PCB, and off-chip effects; straight interconnects >1 GHz
3D field solution	Modeling of complex 3D structures	Very accurate	Very slow; often difficult to set up	>10 GHz Used for highest-frequency on-chip structures, and complex structures such as vias, packages

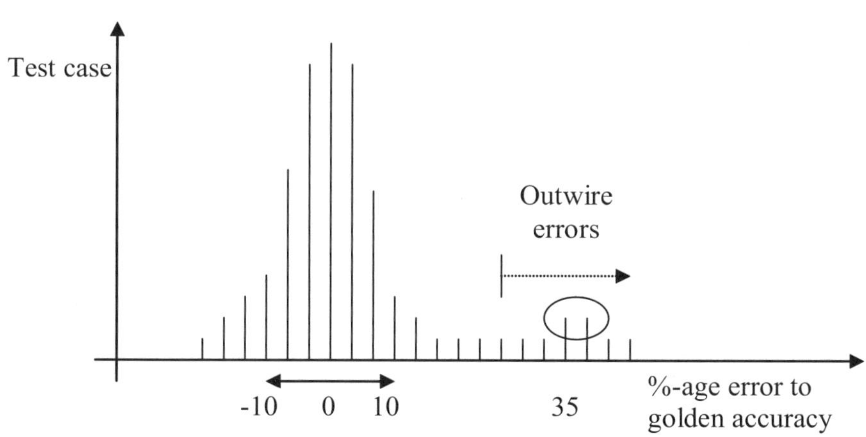

Figure 3.29 A histogram depicting the occurrence of outwire errors in interconnect extraction and modeling.

[6]. Examples of application of FDTD and MOM methods to transmission-line analysis can be found in [7,8]. FDTD is based on the direct discretization of Maxwell's equations in a time domain. It can be easily implemented for parallel computation using a few workstations at the same time. FDTD also can be modified to handle nonlinear and lumped-element circuits. MOM is the integral equation technique, which is mostly used for parameter extraction. It is sometimes computationally expensive and faces all problems of accurate numerical estimation of integrals with singularities, which arise from Green's functions. The MOM system matrix is usually a dense matrix. This prohibits the use of fast sparse matrix computation algorithms.

In the case of the FEM approach, the discrete approximation of the problem under question is transformed into a matrix form. Due to the local support of the interpolating functions, the resulting FEM system matrix is a sparse matrix. Usually, in the FEM method, the entire space—metal and dielectrics—has to be discretized. This results in a large system matrix. The opposite is true for MOM. Despite all of that, due to matrix sparsity in FEM, it is still one of the best techniques to simulate complicated structures, which MOM cannot handle. For example, structures having at the same time different materials can be easily and accurately modeled by using the FEM method. Even electromagnetic problems with a few thousand unknowns in the FEM method can be solved within a reasonable time using modern numerical techniques in combination with powerful workstations. There were a few efforts made to introduce new fast integral equation solvers for interconnect modeling. We should mention here the existence of the adaptive integral methods (AIM) and fast multiple methods (FMM) algorithms, which are implemented for capacitance [9] and inductance [10] extractions. They are fast and accurate techniques, which are widely used for the development of new interconnect models. Another approach is based on random walks, and has been applied with great success for interconnect capacitance extraction [11]. The preceding approaches sometimes can be really slow and difficult to program so they can be implemented as a part of commercial IC design systems. Moreover, the suitability of the mentioned techniques for the true transient time-domain analysis has not been fully demonstrated yet. The techniques mentioned in the earlier discussion are also sensitive to aspect-ratio problem and difficult to set up, especially when one tries to introduce loading ports (sources) into a structure to be modeled.

Measurement-based transmission-line characterization has been reported in [12]. This technique uses measured S-parameters to extract transmission-line electrical characteristics. The same technique can be also implemented if S-parameters are obtained using a full-wave EM solver. The S-parameter-based approach is sometimes not accurate, especially when resonance exists in the interconnect. A slight deviation in S-parameter results can also cause a large error in extracted parameters. We suggest that S-parameters should be used mainly for the verification purposes at a final stage of interconnect modeling. Another method, which can be employed to model transmission-line behavior, consists of solving Maxwell's [13] or *Telegrapher's* [14] equations in the time domain. Such techniques need to be introduced as a part of the CAD-oriented IC design software. This often requires substantial programming efforts. In addition, numerical stability, convergence, and ease of use are not guaranteed.

A suitable methodology is to characterize transmission lines through the development of SPICE-compatible lumped-element models, preferably passive by construction [15]. In general, lossy transmission lines can be represented as a cascade of an infinite number of small RLCG blocks. The SPICE lumped-element models consist of a finite number of cells and are suitable for efficient time- and frequency-domain simulations (including nonlinear, transient, AC, frequency-domain analyses, mixed-signal analysis, and S-parameters). These models should be able to model frequency-dependent skin and proximity effects over a wide frequency range of interest. The simplest way to incorporate frequency-dependent effects is to use ladder networks. Another approach considers the introduction of effective current loops in combination with simple RLCG blocks. After modeling has been completed, S-parameter results are used to correlate models versus EM solver and measurement results. The CAD-oriented examples of equivalent-circuit modeling of transmission-line structures can be found, for instance, in [16]. This approach models interconnects on lossy silicon substrates (CPWs), which becomes important for the higher densities required from CMOS integration.

One promising interconnect modeling methodology has been reported in [17]. This approach currently supports only single and two coupled-wire microstrip structures with an optional side shielding, which are shown in Fig. 3.30. The PCELL layout for the single-wire structure is shown in Fig. 3.31. The PCELL implementation is important, as it leads to efficient instantiation of the transmission lines in the layout, and ensures layout to schematically match the simulations. Both structures are effectively isolated from a lossy substrate by a ground plane. This also reduces losses in the silicon substrate. The equivalent models consist of a eries of *RLC* lumped segments in combination with ladder networks. Ladder networks

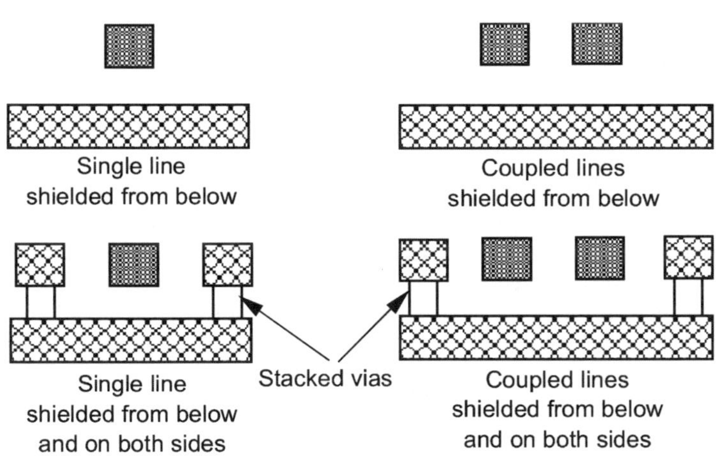

Figure 3.30 Microstrip lines supported in integrated design flow.

Figure 3.31 A single signal-wire parameterized cell over a ground plane, with and without side shielding.

are used to model the frequency-dependent skin effect. They consist of passive, frequency independent elements and are relatively easy to netlist and simulate in commercial SPICE-like simulators. Examples of a ladder network can be found in [18]. As far as *RLC* blocks are concerned, *C* and *L* parameters have to be calculated carefully to take into account mutual electromagnetic coupling. The number of needed *RLC* cells is usually calculated in a way that there are at least 10 segments per shortest on-chip wavelength [1,3,17].

The approach described in [17], adopted in this design flow, uses quasi-static TEM approximation to calculate the low-frequency capacitance and inductance of an interconnect. The quasi-static approach has good accuracy due to the fact the cross-section dimensions of the on-chip interconnects are very small in comparison with a shortest on-chip wavelength, even for high frequencies of operation. After low-frequency capacitance (*C*) is estimated, it is used to calculate high-frequency inductance (*L*) limit. The next step combines low- and high-frequency inductance values to construct ladder networks. The final transmission line model has strong frequency dependencies of resistance (*R*) and inductance (*L*). At the same time, capacitance (*C*) is almost constant over a wide frequency bandwidth.

To show the accuracy of this approach, we simulated a single-wire microstrip without side shielding (see Fig. 3.30) using the adopted lumped-element model in combination with Spectre simulation tool. Then we implemented the method from [12] to extract electrical characteristics of the interconnect under question. Figure 3.32 shows the results over a frequency range of 0.1–100 GHz. For correlation of the models versus full-wave EM solver (using Ansoft's HFSS modeling tool) results, we performed two-port numerical simulations. S-Parameter results normalized to 50-Ω ports are shown in Fig. 3.33. Again, a single-wire microstrip without side shielding is modeled. We also compared the interconnect transmission-line model with measurement results. They can be found in Fig. 3.34. One can see that they are accurate within acceptable levels with S21 magnitude differing by less than 1 dB across a broad frequency range.

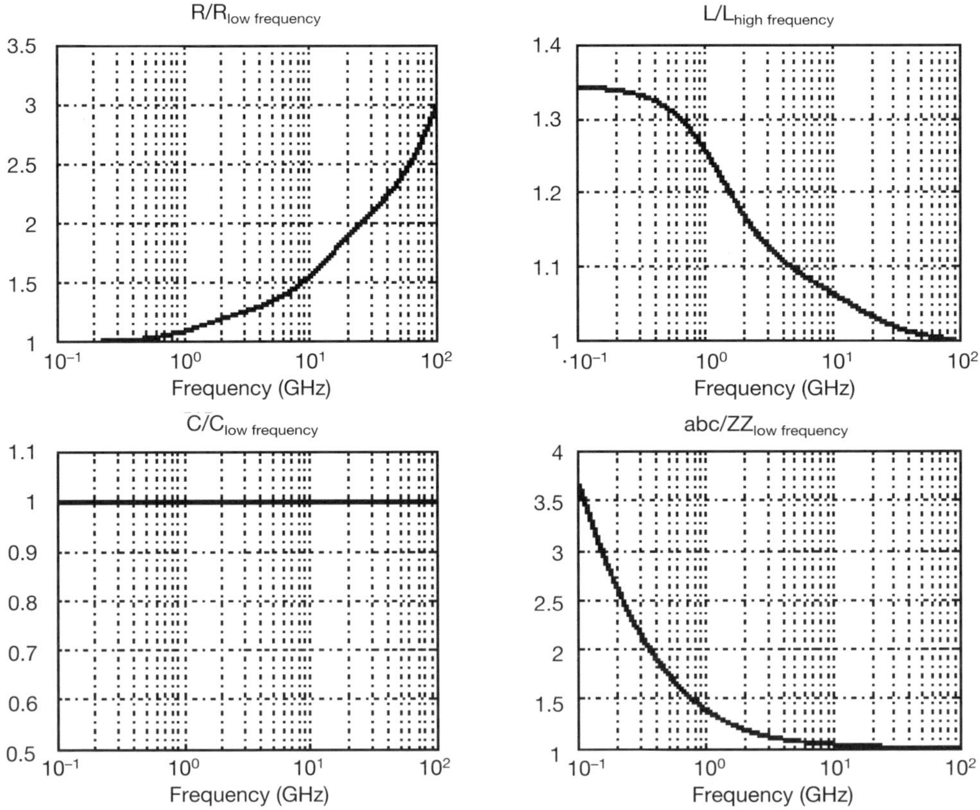

Figure 3.32 Extracted transmission-line parameters, based on the lumped *RLC* model.

3.3.2 Field-Solver Solutions

The availability of fast and accurate electromagnetic-field solvers and parasitic extraction tools is very important for the development of modern IC design and manufacturing technologies. At low frequencies, standard *RLC* parameter extraction tools are of great interest. Based in their results, a large number of passive and active devices are modeled. These tools enable compact device models to be created and incorporated into well-known IC design programs. Low-frequency *RLC* extraction tools usually can be easily scripted to run automatically for a substantial number of test examples within a reasonable time. This greatly simplifies device model development and verification. Some of these tools are based on the new fast integral-equation solvers. Others use finite-element or various finite difference algorithms as their computation engines. Lately, the fast multipole approach as well as method, which uses the stochastic technique, became quite popular. Based on these techniques, a few very efficient *RLC* extraction tools have been developed. Some of them can extract frequency-dependent *R* and *L* parame-

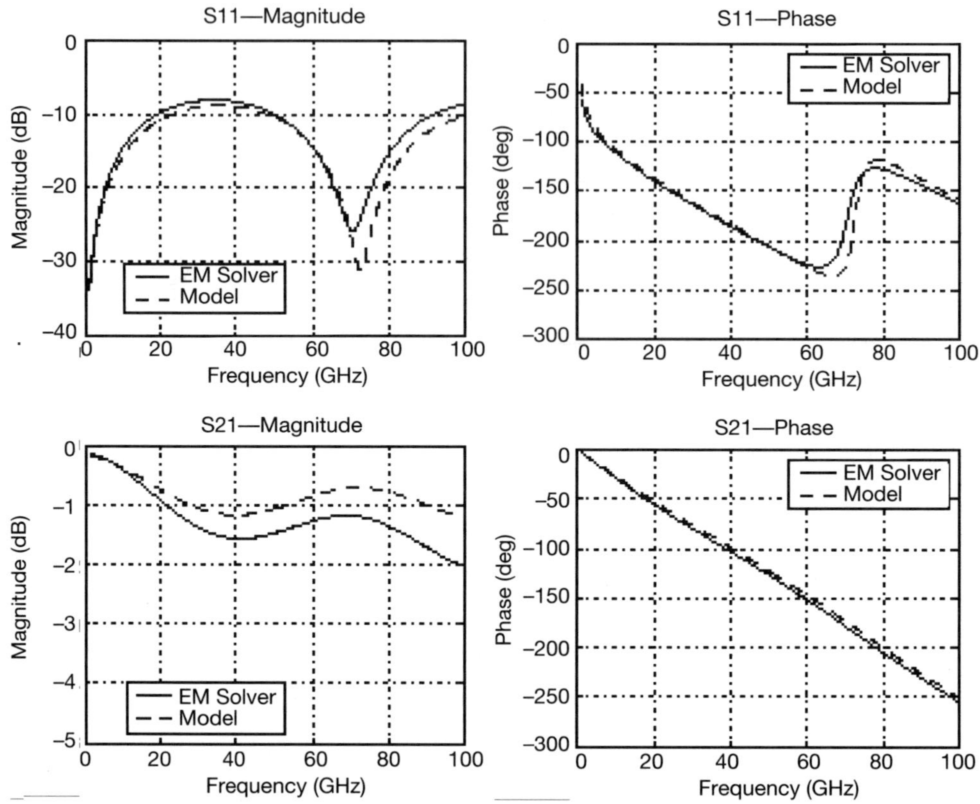

Figure 3.33 Model vs. EM solver: single line, no side shielding.

ters. They also can take into account the presence of a lossy substrate in the modern ICs.

There are numerous two-dimensional electromagnetic-field solvers available in both industry and academia that also can be used for device modeling and circuit simulations. Some of them calculate low frequency electrical parameters of a given circuit with a very good accuracy. They are also easy to set up and run. The most accurate of these tools are based on the finite difference algorithm. Nevertheless, not all of them can be run in a script mode, which prohibits their use for the large-device structure database verification.

The true 3D-field solvers find their application at high frequencies. They also can be used to model complicated circuit elements and are very useful if a compact device model does not exist in a frequency range of interest or it is not accurate enough. The 3D-field solvers or full-wave solvers are usually based on a finite-element technique and can have a few thousand unknowns during simulations. Usually, these tools perform S-parameter simulations, which are more natural for high frequencies. Simulation results can be used to model the device in question as a

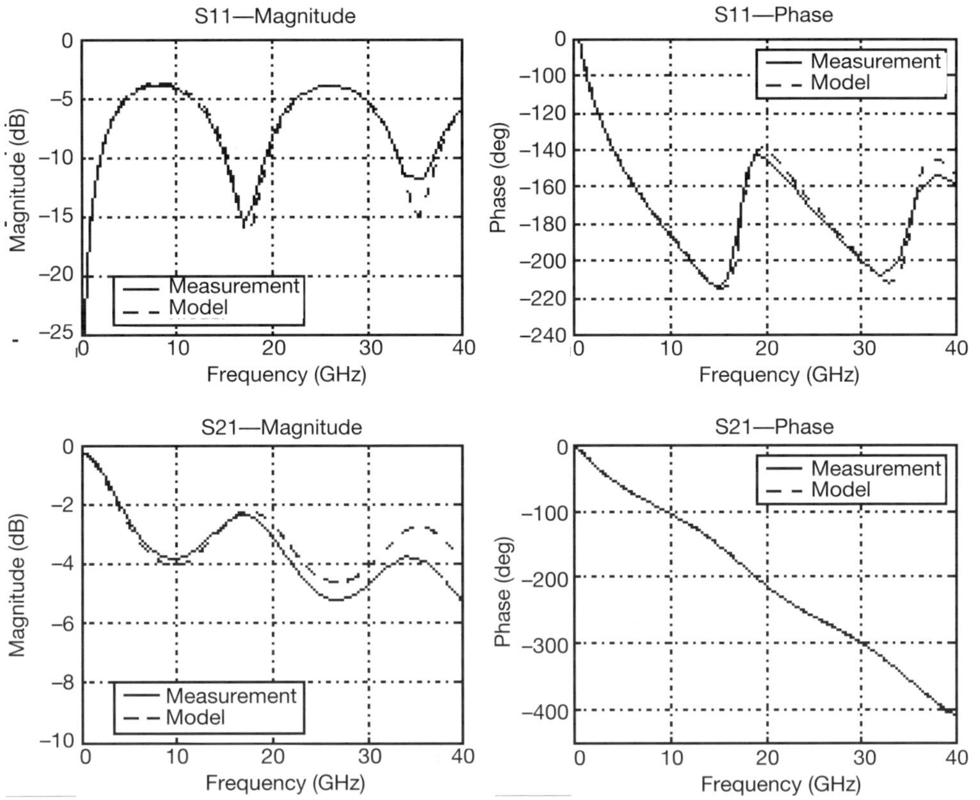

Figure 3.34 Model vs. hardware: single line, no side shielding.

black-box device having only an attached S-parameter data file. The modeler or designer has to be careful in using 3D solvers, especially when setting up a structure to be modeled. Some caution also has to be taken in setting up convergence goal and mesh size to be used during actual simulations. A few of the high-frequency 3D solvers can also perform a parameter extraction frequency sweep. This is very important in modeling skin-effect impact on a device or circuit element electrical properties. Still, it has been found that sometimes low-frequency results obtained with 3D-field solvers are not accurate enough. In that case, the use of *RLC* parasitic extraction tools or 2D field solvers is suggested.

3.3.3 Interconnect-Over-Substrate Modeling

As operating frequencies increase in state-of-the-art wireless designs, highly accurate modeling of critical interconnect paths routed over silicon is crucial for first-pass design success. With this in mind, the interconnect stack of IBM's typical SiGe processes was modeled with highly predictive accuracy as the goal.

Several literature sources are available concerning the modeling of interconnect over silicon. One can readily find predictive equations in the literature for line inductance [19], line capacitance [20], and line conductance (through the substrate) for an interconnect line over Si. However, these equations can substantially deviate from those values predicted by commercial EM solvers such as High-Frequency Structure Simulator (HFSS) [21] and Momentum RF [22]. Appropriate equations were produced to adjust the literature equations to the IBM SiGe process stacks. Equations (3.1)–(3.4) show the initial and fitted-line inductance and line through-silicon conductance equations derived for the lowest (M1) metal layer of a typical IBM SiGe process:

$$L_{cf} = \mathrm{Re}\left(\frac{\mu_0}{4\pi} \ln(1 + 32\left(\frac{h_{eff}}{w} \right)\left[1 + \sqrt{1 + \left(\frac{\pi w}{8h_{eff}} \right)^2} \right] \right) \tag{3.1}$$

where h_{eff} refers to the complex effective height of the interconnect line derived from complex image theory [23]; L_{cf} represents a closed form expression for the external inductance per unit length of an ideal microstrip line [24]. The fitting coefficient derived to fit Equation (3.1) with scattering parameters predicted by Momentum RF [22], α, is given in Equation (3.2):

$$\alpha_{M1} = 0.065 \ln\left(\frac{h}{w} \right) + 0.797 \tag{3.2}$$

In Equation (3.2), h is the height (not effective height) of the interconnect line over the backside of the Si die, and w is the width of the line. Equation (3.1) is multiplied by Equation (3.2) to produce the technology-specific line inductance estimate:

$$G_{Si} = \frac{\sigma_{Si} w_{eff}}{h_{Si}} \tag{3.3}$$

In Equation (3.3), w_{eff} refers to the effective width of the slab of silicon between the interconnect line and a grounded perfect electrical conductor (PEC) on the backside of the wafer such that Equation (3.3) is an accurate representation of simulated through-substrate conduction. Also present in Equation (3.3), σ_{Si} refers to the conductivity of the silicon substrate. The derived equation for this effective width is given in Equation (3.4).

$$w_{eff(M1)} = 3.2w + 2(h_{OX} + h_{line}) + 195 \tag{3.4}$$

where w and h_{line} refer to the width and thickness (height) of the interconnect line, respectively (in μm), and h_{OX} is the vertical separation between the interconnect line and the silicon surface in micrometers.

Using the derived/fitted results just given as well as other fitted circuit-model component value equations for interconnect line capacitance and passive skin-effect network components, modeling accuracy of isolated interconnect lines over silicon in the SiGe 7HP technology (σ_{Si} = 7.41 S/m) was increased to within 1% of S-parameter results predicted by the commercial solvers in the frequency range from 0.5 GHz to 10 GHz, (see excellent matching in Fig. 3.35).

In practical design examples of on-chip interconnect lines over silicon, arbitrary ground terminations will be encountered along the routed path from source to receiver. As a starting point to modeling substrate-to-interconnect coupling accurately for arbitrary termination conditions, a uniform cross-section geometry is first modeled: an isolated interconnect line with parallel heavily implanted, grounded, substrate contacts routed on both sides of the interconnect line. Figure 3.36 shows a cross section of this configuration.

In Fig. 3.36, the implanted substrate contacts are 10 μm wide. The configuration depicted in Fig. 3.36 is sometimes employed to improve noise isolation of interconnect lines without ground metal beneath them (microstrip configuration), and will provide a starting point for studying critical interconnect–substrate coupling in practical wireless designs.

Substrate resistance between the p^+ implanted guard lines was measured and modeled for the SiGe 7HP process, and for this work, converted to the SiGe 5HP process with small changes reflecting the differences between the 5HP and 7HP

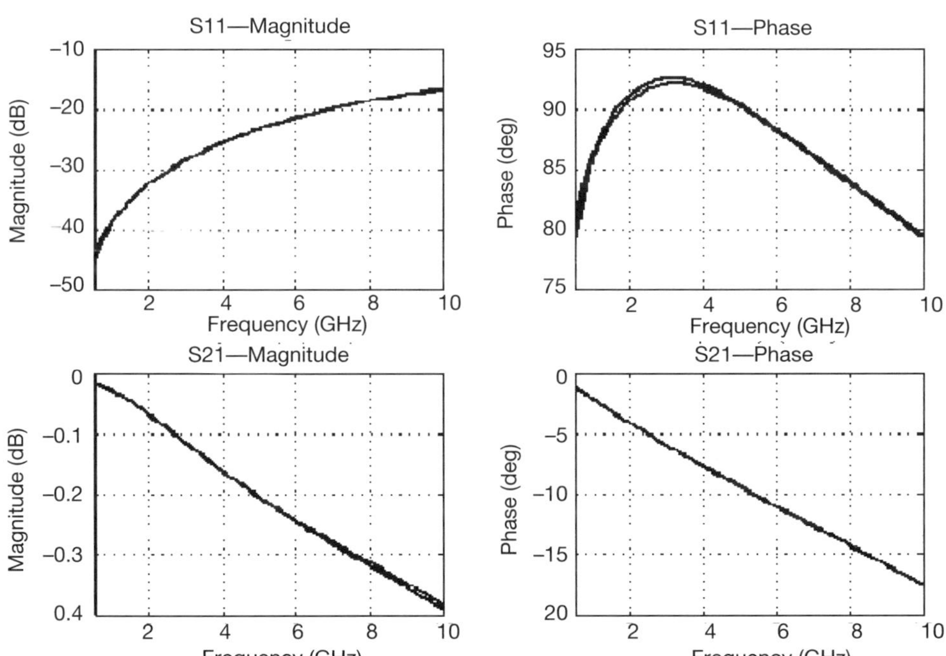

Figure 3.35 S-Parameter agreement between Momentum RF and fitted passive circuit model of a 500-μm-long, 15-μm-wide upper metal layer interconnect line over Si in the SiGe 7HP process.

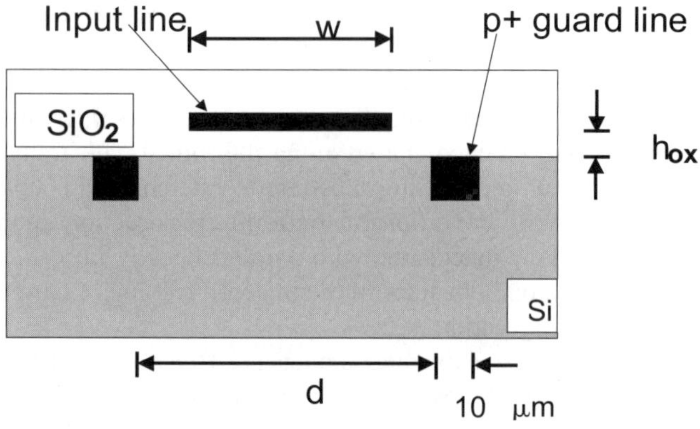

Figure 3.36 Interconnect over silicon with guard lines.

substrate doping profiles. Equation (3.5) shows the fitted empirical relationship for the resistance between two heavily doped substrate contacts.

$$R_{\text{sub}} = 2\frac{R_{sv}}{wl} + A\log_{10}(d + 0.4w)||R_{\text{sh}}\frac{(d + 0.4w)}{l} \tag{3.5}$$

where R_{sv} is the vertical resistance of the contact (1725 Ω-μm^2); w is the width of the substrate contacts in microns; l is the contact length in micrometers; d is the contact separation distance in micrometers; and R_{sh} refers to the sheet resistance of the silicon at the surface (could be epi, well, or intrinsic). The factor A is an empirical offset factor and is weakly related to the contact width (~170 for 10-μm-wide contacts). Note Equation (3.5) represents a numerical fit to measured data. Equation (3.5) was employed to calculate the resistance between the two parallel guard lines shown in Fig. 3.36. This was done by viewing the long implanted guard lines as many substrate contacts connected in parallel.

In this work, the configuration of the guard lines in relation to the interconnect line is assumed to be symmetric. So the guard lines will always be located equidistant from the edges of the interconnect line. This assures that the electrical coupling between the interconnect line and the two grounded p$^+$ implanted guard lines will be equal. Figure 3.37 shows a superimposed circuit schematic on the interconnect over silicon/guard line cross section of Figure 3.36. The capacitance between the interconnect line and the surface of the silicon is represented by c_{OX}; c_{sub} is the capacitance through the silicon die, which is assumed to be 300 μm thick, and G_{sub} is the conductance through the substrate as seen from the line. These circuit-model elements were fitted to within acceptable tolerance of simulated data, as mentioned earlier.

SWCAD Based on the background and theory outlined above, we have developed an effective utility called SWCAD. This tool, which is currently implemented

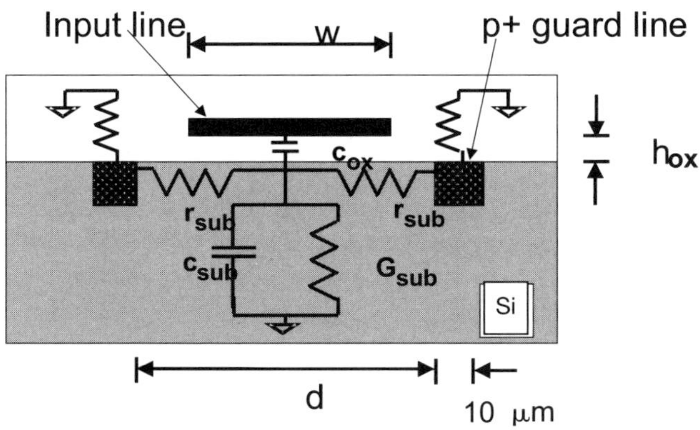

Figure 3.37 Circuit elements of interconnect over Si with neighboring guard lines.

in MATLAB and C code, combines the accuracy of the substrate and interconnect models. The user/designer can use this tool to visualize the substrate parasitics being modeled (Fig. 3.38). Importantly, the tool is designed to fit directly into the designer's CAD environment, and as such outputs SPICE-compatible subcircuits (Fig. 3.39).

The SWCAD tool incorporates the fitted equations mentioned earlier and produces circuit models that accurately model interconnect over silicon in specific IBM technologies. The tool produces stand-alone Spectre (and SPICE) netlists for the electrical representation of interconnect over a silicon substrate with user-defined electrical termination conditions. Circuit topology is automatically changed by the tool based on the frequency of signals in the interconnect and the geometry of interconnect and ground terminations. For example, at high frequencies, extra line sections and skin-effect approximation circuit sections are added to the line netlist. Figure 3.38 shows the 3D circuit topology representing the electrical network of a single interconnect line over silicon with two nearby grounded substrate contacts. The stand-alone Spectre code that the SWCAD tool outputs allows wireless designers to quickly obtain a circuit model representation of the interconnect over silicon and include it in an overall system design netlist. Also, the tool can be used to identify which interconnects may create significant loss within a wireless design.

The text shown in Fig. 3.39 represents one line section of an interconnect line over silicon with symmetrically positioned, grounded substrate contacts on both sides of the line (r_sub_con11 = r_sub_con12). Note that there are two skin-effect approximation circuit sections per line section (l_skin_11, r_skin_11, l_skin_12, and r_skin_12). This sample Spectre netlist output from the SWCAD tool is a numerical netlist, meaning that it represents one specific interconnect-over-substrate condition only. SWCAD can also produce netlists consisting of variables that the designer can freely change at the beginning of the netlist file. The capability to change variables within the stand-alone netlists allows designers to insert the output

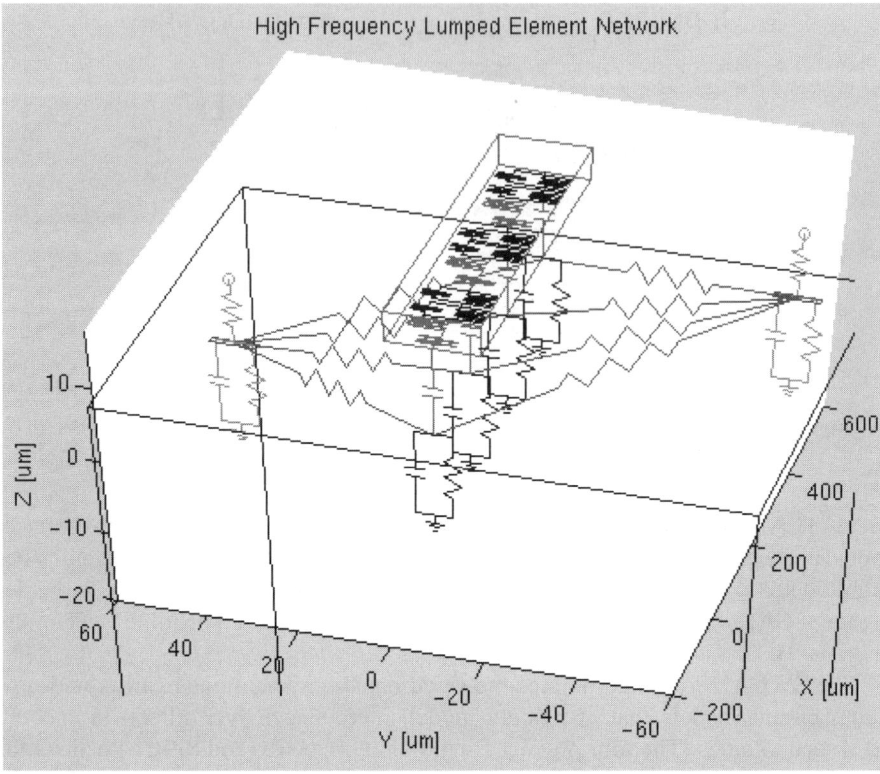

Figure 3.38 SWCAD visualization capability allows designers to investigate significant parasitics in model.

circuit models within a practical design in a scripted or automated fashion. This is particularly important when many of the same type interconnect-to-substrate conditions are present in a high-speed design.

Design Application Example In this work, a test LNA that operates between 2.11 and 2.17 GHz will be used to explore the effects of input line noise management/isolation methodology on overall circuit performance. In the 2.11–2.17-GHz frequency range, displacement current is still being dominated within the substrate, and hence no lateral capacitive elements are included in the model (relaxation frequency for this substrate is around 11.2 GHz). To study the effects of intrasubstrate coupling between the line to the guard lines, inductance and capacitance coupling circuit elements are not included above the substrate (with the exception of c_{OX}). The guard lines are assumed to be contacted every 20 μm with a 10-μm by 10-μm M1 to p^+ contact. Also, the cross section in Fig. 3.36 is kept uniform for the entire interconnect line length.

```
*
* line section 1:
*
coxide1 in net3 capacitor c=1.7013e-14
csub1 net3 gnd capacitor c=1.6341e-14
rsub1 net3 gnd resistor r=1048.9916
rline1 in net1 resistor r=0.38333
lline1 net1 net2 inductor l=1.5747e-10
l_skin_11 net2 net_skin_11 inductor l=4.6684e-12
r_skin_11 net2 net_skin_11 resistor r=0.75191
l_skin_12 net_skin_11 net4 inductor l=1.4763e-12
r_skin_12 net_skin_11 net4 resistor r=2.3777
r_sub_con11 net3 sub_net1 resistor r=704.1418
r_sub_con12 net3 sub_net2 resistor r=704.1418
*
```

Figure 3.39 Example SPICE-compatible output netlist, generated by SWCAD.

To study the effect of interconnect-to-substrate coupling and noise management in a practical circuit design, an LNA input line routed in the lowest metal layer (M1) with parallel p^+ implanted guard lines (as depicted in Fig. 3.36) was chosen to be modeled. Routing the input line in the lowest metal layer maximizes interconnect line capacitance to the substrate (for a given linewidth) while minimizing the lateral extent of the fringing electric fields to the substrate. A two-section line model was used to represent the electrical characteristics of the sensitive input line and is adequate for the target line length of 200 μm and maximum operating frequency of 2.17 GHz (much less than $\lambda/10$ per line section). The circuit model representation of the input line and the guard lines was inserted into the netlist of the test LNA at the LNA's input terminal. Cadence's 446 Analog Environment was then used to model the change in NF, input matching, etc., of the test circuit's overall performance as a function of input linewidth and guard-line separation distance.

A test circuit was employed to explore the effects of critical interconnect modeling and design methodology in a practical wireless design. The test circuit, an LNA, meets typical target specifications in wireless design (gain > 15 dB; NF < 2 dB; reverse isolation > 20 dB; operating frequency: 2.11–2.17 GHz). There are many existing LNA topology examples in the literature, such as cascade, cascode, and folded cascode, that can be employed. Cascade configurations provide increased gain at the expense of higher NFs. Folded cascode configuration examples exhibit high linearity, but require more on-chip inductors than the conventional cascode types [25]. The test LNA circuit has a cascode configuration that provides adequate gain, high-stability, low-noise, and good reverse-isolation characteristics [26].

Figure 3.40 shows the schematic diagram of the LNA and proportional-to-absolute-temperature (PTAT) bias circuits employed in the test circuit design. The LNA portion of the schematic is surrounded by the black box. There are two inductors shown in the LNA design: an emitter degenerator that simultaneously allows minimization of the NF and maximization of linearity [27], and a collector inductor that aids in output impedance matching. The simple PTAT design proved reliable and easy to set to the appropriate bias current and voltage by calculating the three resistor values needed to establish the appropriate cascode collector current.

Using Cadence's analog environment [28] to simulate the Spectre netlist for the test circuit, it was clear that, after optimizing values of the passive and active circuit components, all desired S-parameter requirements were met except for the input matching (S11), which showed a maximum of -4.83 dB (see Fig. 3.41). The desired value should be less than -10 dB. Figures 3.42, 3.43, and 3.44 show the LNA's simulated S22, S12, and S21 performance over the target operating frequency range.

Figure 3.45 shows the calculated effects that the input linewidth and guard-line separation have on the overall NF of the test circuit. Simulated NF estimates the range between 1.533 and 1.895 dB over the design space considered. Input linewidths of 10, 15, 20, and 25 μm were modeled. Guard-line separation (d) was varied between 0, w, (w + 10) μm, (w + 20) μm, (w + 30) μm, and (w + 40) μm (where w refers to the linewidth in Figs. 3.36 and 3.38). The case of $d = 0$ μm separation between the guard lines refers to the special case when the entire area under the input line is assumed to be heavily implanted (grounded guard lines contiguously routed fully under the input interconnect line). In Fig. 3.45, this special case when $d = 0$ μm is revealed by the rapid drop-off in NF as separation distance is set to zero.

Using the $d = 0$-μm case as a baseline for each modeled input linewidth, the percentage increase in NF as a function of guard-line separation distance is documented in Table 3.4. For each line width (w) in Table 2, there is an abrupt difference between the baseline ($d = 0$ μm) NF and the NF when $d = w$, particularly for the larger linewidths. This is expected because the highly conductive p^+ implanted region directly underneath the input line is only present in the $d = 0$-μm baseline case for each linewidth. Table 3.4 shows that a substantial increase in the overall test circuit NF can occur through improper routing of the 10-μm-wide p^+ implanted guard

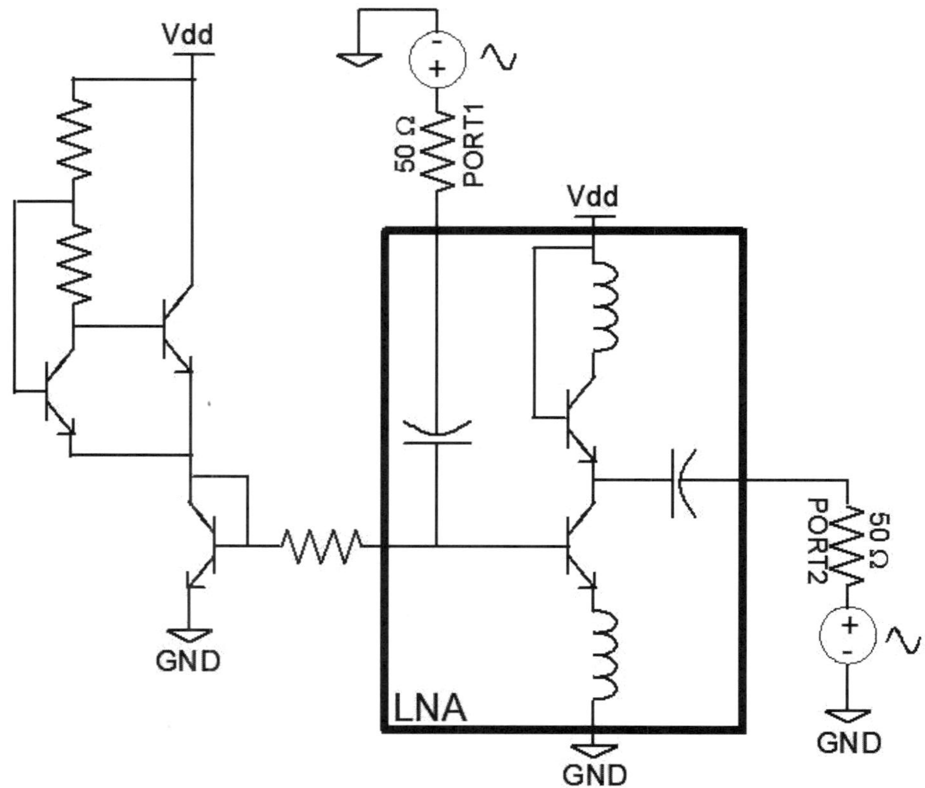

Figure 3.40 Complete test LNA with PTAT schematic. (Test LNA circuit is contained within the box.)

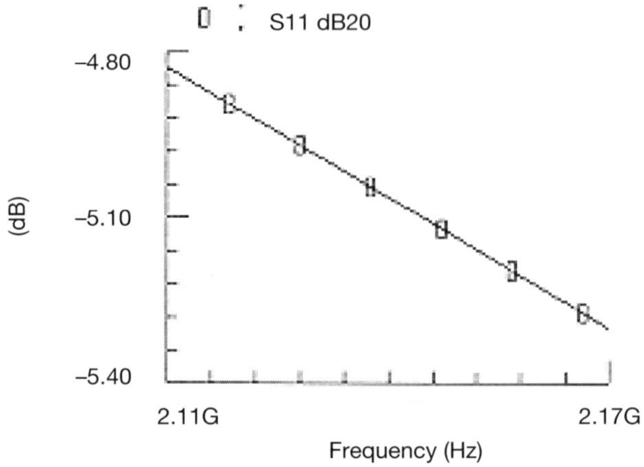

Figure 3.41 Input matching (S11) (desired < –10 dB).

lines. For example, Table 3.4 reveals an 11.41% increase in NF for the 25-μm-wide input line when $d = w + 40$ μm = 65 μm. The percentage increases in NF for the thinner input lines are not as dramatic (due to decreased capacitive coupling to the substrate). For $w = 10$ μm, the maximum percentage change in NF is 2.87%. An increase of up to 2.87% to 11.41% in NF (depending on linewidth and guard-line separation) can cause substantial problems in high-performance/low signal-to-noise input environments. For example, digital conversion bit error rate (BER) restrictions may not be met if several cascaded circuits with increased NFs are present in a data

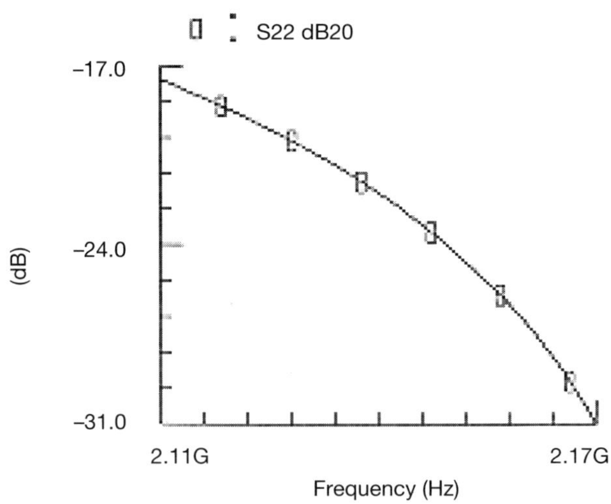

Figure 3.42 Output matching (S22) (desired < –10 dB).

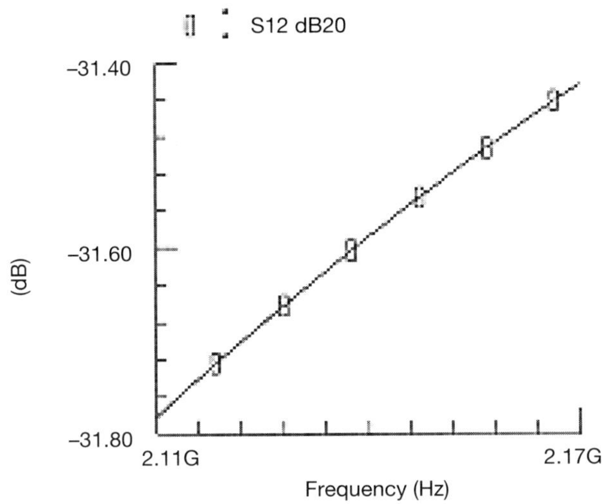

Figure 3.43 Reverse isolation (S12) (desired < –20 dB).

path. Also, the problem is exacerbated at higher frequencies. Calculated NFs and the respective percentage changes observed from the baseline cases ($d = 0$ μm) increased between 2.11and 2.17 GHz. The extrapolated frequency scaling behavior predicts an increase of about 2.7% at 5 GHz over NF estimates calculated at 2.11 GHz for the $d = w + 40$-μm case in the $w = 25$-μm line (so 13.6% maximum deviation instead of 11.4%).

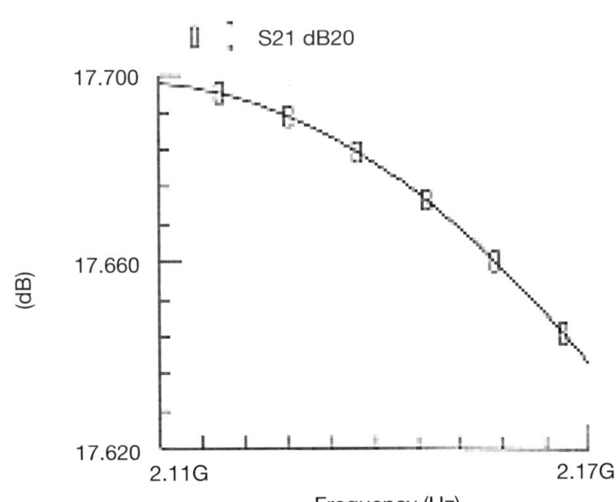

Figure 3.44 Gain (S21) (desired > 15 dB).

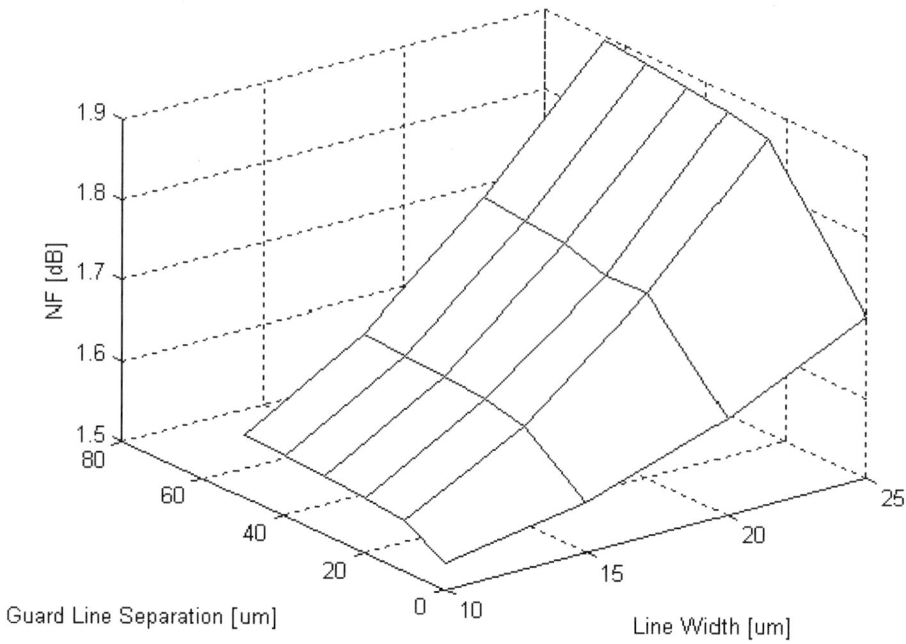

Figure 3.45 Simulated NF change as a function of input and guard-line geometry.

Figure 3.45 and Table 3.4 indicate that NF is highly sensitive to linewidth. As linewidth is increased, the simulated NF increases rapidly. This reflects the roughly proportional increase in line-to-substrate capacitance with linewidth. Even though line inductance and resistance are proportionally decreased with increasing linewidth, the capacitance to the silicon substrate still dominates for noise generation over the input line design space considered. By increasing the capacitive coupling to the substrate, more electromagnetic energy is channeled into the lossy Si substrate. Thus, to minimize overall NF, the input linewidth should be kept small and the guard line should be kept as close to the edges of the input line as possible

Table 3.4 Overall Test Circuit NF Change as a Percentage Difference from the $d = 0$-μm Baseline Case for Each Linewidth (w) vs. Guard-Line Separation Distance (d)

Linewidth (w) [μm]	$d = w$	$d = w + 10$	$d = w + 20$	$d = w + 30$	$d = w + 40$
10	1.89	2.28	2.54	2.74	2.87
15	3.90	4.47	4.80	5.05	5.31
20	6.04	6.54	7.58	7.95	8.26
25	9.52	10.17	10.64	11.05	11.41

(preferably extending the guard lines under the input line as far as possible). Decreasing the input line width to increase noise performance sounds counterintuitive to circuit designers because inductance and line resistance are increased. However, when routing interconnect lines over silicon, especially in the lower metal layers, the strength of the interconnect lines' coupling to the silicon surface is what matters in noise generation.

3.3.4 Summary

In this section, we have discussed in detail the issues in interconnect extraction and modeling. Interconnect parasitic noise effects are critical to the performance of typical SiGe designs, with frequencies ranging up to 60–70 GHz. For many applications, accurate *RLC* extraction tools suffice. However, for higher frequency applications, transmission lines are recommended.

We have also provided an in-depth summary of an effective approach to interconnect-to-substrate modeling. This methodology has been implemented in MATLAB and produces SPICE netlists. This approach has been shown to be effective, when applied to a 2.1-GHz LNA. Looking forward, this methodology is currently being used to investigate numerous subtle effects in RF IC layout design, which are not apparent when using standard EDA parasitic extraction tools.

3.3.5 References

1. R. Singh, *Signal Integrity Effects in Custom IC and ASIC Designs,* IEEE Press and Wiley, New York, 2001.
2. A. Deutsch, P.W. Coteus, G. V. Kopcsay, H. H. Smith, C. W. Surovic, B. L. Krauter, D. C. Edelstein, and P. J. Restle, "On-Chip Wiring Design Challenges for Gigahertz Operation," *IEEE Proc.,* vol. 89, no. 4, pp. 529–555, April 2001.
3. A. E. Ruehli, and A. C. Cangelaris, "Progress in the Methodologies for the Electrical Modeling of Interconnects and Electronic Packages," *IEEE Proc.,* vol. 89, no. 5, pp. 740–771, May 2001.
4. J. Jin, *The Finite Element Method in Electromagnetics,* Wiley, New York, 1993.
5. K. S. Yee, "Numerical Solution of Initial Boundary Value Problems Involving Maxwell's Equations in Isotropic Media," *IEEE Trans. Antennas Propagat.,* vol. 14, pp. 302–307, May 1966.
6. R. F. Harrington, *Field Computation by Moment Methods,* IEEE Press, New York, 1993.
7. Y. Kim, A. Tripathi, R. K. Settaluri, A. Weisshaar, and V. K. Tripathi, "Extraction of Multiple Coupled Line Parameters Using FDTD Simulation," *IEE Proc. Microwave Antennas Propagat.,* vol. 146, no. 6, pp. 443–446, December 1999.
8. M. J. Tsuk and J. A. Kong, "A Hybrid Method for the Calculation of fhe Resistance and Inductance of Transmission Lines with Arbitrary Cross Sections," *IEEE Trans. Microwave Theory Tech.,* vol. 39, pp. 1338–1347, August 1991.
9. J. R. Phillips and J. K. White, "Efficient Capacitance Extraction of 3-D Structures using Generalized Pre-Corrected FFT Methods," in *IEEE 3rd Topical Meeting on Electrical Performance of Electronic Packaging,* pp. 253–256, 1994.

10. M. Kamon, M. J. Ttsuk, and J. K. White, "Fasthenry: A Multipole-Accelerated 3-D Inductance Extraction Program," *IEEE Trans. Microwave Theory Tech.,* vol. 42, no. 9, pp. 1750–1758, September 1994.
11. Y. L. Le Cox and R. B. Iverson, "A Stochastic Algorithm for High Speed Capacitance Extraction in Integrated Circuits," *IEEE Trans. Microwave Theory Tech.,* vol. 40, pp. 1507–1516, July 1992.
12. W. R. Eisenstadt, E. Yungseon, "S-Parameter-Based IC Interconnect Transmission Line Characterization," *IEEE Trans. Components, Hybrids, and Manufacturing Tech.,* vol. 15, no. 4, pp. 483–490, August 1992.
13. A. E. Ruehli, W. P. Pinello, and A. C. Cangelaris, "Comparison of Differential and Common-Mode Response for Short Transmission Line Using PEEC Models," in *IEEE 5th Topical Meeting on Electrical Performance of Electrical Packaging,* pp. 169–171, 1996.
14. L. Corti, M. Magistris, A. Maffucci, and G. Miano, "Efficient Time-Domain Simulation of Lossy Multiconductor Lines with Nonlinear Loads," in *IEEE Intl. Symp. on Electromagnetic Comp.,* vol. 1, pp. 440–445, 1999.
15. T. Dhaene and D. De Zutter, "Selection of Lumped Element Models for Coupled Lossy Transmission Lines," *IEEE Trans. Computer-Aided Design,* vol. 11, no. 7, pp. 805–815, July 1992.
16. J. Zheng, Y. C. Hahm, V. K. Tripathi, and A. Weisshaar, "CAD-Oriented Equivalent-Circuit Modeling of On-Chip Interconnects on Lossy Silicon Substrate," *IEEE Trans. Microwave Tech.,* vol. 48, no. 9, pp. 1443–1451, September 2000.
17. D. Goren, M. Zelikson, T. C. Galambos, R. Gordin, B. Livshitz, A. Amir, A. Sherman, and I. A. Wagner, "An Interconnect-Aware Methodology for Analog and Mixed-Signal Design, Based on High Bandwidth (over 40 GHz) On-Chip Transmission Line Approach," *Design Automation and Test in Europe 2002,* Proceedings, pp. 804–811, March 2002.
18. B. K. Sen and R. L. Wheeler, "Skin Effects Models for Transmission Line Structures Using Generic Spice Circuit Simulators," in *EPEP'98—7th Topical Meeting on Electrical Performance of Electronic Packaging,* pp. 128–131, October 1998.
19. A. Weisshaar and H. Lan, "Accurate Closed-Form Expressions for the Frequency-Dependent Line Parameters of On-Chip Interconnects on Lossy Silicon Substrate," *IEEE MTT-S Digest,* vol. 3, pp. 1753–1756, 2001.
20. Y. Eo and W. R. Eisenstadt, "High-Speed VLSI Interconnect Modeling Based on S-Parameter Measurements," *IEEE Trans. Components, Hybrids, and Manufacturing Technology,* vol. 16, no. 5, pp. 555–562, August 1993.
21. Ansoft Corporation, *High-Frequency Structure Simulator User's Guide,* 2002.
22. Agilent Technologies, *Advance Design System User's Guide,* September 2001.
23. P. R. Bannister, "Applications of Complex-Image Theory," *Radio Science,* vol. 21, no. 4, pp. 605–616, August 1986.
24. D. M. Pozar, *Microwave Engineering,* Second Edition, Wiley, New York, 1998.
25. B. Ray, T. Manku, R. D. Beards, J. J. Nisbet, and W. Kung, "A Highly Linear Bipolar 1V Folded Cascode 1.9 GHz Low Noise Amplifier," in *Proceedings of 1999 BCTM,* pp. 157–160, 1999.
26. G. Gonzalez, *Microwave Transistor Amplifiers Analysis and Design,* Prentice Hall, Upper Saddle River, NJ, pp. 95–102, 1996.

27. C. D. Hull, Analysis and Optimization of Monolothic RF Downconversion Receivers, Ph.D. dissertation, University of California at Berkely, 1992.
28. *Cadence Design Systems Version 4.4.6 User's Guide* Cadence, San Jose, CA, June 2000.

3.4 SUBSTRATE NOISE ISOLATION AND MODELING

Since the early 1990s, there have been numerous research papers published [1] attempting to offer practical and usable approaches to model substrate coupling. This surge of research, in the areas of RF and mixed-signal designs, has been triggered by a realization in the chip-design industry that substrate coupling is an important issue for designers. This realization has in main arisen by the adoption of single IC solutions, where smaller gate lengths have lead to higher frequencies of circuit operations and to high levels of integration. Given these trends, it is somewhat surprising that substrate modeling methodologies are not yet common in commercial verification design flows. This is partly because substrate modeling is not simple and inefficiencies can lead to impractical extraction and simulation time and memory requirements.

Substrate coupling can degrade performance in small RF designs, as shown in Fig. 3.46. In this direct conversion circuit local oscillator (LO) leakage couples through the substrate to the RF signal path and self-mixes with the LO signal (in the mixer) to form DC offset, which is very hard to design against in such designs. Subharmonic mixer architectures now exist that use LO frequencies at different frequencies to the signal, helping to alleviate this problem.

Substrate coupling reduction is an important area for designers. A handful of cost-effective techniques include:

- Guard-ring structures around the noisy and sensitive circuitry, usually tied to dedicated package pins for closer AC grounding. In RF designs, a dedicated pin is not always available, especially when being integrated into a larger design.

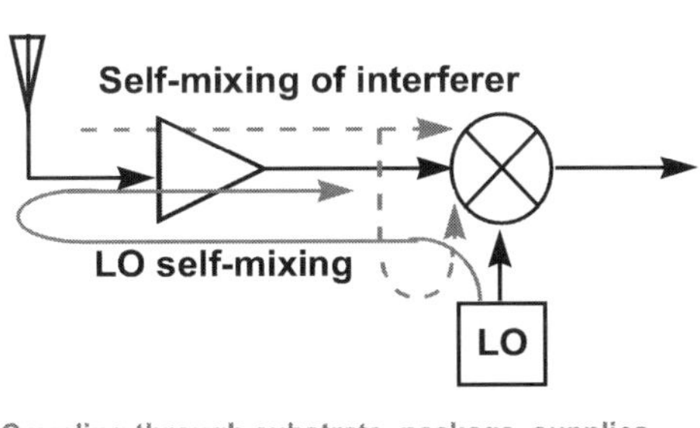

Figure 3.46 Substrate coupling leading to LO feedthrough, in a direct conversion receiver.

- N-Well trenches and DTs between the noisy and sensitive circuitry to block the substrate current flowing near the surface of the substrate.
- Differential circuitry—a favorite for RF designers, but sometimes not as effective as expected with nondifferential bias circuitry affected by substrate coupling leading to performance problems in the circuit.
- Lower package parasitic inductance (for the pins). This is a very effective approach, but one that is sometimes expensive, due to the extra package costs. Lower inductance is also accomplished through multiple- or shorter-bond wire connections, or flip-chip area input/output (I/O) packages.
- Careful floorplanning can also be used. The idea is that the further the sensitive and noisy circuits are apart, the less substrate coupling will affect the circuit's performance.
- Use of low-impedance external package paddles, to effectively "short" the ground pins to the printed circuit bound (PCB) ground network.

3.4.1 Substrate Isolation Evaluation with Technology CAD

TCAD is described in detail in Section 3.1. The motivation for TCAD development and application over the last two decades has primarily been resolution of the critical issues in CMOS scaling, in particular the transient-enhanced diffusion effects that dominate CMOS short-channel effects [1,2], and the inversion-layer mobility [3] and carrier quantization effects [4] that determine current drive and threshold voltages in scaled CMOS devices. In support of the industrial application of TCAD, significant development work in the practical aspects of mesh generation [5] and large-scale solution of the equations solved in process and device simulation [6] has been undertaken. When augmented by the computational infrastructure required to address the large problem sizes and electrical characterization arising from substrate isolation structures, the computational capability represented by modern TCAD tools provides a powerful tool for substrate noise and isolation analysis. TCAD provides a physically based computational connection between process, device-design and layout options, and detailed substrate isolation characteristics. In particular, no assumptions about electrical connectivity between different regions of the structure, or the nature of the electrical interaction between different regions, need to be made. P-N junction and semiconductor–insulator–semiconductor capacitances, and their frequency dependence, as well as the dependence of conductivity on doping and material thickness, are automatically accounted for by standard device simulation approaches.

Application of TCAD to Substrate Isolation Two approaches to applying TCAD to substrate signal propagation appear in the literature:

1. Small-signal S-parameter simulations of large and fairly simple structures [7–10]
2. Large-signal transient and DC simulations of more complex structures [11–15]

The small-signal AC formalism in device simulation provides a way to duplicate AC hardware measurements [16], and has been shown to reflect experimental S-parameter measurements, with quantitative accuracy, on complex semiconductor devices, at frequencies greater than 100 GHz [17]. However, for the large structures characteristic of substrate isolation modeling, high-frequency AC small-signal simulations can represent a significant computational burden, particularly in three dimensions. An example of the correspondence between 2D and 3D simulation can be seen in Figure 3.47, in which 3D and 2D simulations are compared to experimental results found in the literature [18] for a simple test structure consisting of two 100 μm × 100 μm contacts on 0.95 μm of oxide, 35 μm apart, over two different substrates. The 3D simulation shows good agreement with the reported experimental values over many decades of signal frequency, while the 2D simulation of the same structure (the 2D results have been scaled to reflect the same contact width as the 3D structure) show the same qualitative relationship between the two structures, but are offset several dB from the 3D results. This set of simulation results indicates that 2D simulation should not be expected to provide precise quantitative evaluation of different substrate isolation schemes, but rather to provide qualitatively insight into the effectiveness of different isolation schemes combined with a rough quantification of the degree of isolation supported.

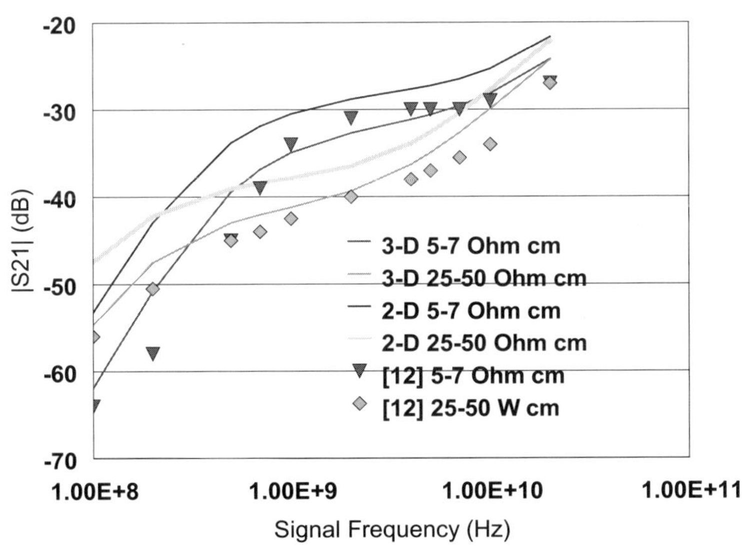

Figure 3.47 Comparison between 2D and 3D TCAD simulation, and experimental data [13] for substrate noise test structure. The test structure consists of two 100-μm × 100-μm contacts on 0.95 μm of oxide, 35 μm apart, over two different substrates Results indicate that 2D TCAD simulations of substrate isolation effects can provide qualitatively accurate evaluations of complex signal propagation effects.

The advantage of large-signal simulations is that detailed noise waveforms can be used in the simulations, and transient semiconductor device simulation typically requires much less computer power than very high-frequency small-signal simulations [16]. Succinctly put, quantifying the isolation characteristics of a structure is less straightforward for a large-signal simulation, requiring either a frequency-domain decomposition of the signal [8] to recover the S-parameters, or choosing a point on the output-node signal characteristics to compare to a point on the input-node signal.

Large-Signal Simulations As an example of the use of TCAD to provide insight into complex substration isolation problems, an extensive set of 2D large-signal transient simulations were performed on a receiver structure consisting of a nominal emitter-size DT-bounded f_T = 47-GHz SiGe HBT [19] biased as a simple common-emitter amplifier, at peak-F_t I_c, surrounded by a 1-μm-wide oxide-lined and polysilicon-filled DT. The structure was created by a detailed 2D process simulation of the manufactured device, and includes a heavily doped buried subcollector and implanted extrinsic base. A 100-mV sinusoidal noise signal at three different frequencies, 1, 10, and 60 GHz, was injected into the substrate at varying distances from the victim structure. In order to assess the most effective way to isolate the HBT from substrate noise, beyond the isolation provided by the DT, different isolation structures, also arising from detailed process simulation, were placed between the noise source and the HBT listener. The general 2D simulation layout is depicted in Fig. 3.48. The resulting signal at the collector of the HBT was recorded for these noise inputs, and is reported as the ratio of the maximum

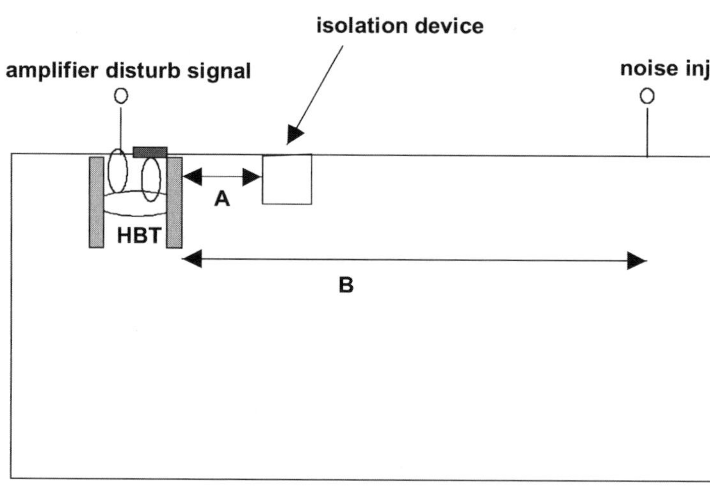

Figure 3.48 Large-signal 2D TCAD simulation structure. A F_t = 60-GHz SiGe HBT device [14], DT bounded, biased at peak-F_t IC as a collector–emitter amplifier, is the listener structure. The p^- substrate is 300 μm thick.

noise voltage to the maximum signal disturbance at the collector of the listener HBT.

Figure 3.49 plots the simulation results for simply spacing the noise signal an increasing distance away from the HBT listener. In this case, the isolation structure is merely the lightly doped p^- substrate. The advantage of increased spacing is observed to saturate for distances above 50 μm. Augmenting the simple substrate isolation with an additional DT, spaced 10 μm from the listener HBT, was found to provide no isolation benefit.

Figsures 3.50 and 3.51 plot the results of a second set of simulations run to evaluate the utility of alternative substrate isolation structures. The often-observed utility of a simple p^+ guard-ring (SX GR), in this technology, consisting of a PFET source/drain diffusion in a p-well, compared to the simple low-doped substrate, is apparent in Fig. 3.51. Adding a DT between the SX GR and the noise source is observed to actually degrade the isolation, presumably because the DT isolates the SX GR from the noise signal. An additional SX GR is observed to add an additional 3.5 dB of isolation. Positioning the SX GR closer to the noise source marginally degrades the substrate isolation, while making the SX GR large marginally improves substrate isolation.

A further enhancement of the SX GR is to make it deeper via a sequence of implants in the MeV range. A process for this type of isolation structure was formulat-

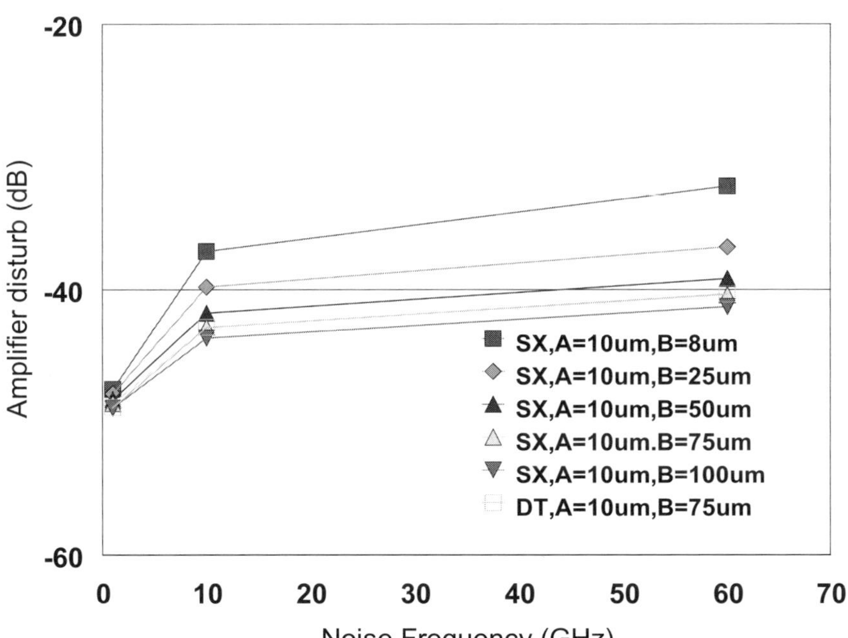

Figure 3.49 Large signal 2D simulation results applied to structure shown in Fig. 3.48. Isolation device is simple p^- substrate (SX).

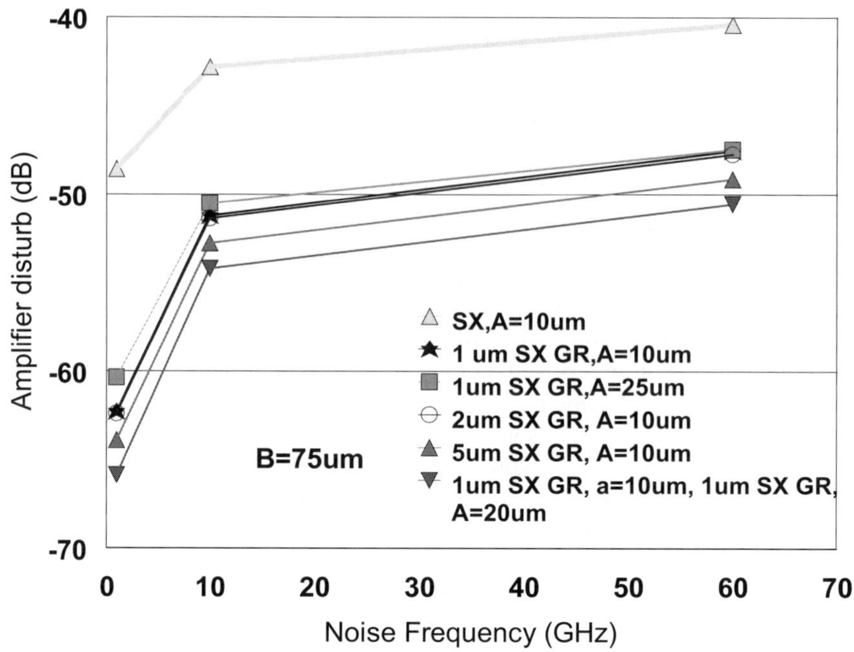

Figure 3.50 Large signal 2D simulation results applied to structure shown in Fig. 3.48. Superior isolation characteristics of a grounded substrate contact (SX GR) are observed.

ed, and simulations of this structure, shown in Fig. 3.52, indicate an additional 6–10 dB in noise isolation could be achieved by significantly extending the p^+ SX GR depth. The utility of increased depth in the SX GR suggests the use of the HBT n-type buried subcollector, contacted by the HBT "reachthrough" or "sinker" diffusion as an isolation device, as it is very heavily doped and lies relatively deep in the substrate. However, the simulations of this device indicate that relative to the SX GR, it is an inferior isolation structure at lower noise frequencies, presumably due to the p-n junction it forms with the p^- substrate, and has no advantage over the simple SX GR at higher frequencies. A final option considered, based on the observed utility of the deeper SX GR, is the polysilicon-filled DT. The polysilicon fill in the DT is floating in the normal process. However, straightforward process changes were devised to allow p-type doping and contacting (to ground) the DT polysilicon fill. When simulated, the isolation characteristics of this device were very similar to those of the HBT buried subcollector.

Small-Signal Simulations The small-signal methodology is illustrated by application to the structure in Fig. 3.52. Two p^+ diffusions in a p^- substrate are separated by large distances, 50 μm–150 μm, and symmetrically placed isolation structures. All contacts are at zero bias. Figure 3.53 plots the results for the case where

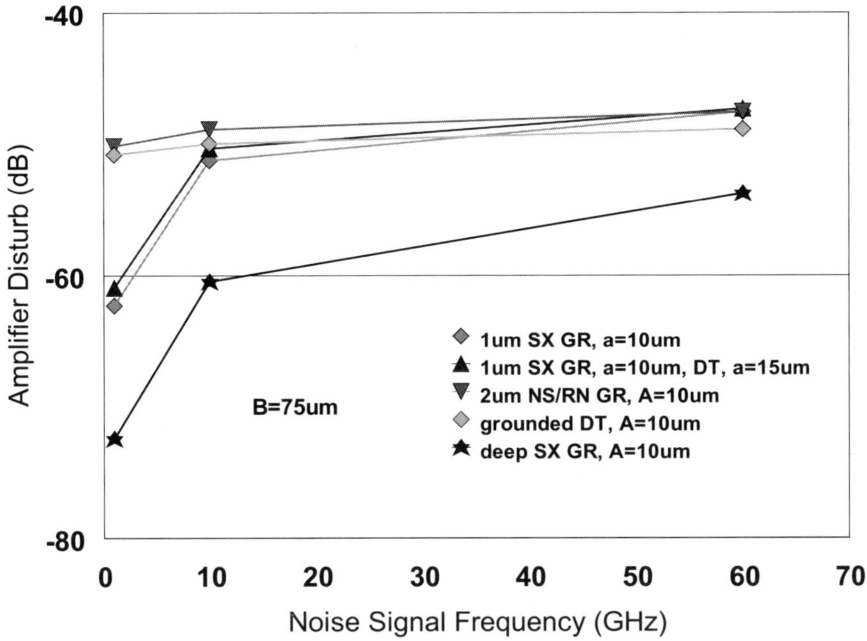

Figure 3.51 Large signal 2D simulation results applied to structure shown in Fig. 3.52. The simulation results for several alternatives to the simple grounded substrate guard ring (SX GR) are shown.

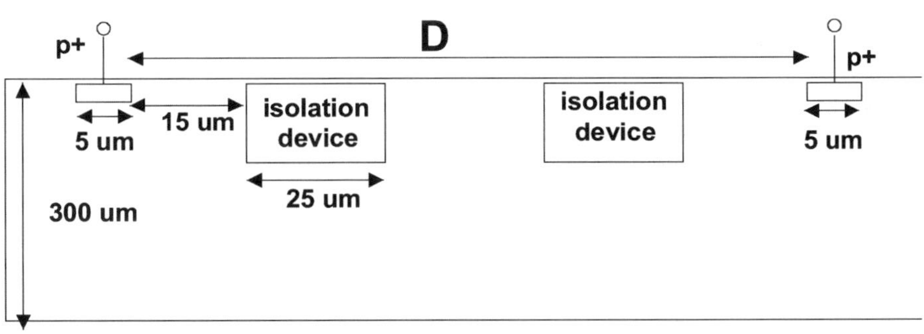

Figure 3.52 Moat structure for 2D small-signal AC simulations. The two signal ports are assumed to be 5 μm × 5 μm. Substrate is doped p^-.

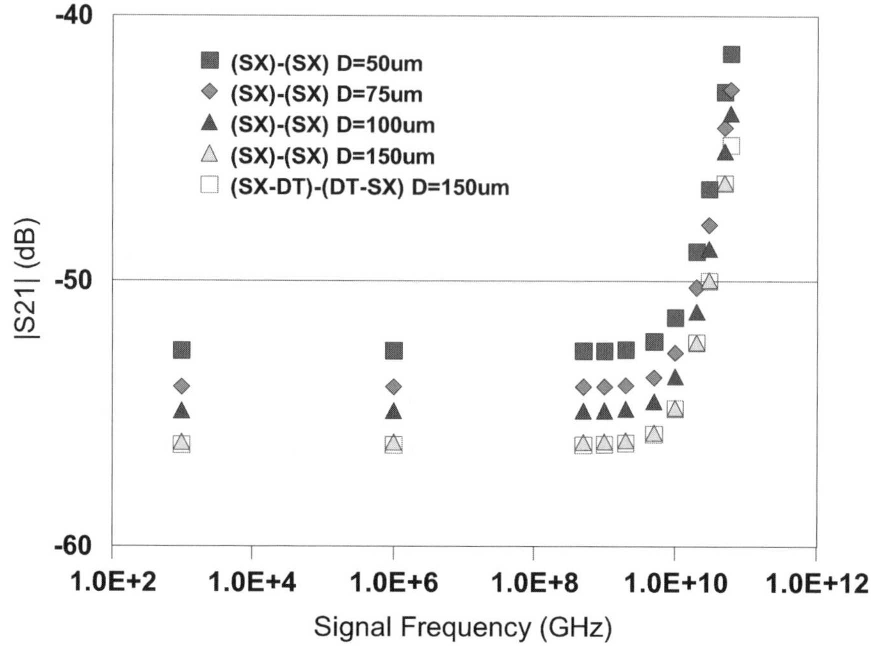

Figure 3.53 Small-signal AC simulation results for structure in Fig. 3.56. Minor marginal substrate isolation improvement is observed for increasing distance between signal and receiver ports.

the isolation structures are simply the p^- substrate. As the large-signal case, for distances greater than 50 μm, the utility of increased injector/listener distance appears to be small, independent of frequency. Note that Fig. 3.53 also indicates the negligible contribution of DT as a substrate isolation device. Figure 3.54 plots the results obtained from using grounded SX GR structures, showing improvements of approximately 10 dB for the case of 75-μm separation and increasing to 15 dB of noise isolation for the 150-μm separation case. Note that the addition of DT to the SX GR noticeably degrades the substrate noise isolation at very low frequencies, and that such structures actually are observed to show improved substrate isolation as noise-signal frequency increases between 1 and 10 GHz, presumably as the DT becomes transparent to the electrical signal and no longer shields the SX GR from it. Additional SX GR isolation structures are observed to improve substrate noise isolation, but at a slower rate compared to the addition of the first one.

3.4.2 Test-Site Results

In high-performance RF and mixed-signal circuits, modeling substrate impedance becomes important for short distances as well as large distances. While, in some cases, it may be important to know the exact value of the substrate impedance af-

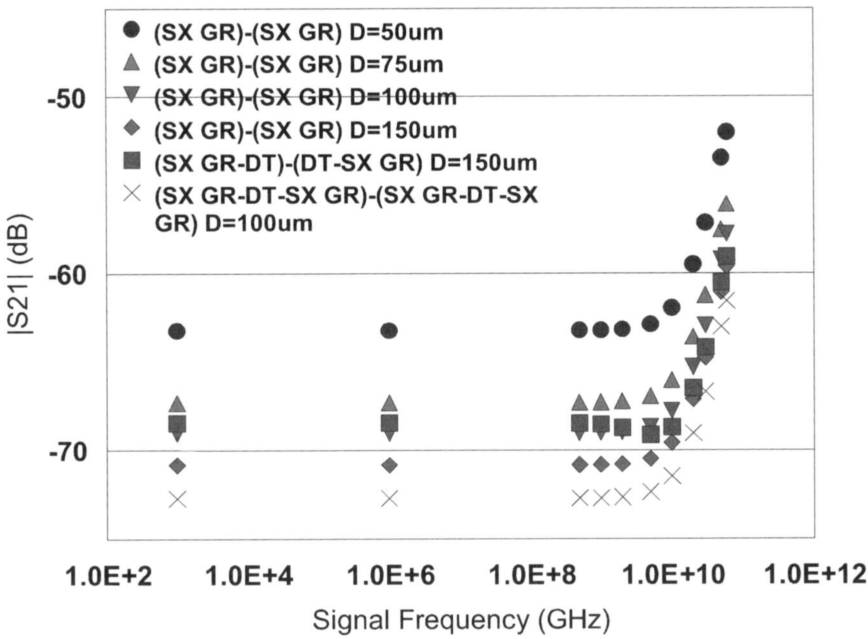

Figure 3.54 Small-signal AC simulation results for structure in Fig. 3.56. Significant substrate isolation improvement is observed with the addition of grounded substrate guard rings between the signal ports.

fecting the device characteristics, in other cases, increasing the isolation between two points in the circuits can be important. A series of systematic test sites were developed, produced, and measured for this work, using IBM's SiGe process. A p^- substrate (10 Ω-cm) was implemented, with a p-well between the devices by default. These results are presented in this section as new, important data for understanding the isolation effects of the substrate.

Based on measurements between identical substrate contacts of different sizes as a function of separation, it was concluded that the total resistance could be broken into horizontal and vertical components, as shown in Fig. 3.55. The vertical component depends on the substrate contact size and the horizontal component on the separation between the contacts. Due to relatively high doping near the surface in the p-well region, the majority of substrate current is expected to flow in the p-well region for short distances relative to the substrate thickness. Figure 3.56 shows the measurement results as a function of distance on a 750-μm-thick wafer. A logarithmic relationship was observed for the horizontal component of the substrate resistance. The empirical formula, shown in Equation (3.6), can be used to estimate substrate resistance up to about 50-μm distance:

$$R_{\text{sub}} = 2*R_{sv}/wl + A*\log(d + 0.4*w) \parallel R_{\text{sh}}*(d + 0.4*w)/l) \tag{3.6}$$

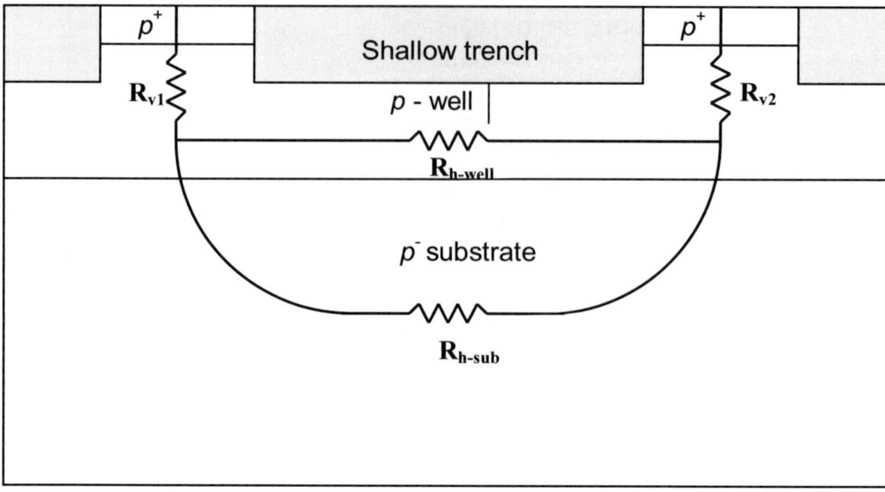

Figure 3.55 Substrate resistance components, with explicit impedance shown for a commonly used extra p-well

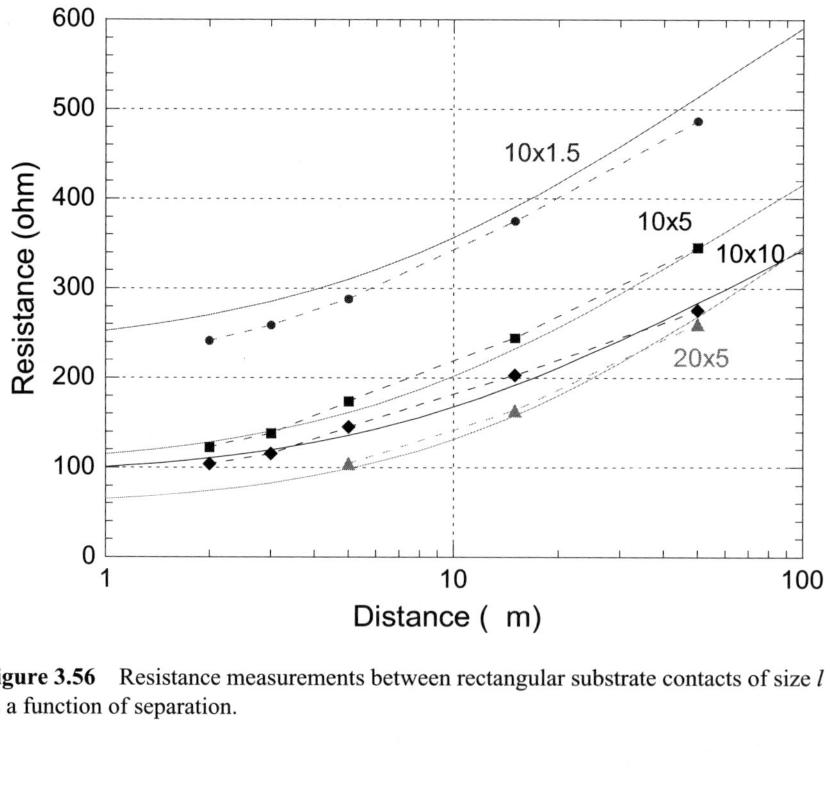

Figure 3.56 Resistance measurements between rectangular substrate contacts of size $l \times w$, as a function of separation.

where

R_{sv} vertical resistance of the contact for unit area (= 1725 W-μm²)

w contact width (in μm)

l contact length (in μm)

d contact separation (in μm)

A weak function of contact width (205, 195, and 170 for 1.5 μm, 5 μm, and 10 μm widths, respectively)

R_{sh} sheet resistance of the p-well

IBM's RF technologies offer layout options for increasing isolation between noisy and sensitive parts of the circuits. The default p-well region can be omitted from selective regions of the layout, giving a lightly doped n-type region near the surface. Since the substrate current flows in the p^- substrate in these isolated regions, higher substrate impedance can be achieved. Another isolation technique available in high performance BiCMOS technologies is the use of a DT to force the current to flow deeper into the substrate (in the high resistive region). Figure 3.57 displays the substrate resistance in the p-well omitted region. Comparing these results with Fig. 3.56, isolation resistance can be increased by orders of magnitude by using such isolation.

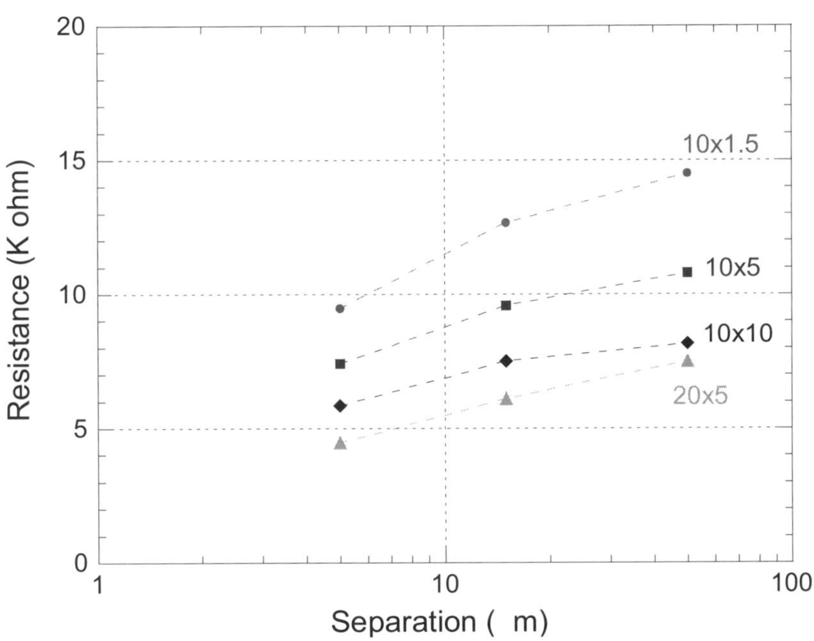

Figure 3.57 Resistance measurement between rectangular substrate contacts of size $l \times w$ in IBM's BiCMOS 7HP technology without p-well between the contacts.

IBM's SiGe processes use 11–16 Ω-cm substrates to minimize substrate losses. An interesting effect of using a lightly doped substrate is that the transition frequency F_t of the substrate moves down to between 10 and 20 GHz. For a substrate model, which consists of a network of resistors, the transition frequency is the frequency at which each of the resistors is shunted by a capacitive reactance, which is equal in magnitude to the value of each of the resistors. Equations (3.7) and (3.8) describe how to obtain values for F_t and C_{shunt}. Equation (3.8) was previously presented in [20]:

$$F_t = \tfrac{1}{2}\pi\rho\varepsilon \tag{3.7}$$

where

F_t substrate transition frequency (Hz)
ρ substrate resistivity (Ω-cm)
ε substrate permitivity (F/cm)

$$C_{\text{shunt}} = \rho\varepsilon / R_{\text{shunt}} \tag{3.8}$$

where:

R_{shunt} substrate resistive element (Ω)
C_{shunt} substrate capacitive element (F)

At higher frequencies, substrate currents may increase due to the ever-decreasing substrate impedance, and substrate power dissipation may decrease, since less current will be flowing through the resistive part of the substrate network. Parasitic capacitance of metal begins to be terminated into a series capacitor instead of a series resistor. There can be a substantial change in the character of the substrate at the transition frequency for designers to ponder.

Finally, *linking the previous TCAD discussion to the test-site discussion,* we present in Fig. 3.58 initial results mapping our 3D TCAD results to test-site results. As can be seen, the results are very accurate for AC and DC measurements/simulations, demonstrating the power of the predictive modeling capability in the substrate modeling area.

3.4.3 Substrate Modeling Techniques

For the designer, it is important to understand some of the most used algorithms in substrate modeling and their applicability to RF and AMS designs. A number of approaches have been proposed, including using the Green's function and finite difference techniques [1]. In this section only the Green's function approach is discussed. This extraction method involves the following generic steps:

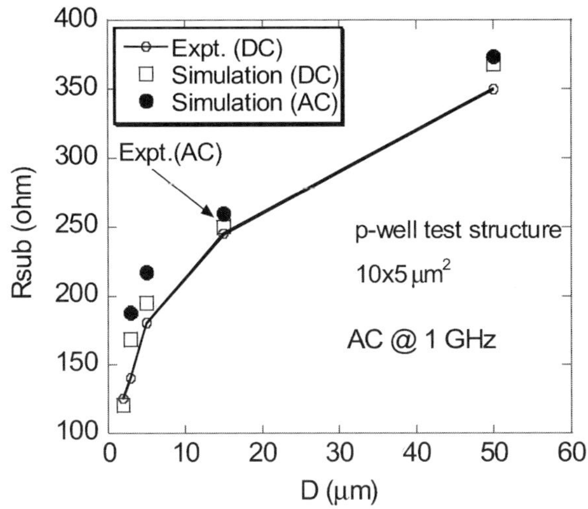

Figure 3.58 Resistance vs. separation (DC and AC) for p-well substrate contact test structure with contact dimension $10 \times 5\ \mu m^2$.

- Identify the areas to be connected to the substrate (termed *substrate ports*)
- Extract a resistive mesh and link the substrate ports
- Netlist and simulate the design, using an analog simulator

Figure 3.58 shows how the Green's function technique can be used to connect the resistive mesh to the substrate ports. The Green's function approach inherently leads to capacity problems. This is due to thc large number of resistors extracted to create an accurate model of the substrate. These resistors cause long extraction and simulation times and large CPU memory requirements.

Using the Green's functions method, a substrate resistor is extracted between every substrate port, as shown in Fig. 3.59. This leads to density in the nodal matrix that the analog simulator solves. Analog-simulation engines, such as SPICE, are designed to solve sparse matrices, and therefore are not efficient in solving dense matrices. Previous tests have shown that up to 300 devices can be connected to a substrate mesh, using this algorithm in the RF parasitic verification flow. With bigger designs, the simulation time and memory requirements have been found to be a major bottleneck.

One approach to increase capacity is to approximate the effect of distant coupling using a *windowing* technique, which effectively creates local substrate models. This has been covered in [1] and does not offer a capacity increase sufficient for large AMS designs, because substrate coupling can be a global effect with the effect of distant coupling often significant. Alternatively, selection layers have been used to allow increased capacity in the verification flow. In this flow, the user can select

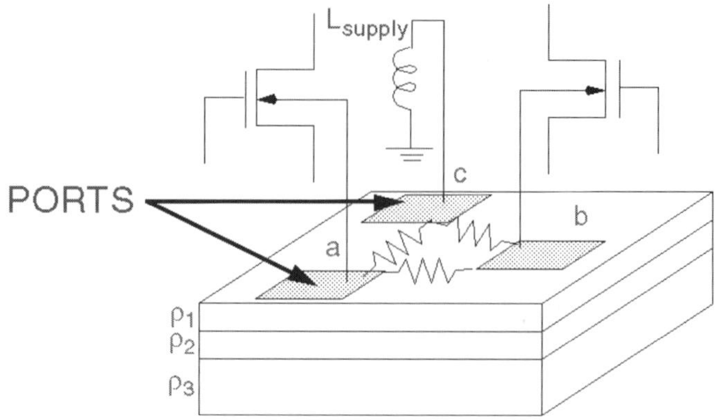

Figure 3.59 Using Green's function for efficient RF IC substrate modeling, through substrate ports.

the areas of the circuit to be connected to the substrate. In this way, the capacity of the substrate modeling tool remains the same, but the circuit can be much larger in the backend verification flow. This approach offers accurate verification of designs up to approximately 2000 devices, depending on the amount of substrate coupling in the circuit. This is sufficient for RF-only circuits, but not for larger AMS designs.

3.4.4 Summary

We have presented TCAD and test-site work in this section, demonstrating the beginnings of a coordinated methodology for achieving two clear goals: (1) to allow for impedance estimation for interblock isolation, for which an initial DC equation has been presented; and (2) to allow optimization of the isolation techniques used in the circuit design. Initial TCAD to test-site correlation results were also presented.

3.4.5 References

1. R. Singh, *Signal Integrity Effects in Custom IC and ASIC Designs,* IEEE Press and Wiley, New York, 2001.
2. C. S. Rafferty, "Front-End Process Simulation," *Solid-State Elec.,* vol. 44, pp. 863–868, 2000.
3. A. Mujtaba, S. Takagi, and R. Dutton, "Accurate Modeling of Coulombic Scattering, and Its Impact on Scaled MOSFETs," in *Proceedings 1995 Symposium on VLSI Technology, Digest of Technical Papers,* pp. 99–100, 1995.
4. M. Ieong, H.-S. Wong, Y. Taur, P. Oldiges, and D. Frank, "DC and AC Performance Analysis of 25 nm Symmetric/Asymmetric Double-Gate, Back-Gate and Bulk CMOS," in *International Conference on Simulation of Semiconductor Processes and Devices, SISPAD 2000,* pp. 147–150, 2000.

5. V. Axelrad and M. Duane, "Controlling Mesh Effects in Integrated Process and Device Simulation," in *Proceedings, International Workshop on Statistical Metrology, IWSM 1998,* pp. 100–103, 1998.
6. B. P. Herndon, A Methodology for the Parallelization of PDE Solvers: Application to Semiconductor Device Physics, Ph.D. dissertation, Stanford University, 1995.
7. K. Joardar, "Substrate Crosstalk in BiCMOS Mixed Mode Integrated Circuits," *Solid-State Elec.,* vol. 39, no. 4, 1996, pp. 511–516, 1996.
8. S. Bharatan, P. Welch, K. H. To, R. Thoma, and M. Huang, "3D Simulation and Modeling of Signal Isolation in RF/IF Circuit," *Modeling and Simulation of Microsystems 2001,* pp. 470–473, 2001.
9. K. Joardar, "Comparison of SOI and Junction Isolation for Substrate Crosstalk Suppression in Mixed Mode Integrated Circuits," *Electronics Lett.,* v. 31, pp. 1230–1231, 1995.
10. K. Joardar, "Signal Isolation in BiCMOS Mixed Mode Integrated Circuits," in *Bipolar/BiCMOS Circuits and Technology Meeting,* pp. 178–181, 1995.
11. D. K. Su, M. J. Loinaz, S. Masui, and B. A. Wooley, "Experimental Results and Modeling Techniques for Substrate Noise in Mixed-Signal Integrated Circuits," *IEEE J. Solid-State Circuits,* vol. 28, pp. 420–430, 1993.
12. J. M. Casalta, X. Aragones, and A. Rubio, "Substrate Coupling Evaluation in BiCMOS Technology," *IEEE J. Solid-State Circuits,* vol. 32, pp. 598–603, 1997.
13. S. Masui, "Simulation of Substrate Coupling in Mixed-Signal MOS Circuits," in *Symposium on VLSI Circuits Digest of Technical Papers,* pp. 42–43, 1992.
14. X. Aragones, F. Moll, M. Roca, and A. Rubio, "Analysis and Modelling of Parasitic Substrate Coupling in CMOS Circuits," *IEE Proc.-Circuits Devices Syst.,* vol. 142, pp. 307–312, 1995.
15. K. Mayaram, "Substrate Noise Coupling Modeling and Applications to RF VCOs," paper presented at IEEE SCV SSC meeting, November 2000. Available at www.ewh.ieee.org/r6/scv/scv_sscnov2000.html.
16. S. E. Laux, "Techniques for Small-Signal Analysis of Semiconductor Devices," *IEEE Trans. Electron Devices,* vol. 32, pp. 2028–2037, 1985.
17. J. B. Johnson, A. Stricker, A. Joseph, and J. A. Slinkman, "A Technology Simulation Methodology for AC-Performance Optimization of SiGe HBTs," *IEDM Tech. Digest,* pp. 489–492, December 2001.
18. W. Jin, Y. Eo, J. I. Shim, W. R. Eisenstadt, M. Y. Park, and H. K. Yu, "Silicon Substrate Coupling Noise Modeling, Analysis, and Experimental Verification for Mixed Signal Integrated Circuit Design," *Microwave Symposium Digest, 2001 IEEE MTT-S International,* vol. 3, pp. 1727–1730, 2001.
19. S. A. St. Onge, D. L. Harame, J. S. Dunn, S. Subbanna, C. C. Ahlgren, G. Freeman, B. Jagannathan, S. J. Jeng, K. Schonenberg, K. Stein, R. Groves, D. Coolbaugh, N. Feilchenfeld, P. Geiss, M. Gordon, P. Gray, D. Hershberger, S. Kilpatrick, R. Johnson, A. Joseph, L. Lanzerotti, J. Malinowski, B. Orner, and M. Zierak, "A 0.24 μm SiGe BiCMOS Mixed-Signal RF Production Technology Featuring a 47GHz ft HBT and 0.18 μm L_{EFF} CMOS," in *Proceedings of 1999 BCTM,* pp. 117–120, 1999.
20. M. Pfost and H.-M. Rein, "Modeling and Measurement of Substrate Coupling in Si-Bipolar IC's up to 40 GHz," *IEEE J. Solid-State Circuits,* vol. 33, no. 4, pp. 589–591, April 1998

4

LEADING-EDGE APPLICATIONS

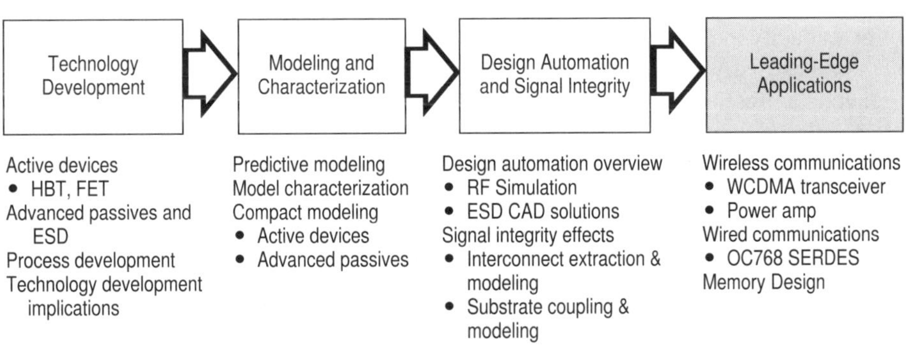

OVERVIEW

This chapter provides an insight into some of the leading-edge applications, which have been enabled using the IBM advanced SiGe process technologies. These examples have been chosen to show not only the leading-edge nature of the applications, but also the breadth of the application space, as initially discussed in the Introduction.

ISBN 0-471-44653-X

The chapter presents implementations of a wired SONET 40-GB/s design, a 3G direct conversion receiver, a PA for cellular radio, and a high-speed memory design.

- Section 4.1 presents an OC768 40–56-GB/s wired serializer/deserializer design.
- Section 4.2 discusses the design of a 3G wideband CDMA (W-CDMA) direct conversion receiver.
- Section 4.3 presents an Ericssom PA design for use in cellular phones.
- Section 4.4 presents an SiGe memory-circuit design for use with high-speed microprocessors.

4.1 WIRED COMMUNICATIONS: SONET DESIGN

The continuing demand for new communication services and higher user-end bandwidth has necessitated the development of Ethernet, Fibre Channel, and SONET standards for transmission at data rates of *f* 10 Gb/s, backed by the development of hardware capable of supporting such data rates. For example, the 10-Gb/s Ethernet standards activity, which targets the broadband local area networks (LAN) and wireless area network (WAN) applications, was completed before the end of 2002. For telecommunication applications, a well-established standard for 10-Gb/s serial links has also been defined by SONET OC-192, which calls for a line rate of ~9.95 Gb/s. Telecommunication equipment for higher data-rate standards, such as SONET OC-768 at 40 Gb/s and its variants with forward error correction (FEC) at 43–50 Gb/s, for high-capacity long-haul applications is currently being commercialized.

As these very high data-rate communication markets continue to mature, they generate a pressing need for higher levels of integration to bring the cost and power dissipation down. However, this must be achieved while still complying with stringent serial-link requirements, such as minimizing jitter and BER. In order to put some of these requirements into more perspective, consider an example based on a realistic serial-link jitter-budget scenario: a clock generator at the transmitter of a 10-Gb/s link typically must have less than 10% peak-to-peak jitter in order to enable a total-link BER better than 1E–12. Assuming that the equivalent noise source is white, one can show that this peak-to-peak requirement translates to a sub–0.7-ps-rms jitter-generation specification, thus indicating a key challenge for circuit designers, namely, how to design high-performance clocking circuits while keeping power dissipation low. In addition, one still has to aim at high levels of integration that may include all the analog and digital functions related to serialization, deserialization, coding, framing, and built-in self-test (BIST). It is well known that whenever analog and digital circuits are used on the same chip, there is potential for deleterious crosstalk and substrate coupling, especially at these speeds. These considerations clearly show that circuit designers targeting such ultra broadband applications will need all the help they can get from process-technology developers [1].

This section is organized as follows. In Section 4.1.1, we describe how SiGe BiCMOS 5HP has been extensively used in our 10–13-Gb/s SERDES work. The design and hardware characterization details of all the SERDES building blocks, such as the VCOs, transmit and receive PLLs, and multiplexer/demultiplexer as well as the fully integrated 10.3125 and 12.5-Gb/s versions of the SERDES chip, are discussed in that section. The more demanding bandwidth and jitter requirements of OC-768 applications require the use of the BiCMOS 7HP technology, as described in Sections 4.1.2 and 4.1.3. SERDES building blocks operating at 40 to 56 Gb/s with half-rate 20- to 28-GHz clocks are reviewed in detail in Section 4.1.2. Some of the challenges of packaging and testing at these speeds are also discussed. Section 4.1.3 gives an overview of two analog front-end functions and their implementation for SONET OC-768 in BiCMOS 7HP, namely, the limiting amplifier and the electroabsorption modulator (EAM) driver. The hardware results from the EAM driver work show that SiGe HBTs can address high-voltage drive applications, contrary to what is commonly assumed. The significance of accurate modeling of both active and passive devices, including interconnection wiring, is also highlighted in that section. Finally, a summary of the current status and potential directions for the future are presented in Section 4.1.4.

4.1.1 SERDES at 10 to 13 Gb/s

Introduction In this section, we describe an integrated SERDES chip implemented in SiGe 5HP, first presenting PLL details in the form of early test-chip hardware and results, followed by full SERDES hardware and results Among the key challenges in such a design is building low-power, high-performance integrated circuits that support very high data rates while complying with stringent jitter specifications.

Figure 4.1 shows a typical block diagram of an optical serial link and associated electronics. The objective of the link is to take parallel data streams at one end of the link, serialize them for high-speed transmission over an optical fiber, then deserialize them at the other end of the link. In the first step for the link, then, the serializer block takes low-speed parallel data and multiplexes it onto a high-speed serial line. These serial data are then provided to a laser-diode driver, which, in conjunc-

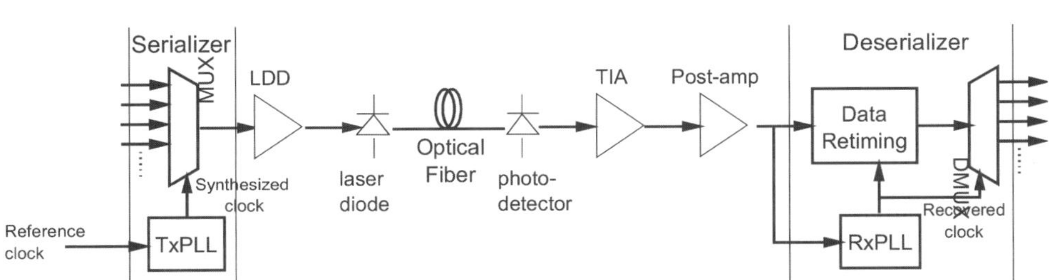

Figure 4.1 Typical block diagram of an optical serial link.

tion with a laser diode, performs an electrical-to-optical conversion and sends the data over an optical fiber. On the receiving end of the serial data stream, the light is sensed by a photodetector, with the resulting signal amplified by a transimpedance amplifier and then a postamplifier to provide sufficient signal amplitude for the deserializer input. The deserializer includes a clock and data recovery (CDR) which recovers the serial data and the clock embedded in that data, and a demultiplexer, which converts the serial data stream to a final parallel output. The overall jitter budget for the link is very tight, particularly at high data rates, and much of this budget is typically claimed by the link's optical and optoelectronic conversion pieces.

Within the SERDES, the transmit PLL (TxPLL) and receive PLL (RxPLL) are thus two critical circuits in the link; they execute key functions and must operate within a fraction of the overall link jitter budget. The TxPLL uses a local low-frequency clock reference to synthesize a low-jitter, high-frequency clock required in the multiplexer stages, and determines the random jitter present on the serializer's output serial data stream. The RxPLL is used to extract the clock signal from the received data stream, which is corrupted by noise due to the physical medium and to the optical and electronic components; it must effectively undo all of the distortion introduced in the data stream from its creation at the remote serializer to its arrival at the RxPLL serial input. The clock it extracts from the data stream is then used for data retiming and demultiplexing.

Stand-Alone PLL Test-site Designs and Results In this section, we describe the core design for the transmit and receive PLLs used in the 10- to 13-Gb/s SERDES implementations described in the paper by Friedman et al. [2]. The PLLs were first implemented in stand-alone test sites to enable independent evaluation; results from these test sites are presented below. These stand-alone PLLs were used, with slight modifications, in the full integrated SERDES implementations described in the next section. Initial designs were executed targeting 12.5-Gb/s data rates, intending to cover the 10-Gb Ethernet standard if 8-b/10-b coding were chosen, as well as SONET rates with significant forward error correction overhead. When the 10-Gb Ethernet standards body chose a 64-b/66-b coding scheme, the target data rate was shifted to 10.3125-Gb/s. This new data rate was primarily addressed in the fully integrated SERDES designs, but did not engender any significant architecture changes; the earlier test site designs generally targeted 12.5-Gb/s operation.

1. *TxPLL Design* Figure 4.2 shows a block diagram of the TxPLL. The main components of the loop, which uses a very standard architecture, are the VCO, the phase and frequency detector, the charge pump, the low-pass filter, and the divider. The TxPLL input is a low-frequency reference clock. The action of the TxPLL creates an output clock at a frequency N times that of the input, where N is the divide ratio of the divider. For 12.5-Gb/s data rates, a full-rate 12.5 GHz clock is generated. Unlike a TxPLL based on a half-rate scheme [3], the full-rate approach implemented here enables a final retiming of the multiplexed data stream, thus minimiz-

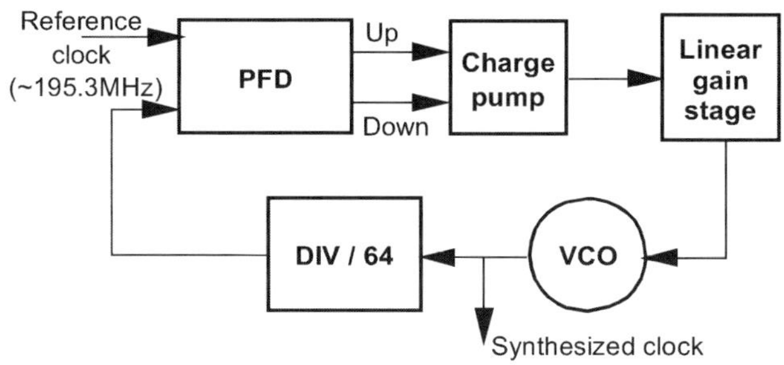

Figure 4.2 TxPLL block diagram.

ing duty cycle variation. Because a low-phase noise low-conduction VCO (LC-VCO) is used in the TxPLL, the TxPLL bandwidth was chosen to be relatively low, ~300 kHz. This bandwidth choice enables the loop to suppress more in-band noise, while using a low-phase noise VCO mitigates the effect of passing more VCO-generated phase noise to the PLL output. All circuits in the TxPLL are bipolar ECL or common-mode logic (CML). Eliminating high-swing CMOS circuits from this block helps reduce noise injection that can degrade circuit performance.

2. *Differential VCO* Figure 4.3 shows the simplified schematic of the differential LC-VCO used in both the TxPLL and the RxPLL. The cross-coupled differential pair of bipolar transistors uses positive feedback to form a negative-resistance cell that restores energy lost by the oscillator in each cycle. The on-chip spiral inductor connected to the collectors of the differential pair devices resonates with the on-chip varactor and parasitic capacitance at that node, setting the base operating frequency of the oscillator. The value of the varactor capacitance is tuned by changing the V_{var} voltage, thus tuning the VCO. A top-level metal 0.34-nH inductor with

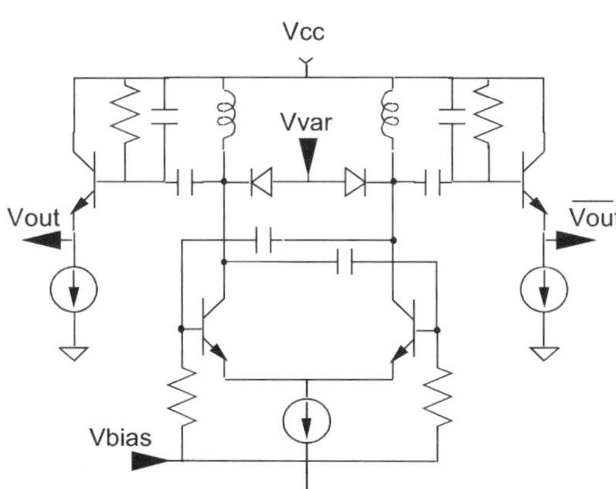

Figure 4.3 Schematic of the LC-VCO.

a quality factor of 11 at 12.5 GHz was used. A pair of emitter–follower stages buffers the VCO output, thus isolating the resonator from external perturbations. The emitter–followers are connected to the core of the VCO using a capacitive divider built using high-Q MIM capacitors. This divider is necessary to step down the high-amplitude signal generated in the tank (which helps reduce VCO phase noise) for use in the ECL circuits connected to the VCO. High-Q MIM capacitors are also used for feedback and bypass capacitors. Local power supply and DC node decoupling was achieved using on-chip MOS capacitors.

3. *Phase and Frequency Detector* A conventional phase-frequency detector (PFD) circuit is used in the TxPLL. This PFD is built from two synchronous set, asynchronous reset flip-flops and a NOR gate. It acts linearly in response to phase errors and provides a saturating output in response to frequency errors. This fully differential circuit takes the reference clock and the frequency-divided VCO output as inputs and produces up and down pulses as needed to adjust the VCO frequency and/or phase. The two PFD flip-flops (RS-DFF) are synchronously set, one by the reference and one by the divided VCO, and are asynchronously reset by the NOR gate output. In the locked condition, narrow, matched up and down pulses are generated. In order to minimize static phase error, the NOR gate includes dummy devices that allow the two flip-flops to drive nominally identical loads. Because of the relatively low operating speed of this circuit, bipolar CML is used throughout.

4. *Charge Pump, Gain Stage, and Low-Pass Filter* Figure 4.4 shows the schematic of the differential charge pump. This circuit includes two differential pair switches, each one connected to the UP and DOWN differential outputs of the PFD. The charge pump current I_{CP} flows through both the capacitor C and the resistor R_Z, forming the transmission zero to stabilize the phase loop. A negative resistor cell is used to compensate the current leakage through the load resistors R_L [4]. Therefore, the loop filter acts ideally as a perfect integrator. Mismatch in V_{BE} between the two transistors that form the negative resistance as well as mismatch between resistors will cause nonideal cancellation of the filter leakage current, meaning that the real charge pump will act as an imperfect integrator. Because the charge pump output

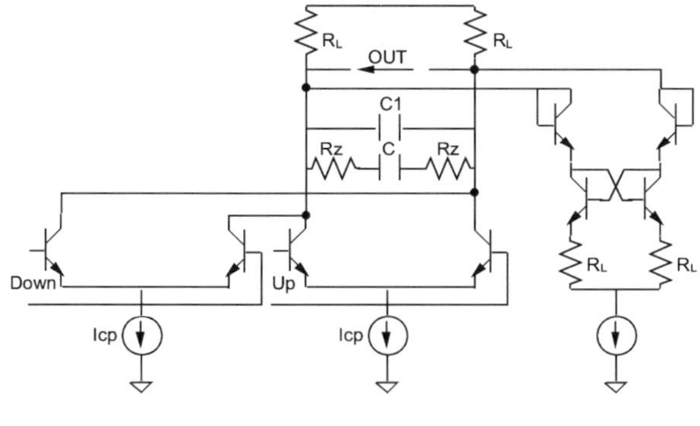

Figure 4.4 Schematic of the negative resistance charge pump.

swing is limited to a relatively narrow range to keep charge pump transistors out of saturation, the low-pass filter output is connected through a linear gain stage to the varactor control input V_{var}. This linear gain stage extends the tuning range of the VCO when it is operating within the loop by amplifying the charge pump output. High-frequency poles are formed using C1, in parallel with C and R_Z, and another capacitor on the output of the gain stage. These poles act to suppress PFD ripple.

5. *TxPLL On-Wafer Measurement Results* Figure 4.5 shows the results of four phase-noise measurements and illustrates both TxPLL performance and the effect of the reference clock choice on that performance. The open-loop VCO phase noise is shown in trace A, while the phase noise of a relatively noisy input reference is shown in trace B. Trace C shows the phase noise of the TxPLL output clock when the noisy input reference is used; trace D shows the phase noise of the TxPLL output when a low-noise reference is used. As expected, when the noisy reference is used, inside the loop bandwidth the clock phase noise (trace C) follows the reference clock input phase noise multiplied by the divider ratio, and outside the loop bandwidth follows the open-loop VCO phase noise. This plot indicates that loop bandwidth is approximately 300 kHz, consistent with bandwidth prediction by linear PLL theory. In this case, jitter generation is still relatively high and is dominated by the input noise. When a low-noise source is used as the reference, in-band noise is dominated by charge-pump and other circuit-element noise, not by input noise (see trace D). The resulting TxPLL clock phase noise yields jitter generation in a 10-kHz–100-MHz bandwidth of 0.4 ps rms. The measured power dissipation of the circuit is 270 mW at 3.3 V (excluding the output testing buffers).

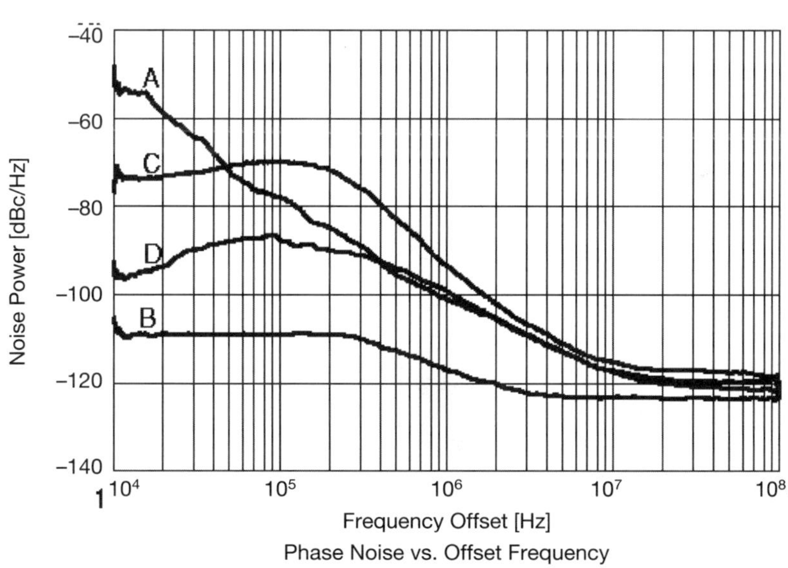

Figure 4.5 Phase noise plots. (A) Free-running VCO, (B) reference, (C) locked VCO, noisy reference, (D) locked VCO, clean reference.

The VCO oscillation frequency can be set between 11 and 13 GHz by varying the tuning voltage between 3.3 V and 0. This is a tuning range of more than 16% around the center frequency. The VCO gain is approximately 300 MHz/V when oscillating at 12.5 GHz. The measured free-running VCO phase noise at 12.5 GHz is –101 dBc/Hz at 1 MHz offset from the carrier (Fig. 4.6, trace A). This represents a ~20-dB phase-noise improvement over a previously reported ring VCO design [5]. Since the TxPLL bandwidth is relatively small, a low phase-noise VCO is mandatory to achieve low clock jitter generation. The effect of the VCO phase noise is less important in the RxPLL because of its much higher bandwidth (>10 MHz). However, the LC-VCO has four times less gain than that of the previously reported ring VCO, and thus is less sensitive to the noise on its input frequency control, which improves RxPLL performance.

6. *RxPLL Design* The RxPLL, shown schematically in Fig. 4.6, uses a dual-loop architecture. The first of these loops is responsible for frequency acquisition, while the second executes data recovery. A frequency-acquisition aid is required, because the pull-in range of a PLL in data recovery mode is too small to guarantee that the PLL will lock to the incoming data stream independent of the initial frequency of the RxPLL VCO. Furthermore, keeping the VCO frequency close to the desired operating frequency will also prevent false locking of the data loop. In operation, the frequency-acquisition loop drives the RxPLL VCO frequency to a multiple of a stable reference clock input, where the reference clock frequency is set at 1/64 of the target data rate. A lock detector automatically switches the circuit to the data recovery loop when the divided VCO clock frequency and the reference clock agree to within ±0.1%. In this mode, the lock-in time of the data loop is very short because the VCO frequency is already very close to the incoming data rate. If lock is lost, the lock detector automatically switches the circuit back to frequency-acqui-

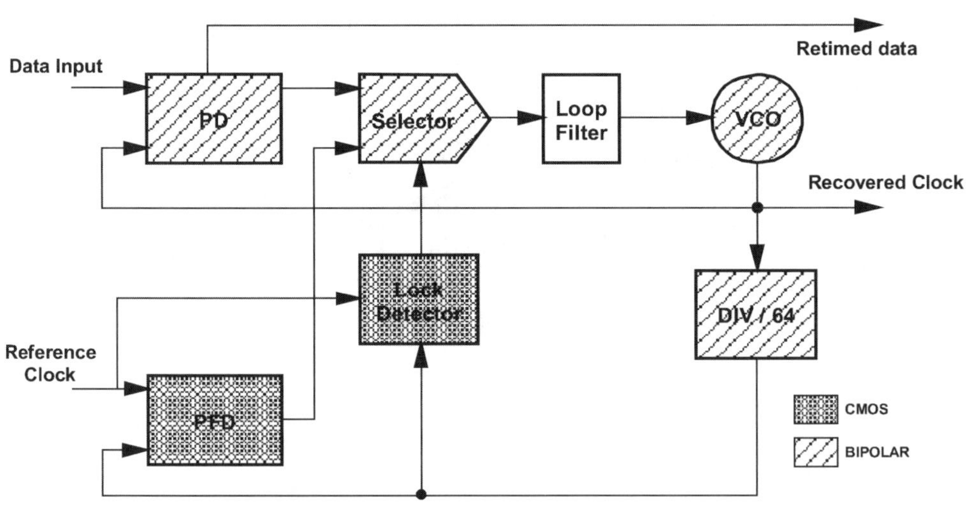

Figure 4.6 RxPLL block diagram.

sition mode. To reduce digital switching noise coupling, the CMOS PFD is automatically turned off by the lock detector while the loop is in data-acquisition mode [5]. A bang-bang phase detector that also acts as a decision circuit for data retiming is used in the data recovery loop [5]. A high DC gain active loop filter is used to reduce the static phase error. The high-speed portion of the circuit uses bipolar ECL.

7. *Frequency Acquisition Loop Enabling Automatic RxPLL Data Acquisition* The frequency-acquisition loop and the data-acquisition loop share many components, with the key extra circuits required by the frequency loop being the PFD and the divider. In this implementation, a conventional CMOS three-state PFD is used in the frequency-acquisition loop [6]; in later designs, this was replaced with a bipolar PFD for consistency with the TxPLL loop and to reduce CMOS switching noise in the design. The up and down signals from the PFD are fed via the selector to the charge pump/loop filter as differential signals after a CMOS-to-ECL conversion.

The CMOS lock detector used to switch the RxPLL between frequency-acquisition mode and data recovery mode consists of two gray-counters, one driven by the reference clock (refclk) and the other by "divclk," the VCO frequency divided by 64. The counters count for a period of 2048 refclk cycles, or 4096 divclk cycles, whichever is less. At the end of this interval the counter values are compared to determine whether the two frequencies agree to within ±0.1%. The counters are then reset before the next counting interval begins. Under ordinary conditions it takes 2048 refclk cycles to count, plus 9 refclk cycles to compare and reset the counters, for a total of 2057 refclk cycles between each measurement update. If the frequencies are within tolerance, the lock detector switches the RxPLL to data recovery mode immediately. Conversely, if the frequencies are out of tolerance for 1000 consecutive measurement intervals, then the RxPLL is toggled back to frequency-acquisition mode. This bias toward data recovery is designed to allow the circuit to tolerate low-frequency, high-amplitude jitter on the data. In later, full SERDES implementations, this feature was eliminated, so that single out-of-tolerance measurements switch the RxPLL to frequency acquisition mode.

8. *RxPLL On-Wafer Measurement Results* Measurements of the stand-alone RxPLL circuit were made on-wafer using 12.5-Gb/s pseudorandom bit stream (PRBS) input data. The data-acquisition loop pull-in range is about ±0.5%, which amply covers the ±0.1% tolerance of the lock detector. The jitter generation (in a 100-MHz bandwidth) of the recovered clock is 0.3 ps rms when the RxPLL is locked to 2^7–1 PRBS input data (relevant to 8-b/10-b coded data, which has a maximum run length of 5 [7]). This jitter increases to 0.6 ps rms when using 2^{31}–1 PRBS input data. The DC power consumption of the chip is 330 mW at 3.3 V, excluding output test buffers, with half of this power consumed in the frequency-acquisition aid circuitry.

To evaluate jitter tolerance of the RxPLL in a real-world environment, a 14-km length single-mode optical fiber link test bench was built. Figure 4.7 shows the RxPLL single-ended input and output data eye diagram and recovered clock when receiving 2^{31}–1 PRBS data. The measured input data jitter is ~42 ps peak-to-peak. The measured BER of the RxPLL under test is less than 10^{-12}, with an input sensitivity of 100 mV.

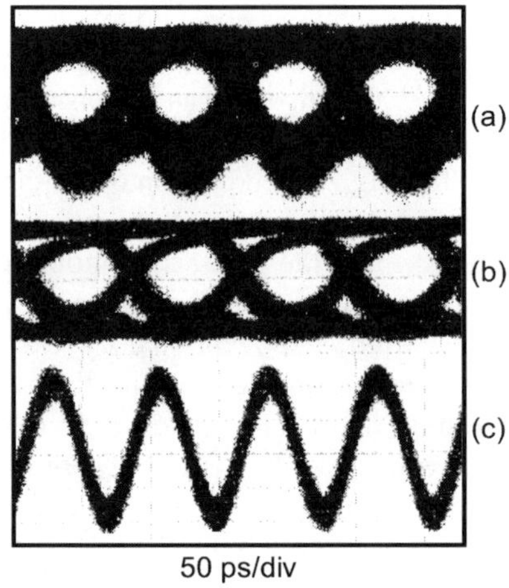

Figure 4.7 Input and output waveforms of the RxPLL. (a) 12.5-Gb/s input eye diagram; (b) 12.5-Gb/s recovered eye diagram; (c) 12.5-GHz recovered clock.

9. *PLL Physical Design* One of the challenges of mixed analog/digital designs such as the one presented in this section is avoiding coupling digital switching noise into sensitive analog nodes through the substrate and power-supply lines. In the physical design of these PLLs, several measures have been taken to reduce the digital noise coupling:

- Since the high-speed part of the IC is fully differential, it has been implemented in a symmetrical way with respect to complementary signal paths.
- Local bias generators are used for each sensitive part of the IC, e.g., VCO, phase detector, etc.
- Separate power supply and grounds for the analog part and digital part within the two loops of the RxPLL are used to avoid coupling. Special care is also taken to reduce the parasitic inductance of supply lines and to improve ground homogeneity.
- A ring of substrate contacts surrounds the VCO in both PLLs and the CMOS logic present in the RxPLL. This ring itself is also surrounded by a ring of DT oxide.

The chip microphotographs for the stand-alone PLLs are shown in Fig. 4.8.

Single-Chip SERDES Design and Results In this section, we describe the design and evaluation single-chip SERDES chips operating from 10 Gb/s to 13 Gb/s [8]. An initial SERDES design targeted 12.5-Gb/s operation, but later designs retuned the VCO to enable 10.3-Gb/s operation as required to meet the final 10-Gb Ethernet standard. Recent related work in this area includes an SiGe BiCMOS/SOI

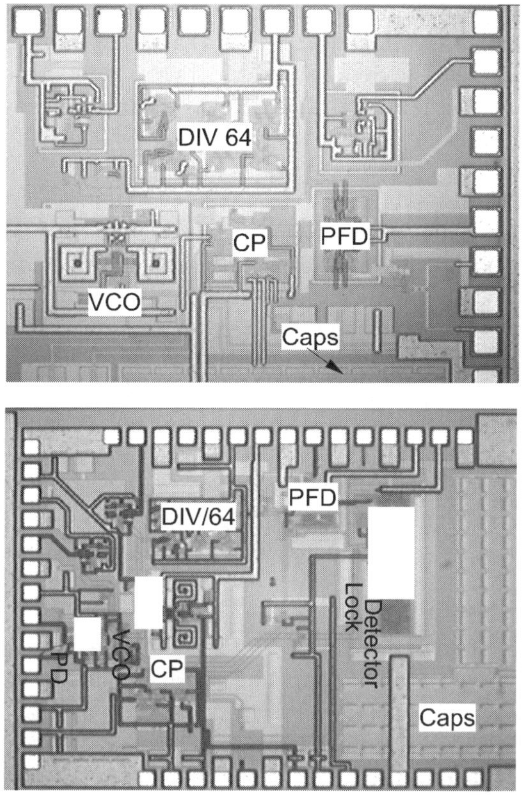

Figure 4.8 Chip photomicrograph of TxPLL (top) and RxPLL (bottom).

full transceiver implementation [9], an SiGe receiver/1:8 demultiplexer chip [10], and a CMOS clock and data recovery circuit [11], and, more recently, CMOS serializers and deserializers [12,13].

The SERDES presented here is a fully monolithic, single-chip 16:1 serializer and 1:16 deserializer, including integrated transmit and receive PLLs with on-chip loop filters, featuring very good jitter and error-rate performance and low power consumption in BiCMOS 5HP; significant power reduction could be realized by porting this design to BiCMOS 7HP. At start-up, an automatic trim operation sets the optimal center frequency of the VCOs used in the chip's two PLLs. This is achieved using available wrap modes and on-chip data checking.

The PLLs use LC-VCOs, one for the TxPLL and one for the RxPLL, that operate over multiple overlapping frequency bands. The chip automatically selects the correct band setting for each VCO at start-up. In this way, trimming operations at test are not required for the circuit to accommodate process, supply voltage, and temperature variations. An automatic phase-selection mechanism enables input parallel data to be correctly latched into the transceiver independent of the relationship between the arrival time of that data and the transceiver output clock edge that times the request for parallel data. Iddq (leakage test) and scan features are built into the design to enhance testability. Serial and parallel wrap modes coupled with circuitry that checks a fixed pattern's successful transit of the transceiver further enable efficient on-wafer test using low-frequency test equipment.

1. *Transceiver Architecture* Figure 4.9 shows a block diagram of the transceiver. In the receive path, the input data stream is fed to the CDR PLL. The recovered data output is then demultiplexed to 16 parallel recovered data streams using the recovered clock. This output is sent to the off-chip drivers and can also be wrapped to the transmit path data input when the part is operated in a parallel wrap test mode. A selectable full- or half-parallel-rate clock accompanies the parallel output data. In the transmit path, 16 parallel input data streams are multiplexed to a single serial-output data stream. A clock multiplier unit (CMU) PLL uses an input reference clock to generate the full-rate clock. This clock is required to create the clock frequencies and phases needed to stagger and multiplex the incoming data and also to execute the final full-rate retiming of the serial output. Finally, the transmit section outputs a full-parallel-rate clock to be used by another chip in the system to time the data sourced to the transceiver for serialization.

Test features as well as configuration and start-up circuitry are controlled by the CMOS block of the transceiver. These features, implemented in a combination of CMOS and bipolar logic, include wrap modes, self-test, automatic VCO coarse tuning, iddq testability, and general scan design testability. Three different reference clock frequencies, the data rate divided by 16, 64, and 80, were supported in the initial 12.5-Gb/s design. A divide-by-66 mode replaced the divide-by-80 mode for the 10.3-Gb/s design..

2. *Transmit and Receive PLLs* As in the case of the stand-alone TxPLL, the transceiver transmit PLL takes a selectable low-frequency reference frequency and

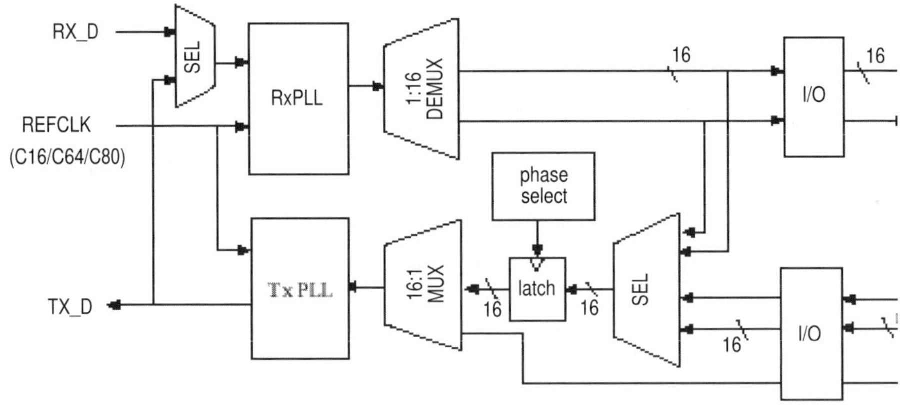

Figure 4.9 Transceiver block diagram.

creates an output clock at the desired serial line rates as well as divided clock frequencies used in the multiplexer. By executing a final retiming of the multiplexed data stream, duty cycle variation in the output is minimized. As in the case of the stand-alone RxPLL, the receive PLL uses a dual-loop structure: a data recovery loop that recovers clock and data from the input data stream, and a frequency-acquisition loop that moves the oscillation frequency of the Rx VCO inside the capture range of the data recovery loop. All elements of both PLLs are implemented in bipolar ECL or CML, with the exception of lock to reference indicator circuits, which are implemented in CMOS. The detailed structure of the transmit and receive PLLs implemented here is closely related to those described earlier (see Figs. 4.2 and 4.4) and in [2] and [5].

Each PLL uses an LC-VCO. Such VCOs typically suffer from relatively narrow tuning ranges, making it difficult to for these VCOs to run at target operating frequencies in the presence of process, supply voltage, and/or temperature variations. Furthermore, even if an LC-VCO with a broad tuning range were available, such a VCO will have high gain, which is generally undesirable due to noise sensitivity considerations, particularly in the transmit PLL.

The VCOs (Fig. 4.10) used in the transceiver's PLLs address these problems by operating in multiple overlapping frequency bands, where the VCO gain in any given band is modest. Band selection is digitally controlled. At start-up, with a valid reference clock applied to the chip, the transceiver will operate in a serial wrap mode, sending a known parallel data pattern from the transmitter parallel input all the way to the receiver parallel output, where the data pattern is checked.

Combinations of band settings for the transmit and receive VCOs are automatically tried, and the transceiver searches for an optimal combination. Each combina-

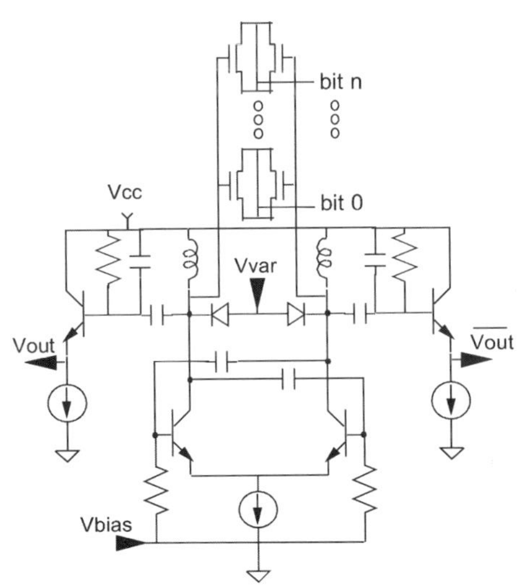

Figure 4.10 Schematic of the LC-VCO used in the PLLs.

tion is evaluated for a fixed number of reference clock cycles. The combination's quality is assessed by checking whether both PLLs successfully locked, whether the known data pattern was successfully recovered, and whether the locked VCOs are tuned near the middle of their tuning range.

The output of the calibration circuit is a set of trim bits that set the coarse tuning of the Rx and Tx VCOs. As the automatic selection progresses, the output trim bits are modified each time a combination yielding a better assessment is identified, until the automatic trim process is complete. During operation, the 10.3-Gb/s implementation also detects whether the VCO control voltage is approaching a band edge, and will automatically shift its coarse tuning to the appropriate adjacent band, provided this update feature is enabled. Automatic band selection and update capability thus enables the use of a VCO with low effective gain when operating within the main PLLs, yet does not require extraordinary measures such as multiple VCOs or fuses that can be blown at wafer test for the transceiver to be robust against process, temperature, and supply voltage variations.

3. *Multiplexer and Demultiplexer* In both the 12.5-Gb/s and 10.3-Gb/s designs, the 16:1 multiplexer is composed of a recursive series of 2:1 multiplexer blocks. The 1:16 demultiplexer, similarly, is composed of a recursive series of 1:2 demultiplexer blocks. In both cases, data and clock paths are intertwined in the macros.

The sampling of the parallel Tx data coming into the transceiver presents a special timing problem. Since the clocking of all elements in the Tx multiplexer macro is derived from the Tx VCO, there is naturally a question regarding how to synchronize the incoming data with those clocks. Typically (as is the case in this transceiver), a clock is fed forward from the transceiver to a framer chip that is responsible for providing the parallel input data to the transceiver transmit path. Since the total delay in this loop, comprising of driver delays, receiver delays, and card delays, is typically not known and is likely to be significant with respect to a baud interval, a mechanism is needed to ensure proper sampling of the incoming parallel data by the transceiver. In the initial transceiver design described here, a source-synchronous clock (half parallel rate) that accompanies the parallel data is used as an input to a phase-detection and -selection circuit; in the follow-on designs, both half-rate and full-rate parallel clocks are supported. In any case, the phase-selection and -detection circuit then selects a desirable phase of the local clock to sample the parallel input. In cases where automatic phase selection is not appropriate, phase selection can be explicitly set using CMOS control inputs. The phase-selection mechanism can operate continuously while the part is processing data to accommodate significant changes in path delay during part operation, although errors will be generated as the phase setting is updated. This implementation is more compact and power-efficient than the first-in-first-out-based (FIFO) approach typically used to solve this problem.

4. *Test Environment and Results* Both the 12.5-Gb/s and 10.3-Gb/s transceivers were first evaluated on-wafer using a standard chip tester connected to a probe card plus cabling with typical I/O bandwidths of 500 MHz. This environment does not allow generic full-rate testing of the part, but does enable a successful demonstration of

the full parallel interface, control circuitry, and test features such as iddq and scan. It also allowed the evaluation of the full range of frequencies at which the transmit and receive PLLs would lock. Successful lock and optimal VCO trim settings were automatically generated at VCO operating frequencies from 11 GHz to 13 GHz in the 12.5-Gb/s design and 9.6 GHz to 11.5 GHz in the 10.3-Gb/s design.

Full-rate testing was accomplished using a surface laminar circuit (SLC) carrier-packaged pair of transceiver chips. SLC packaging supports finer linewidths than typical board-level technologies. This characteristic enables the transceiver chips, which have solder bumps on their I/O pads, to be direct flip-chip attached to the SLC; a ceramic or other space-transforming carrier is not required. This approach not only saves the cost of a package, but also reduces the number of transitions through which high-speed signals must pass, improving high-speed signal integrity.

The test configuration most like the expected operating mode of the transceiver is shown in Fig. 4.11. In this case, a 12.5-Gb pulse pattern generator (PPG) feeds a $2^{31} - 1$ PRBS data stream to the serial Rx input (RxS) of the first transceiver (SERDES1). These data are demultiplexed, passed over the parallel interface from the SERDES1 parallel output (RxP) to the SERDES2 parallel input (TxP), multiplexed in SERDES2, and is output from the SERDES2 serial output (TxS). The SERDES2 output is compared with the SERDES1 input using a BER tester. Simultaneously, a second 12.5-Gb PPG feeds a second $2^{31} - 1$ PRBS data stream at a nominally identical (but not synchronized) data rate to the RxS input of the second transceiver, SERDES2. Again, these data are demultiplexed, passed over the parallel interface from the SERDES2 RxP parallel output to the SERDES1 TxP parallel input, multiplexed in SERDES1, and are output from SERDES1 TxS. In this configuration, then, each transceiver effectively acts as a framer for the other transceiver. Furthermore, independently controllable data streams at noncommensurate frequencies run in opposite directions through both transceivers, enabling chip performance under worst-case crosstalk conditions to be evaluated. No performance degradation and no injection locking is observed under these conditions, even when the operating frequency of one data path is dragged through that of the second. This test configuration was successfully exercised for both the 12.5-Gb/s and 10.3 Gb/s versions of the transceiver.

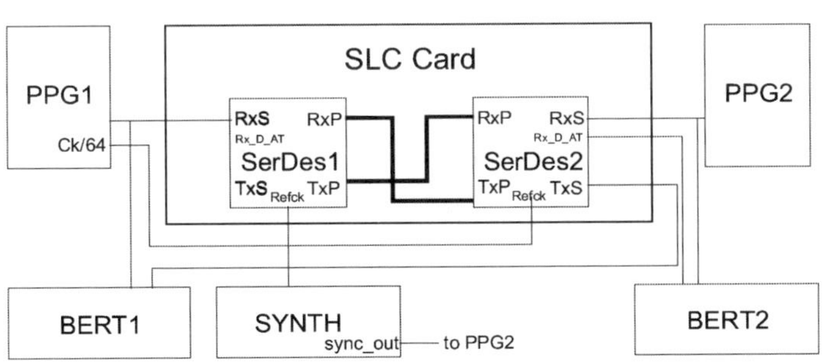

Figure 4.11 Back-to-back SERDES test setup.

In the operating condition described earlier, the card using 12.5-Gb/s transceivers performed error-free overnight at data rates from 11 Gb/s to 11.9 Gb/s. At 12.5 Gb/s, error rates below 5E–14 were measured. The degradation in error rate is due to a noise-coupling path that was eliminated in the follow-on design, resulting in error-free operation across the entire operating band of the part. Both Tx and Rx jitter generation at 12.5 Gb/s was below 0.5 ps rms, measured with a spectrum analyzer phase noise utility with integration limits from 10 kHz to 100 MHz. The TxS output eye diagram at 11.1 Gb/s is shown in Fig. 4.12. The TxS output eye diagram at 10.3 Gb/s from the 10.3 Gb/s design is shown in Fig. 4.13. Supply voltage variations of ±5% about the nominal 3.3-V level did not affect performance. In both the 12.5-Gb/s and the 10.3-Gb/s designs, operation was also demonstrated on-wafer at ambient temperatures ranging from 0 to 100°C. Total power dissipation at nominal operating conditions was 3.3 W, with significant reduction possible by porting the design to a more advanced process.

5. *Transceiver Physical Design* In high-speed circuits, significant attention must be paid to physical design both to avoid speed degradation and to minimize noise coupling [14]. In this transceiver design, the target operating mode of the part, in which transmit and receive paths operate simultaneously at noncommensurate yet almost identical frequencies, poses a particularly challenging problem with respect to noise coupling. In addition, switching noise from CMOS logic and digital drivers must not significantly degrade PLL performance. Several key physical design strategies were followed in the transceiver design reported here:

- The Rx and Tx VCOs were placed far apart, with a ring of substrate contacts and DT isolation surrounding each VCO. The Rx and Tx PLLs were also isolated from each other and from the rest of the circuit with substrate contact regions and DT rings. The CMOS section was placed on the opposite side of the chip from the PLLs.

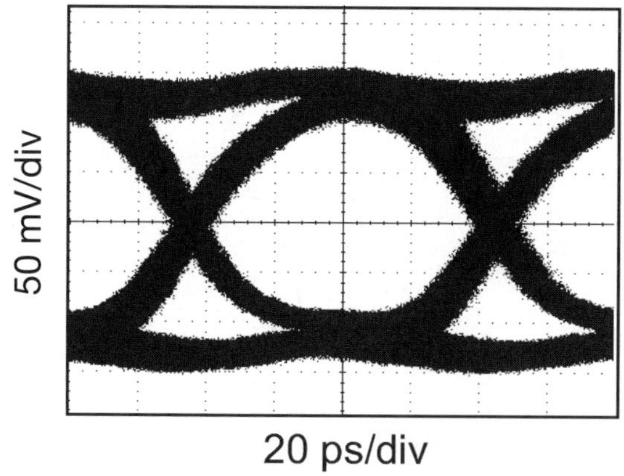

Figure 4.12 Sample serial output data eye diagram at 11.1 Gb/s.

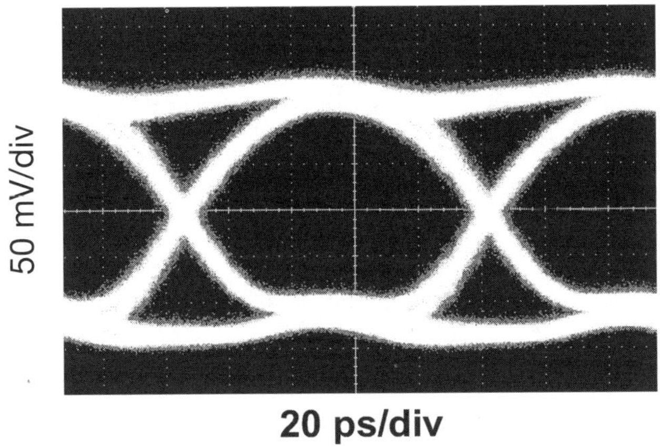

Figure 4.13 Tx serial output @ 10.3 Gb/s $2^{31} - 1$ PRBS.

- Three power-supply domains were maintained on-chip, one for the PLLs, one for the multiplexer, demultiplexer, and parallel I/O, and one for the CMOS section. Low inductance connections to power-supply pads were implemented throughout.
- Substrate contacts outside of the CMOS region were not tied to the analog ground; these contacts nominally carry no current.
- Separate bias generators were used for key blocks.

The chip microphotograph for the 12.5 Gb/s design is shown in Fig. 4.14. The design integrates approximately 8400 HBTs and 30,000 FETs.

4.1.2 SERDES Circuits at 40 Gb/s

Introduction A 40-Gb/s serial data communication link looks much like a 10-Gb/s communication link with everything running 4 times faster. In the case of the 40-Gb/s link, the laser-diode driver used in a typical 10-Gb/s data link is replaced by an EAM driver; for the 40-Gb application, the laser is operated in a continuous-wave (CW) mode and an EAM is used to modulate the effective power coupled to the fiber. A generic 40-Gb/s link is shown in Fig. 4.15. Beyond serial data communication links, the core multiplexing elements of serializers and core demultiplexing elements of deserializers are also necessities for test equipment operating at 40-Gb data rates such as pattern generators and error detectors.

Two of the key elements of this link are the serializer and deserializer. In this section, building blocks for serializers and deserializers that support 40-Gb/s nonreturn-to-zero (NRZ) serial data rates will be described. In Section 4.1.4 some details regarding front-end chips will be provided.

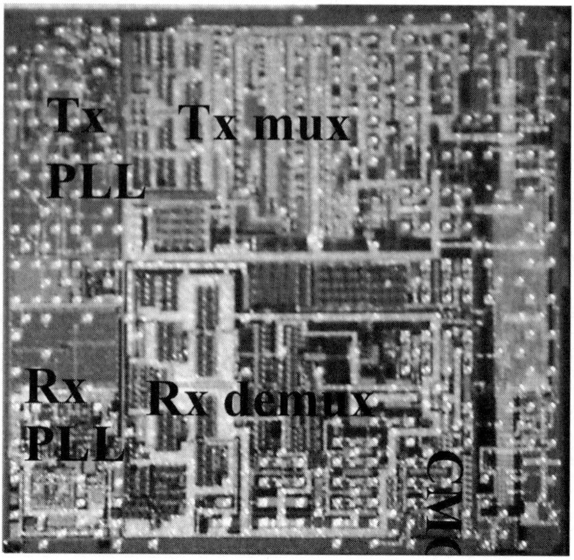

Figure 4.14 Die photo of transceiver chip.

The main building blocks for the serializer and deserializer are the multiplexer, the demultiplexer, the CMU PLL, and the CDR PLL. The number of signals multiplexed in the serializer and demultiplexed in the deserializer is implementation-specific, but typically might be on the order of 16. The most critical pieces of these designs are their highest-frequency sections, namely, the 4:1 multiplexer stage and the 1:4 demultiplexer stage. Implementing and demonstrating these core circuits is thus an excellent first step in developing a working 40-Gb/s serializer and deserializer, as is shown by the CMU and the CDR. In the remainder of this section, we will present details of the core circuits needed to build serializers and deserializers that operate at 40-Gb/s data rates as implemented in SiGe 7HP.

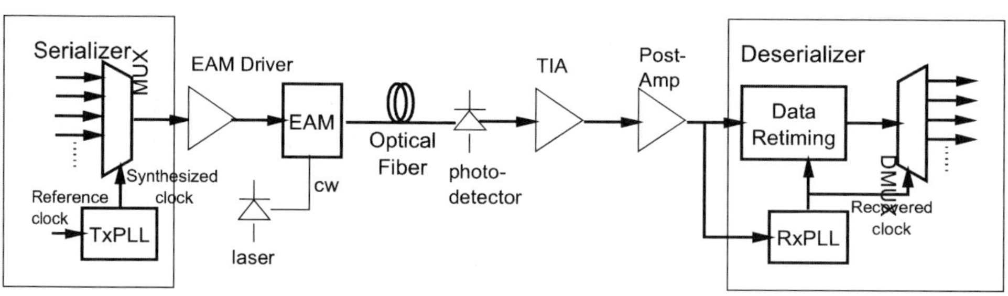

Figure 4.15 Optical link at 40 Gb/s.

4:1 Multiplexer and 1:4 Demultiplexer There are several key challenges associated with the design of the 4:1 multiplexer and 1:4 demultiplexer circuits [15]. The first of these challenges is, of course, the high data rate these circuits must handle. The base SONET OC-768 data rate is just below 40 Gb/s. Once forward error correction schemes are added, however, the line rate will exceed 40 Gb/s, with the actual line rate depending on the chosen scheme. SiGe 7HP does not support a 40-Gb/s full-rate latch at moderate power with margin. This is why the multiplexer, demultiplexer, CDR, and CMU PLLs use a half-rate architecture. The high operating speeds of the circuits also present significant additional challenges in layout, packaging, and test. We illustrate how these challenges can be met below, through a description of 4:1 multiplexer and 1:4 demultiplexer designs that were implemented in SiGe BiCMOS7HP, tested on-wafer to data rates above 50 Gb/s, packaged, and then demonstrated as packaged parts to data rates above 50 Gb/s.

The block diagram of the 4:1 multiplexer circuit is shown in Fig. 4.16. The multiplexer is implemented using a tree architecture, where the base elements in the tree are a recursive series of 2:1 multiplexer stages. The logic family chosen for this design is emitter-coupled logic, which offers sufficient performance while enabling the use of supply voltages in the –3.3-V range. Single-ended internal signal swing levels were chosen to be 300 mV throughout the design, providing adequate SNR without demanding a larger supply voltage.

The multiplexer IC takes four parallel single-ended data streams at its input and a half-rate input clock, e.g., 20 GHz for 40-Gb/s operation, to create the final multiplexed output. The parallel data are received by single-ended to differential conversion buffers with on-chip 50-Ω matching resistors. These buffers are built with two differential pairs in order to achieve a good common mode rejection at the input of the two first multiplexing stages. The input clock is internally divided by two with a static divider (built with a toggle flip-flop) in order to latch the four parallel data inputs and perform the first multiplexing operation. The clock input is received with a double-stage-wide bandwidth Cherry-Hooper amplifier [16] in order to speed up clock rise and fall edges, since sine-wave synthesizers are used for testing at these frequencies. These amplifiers enable the use of very low power-level input clocks. In each 2:1 multiplexer stage, proper data timing is achieved by using latches. These latches act to offset the two input data streams by 90° with respect to each other and need to operate with a clock to data output time delay less than half the bit time of the output multiplexed data. The current consumption of the multiplexer is 410 mA from a nominal –3.6-V supply voltage. Thirty-seven percent of this current is consumed in the input and output buffers.

The demultiplexer IC receives a full-rate differential data stream and outputs four parallel single-ended quarter-rate data streams. The demultiplexer circuit block diagram is shown in Fig. 4.17. Like the multiplexer, it uses a tree architecture, here with a recursive series of 1:2 demultiplexer stages. It uses a half-rate input clock for the first 1:2 demultiplexing stage. This clock is then divided by two with a static divider in order to perform the last 1:2 demultiplexing operation. Each 1:2 demultiplexer stage uses latches for demultiplexing and data alignment. The serial input data and input clock are received with double-stage-wide bandwidth Cherry-Hoop-

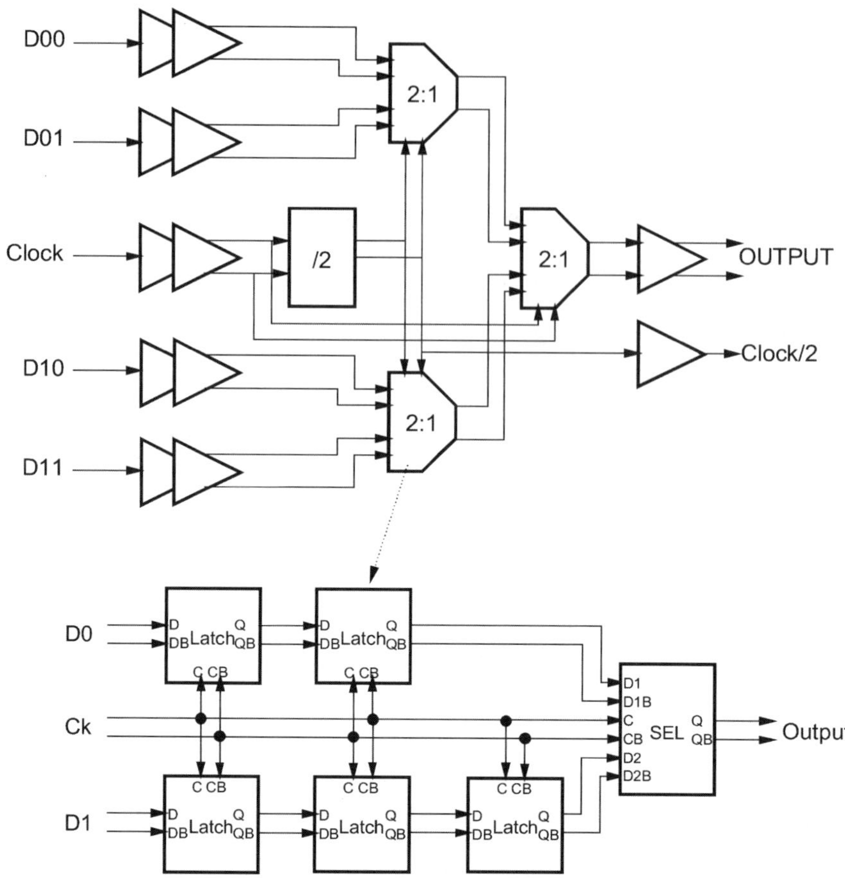

Figure 4.16 Simplified block diagram of the 4:1 multiplexer circuit.

er amplifiers with on-chip 50-Ω matching resistors. A "bit-skip" DC signal is XORed with the divided clock to allow bit rotation control at the parallel outputs of the demultiplexer. The chip consumes 430 mA from a –3.6-V supply voltage, with 40% of this current consumed in the input and output buffers.

At the high data rates intended for the multiplexer and demultiplexer, attention to detail in the physical design is crucial in order to maximize circuit performance. Layout parasitics must be minimized, particularly on high-speed lines. In order to preserve signal integrity, transmission lines are used on long runs. Finally, in order to support at-speed evaluation prior to packaging, the layout needs to be compatible with both wafer-level probing and packaging.

The multiplexer and demultiplexer ICs were mounted in a custom-designed package that enabled high-frequency operation of the chips [17,18]. The package includes a housing, a ceramic substrate, and the IC itself. The Al/Gr composite

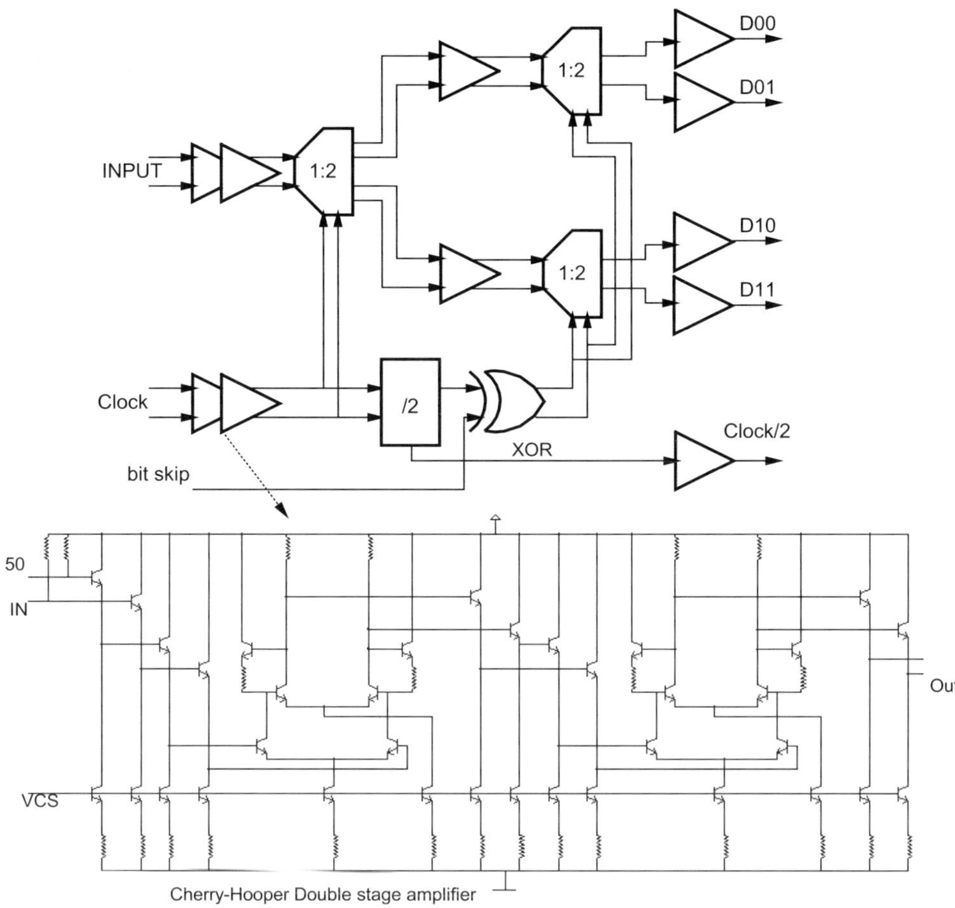

Figure 4.17 Simplified block diagram of the 1:4 demultiplexer circuit.

housing provides support for high-speed V-connectors, protects the ceramic and chip, and serves as a heat sink. The ceramic substrate measures 1.75 in. by 1.95 in. On the ceramic, a finite ground-plane coplanar waveguide was used for signal lines to limit the number of propagation modes and achieve better isolation between signal traces. Ribbon bonds were used for chip-level interconnection, and the bonding lengths were minimized and well controlled to minimize parasitic inductance. Feedthroughs with internal decoupling capacitors were used for power and bias supplies, and surface-mount capacitors were used on selected power traces near the chip within the package to further reduce supply noise. An assembled package and its chip-level interconnections are shown in Fig. 4.18.

Testing the multiplexer requires that four quarter-rate data streams and a commensurate half-rate clock be provided to the circuit. In test, the data streams and

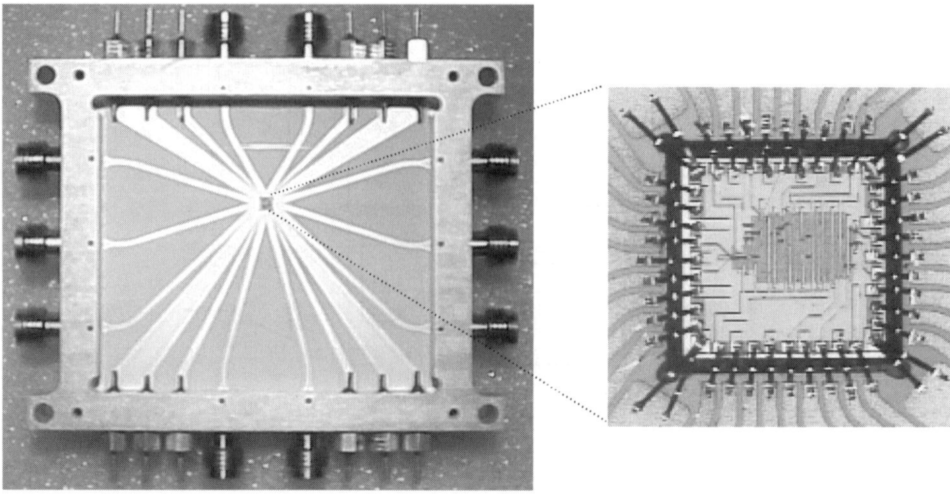

Figure 4.18 An assembled package and chip-level interconnections.

clock come from test equipment through cables to the part; delay lines are used to match the arrival times of the data and to position the clock appropriately for successful multiplexing. When the clock and data frequency are changed, some delay adjustment must generally be repeated if cable lengths do not match. The output signal is observed using a high-frequency oscilloscope and the quality of the output eye evaluated. Testing of the demultiplexer requires that full-rate data and a commensurate half-rate clock be provided to the circuit. Again, the data stream and the clock come from test equipment through cables to the part so that initial delay adjustment between clock and data is necessary, as is follow-on adjustment for significant input frequency changes. The four demultiplexed outputs can then be observed using an oscilloscope and the quality of the resulting four output eyes evaluated. Once packaged parts are available, testing of the multiplexer and demultiplexer in tandem is possible, either by connecting the parallel ports or the serial ports of the devices. Error-rate testing of the multiplexer and demultiplexer in isolation can also be executed.

Both the multiplexer and demultiplexer circuits were first tested on wafer to check functionality. On wafer, the demultiplexer initially could not be evaluated at target data rates due to test equipment limitations, but basic operation was demonstrated up to 12.5 Gb/s. However, in early testing we were able to demonstrate multiplexer operation on wafer up to 56 Gb/s at a supply voltage of –3.3 V by feeding it 4×14 Gb/s $2^7 - 1$ PRBS inputs. At the time, 14 Gb/s constituted the maximum test equipment speed available to us. In later testing, 60-Gb/s operation of the multiplexer and demultiplexer were both demonstrated. The input amplitude was 500-mV single-ended (the inputs are referenced to a DC level of –250 mV). Once functionality was checked, both ICs were packaged. This enabled us to perform BER testing on both circuits in order to confirm their functionality at full speed. Figure 4.19 gives a qualitative comparison of the multiplexer performance at 40 Gb/s when tested on-wafer and in packaged form. The IC was tested with $2^{31} - 1$ PRBS, –3.3-V supply voltage and 100°C chip temperature (during on-wafer testing). The modest performance degradation observed in the packaged multiplexer eye as compared to the on wafer multiplexer eye is probably due to return loss at the serial output in the packaged part.

The final test for the packaged parts was evaluation in a back-to-back configuration connected through their serial interface. When testing in this configuration, shown in Fig. 4.20, error-free operation ($<10^{-15}$) at 40 Gb/s was achieved on all four parallel outputs of the demultiplexer with $2^{31} - 1$ PRBS parallel inputs at the multiplexer. At that speed, consistent error-free operation was observed when varying the supply voltage from –3.3 V to –3.9 V. The clock phase margin of the demultiplexer is approximately 11 ps, or 44% of a unit interval. Error-free operation ($<10^{-15}$) was also achieved at 50 Gb/s with a supply voltage of –3.5 V and $2^7 - 1$ PRBS data at the multiplexer parallel inputs. At this data rate, the clock phase margin of the demultiplexer was reduced to 6 ps, or 31% of a unit interval. Corresponding 50-Gb/s eye diagrams are shown in Fig. 4.21. Error-free operation up to 52.2 Gb/s was also achieved under the same supply voltage conditions.

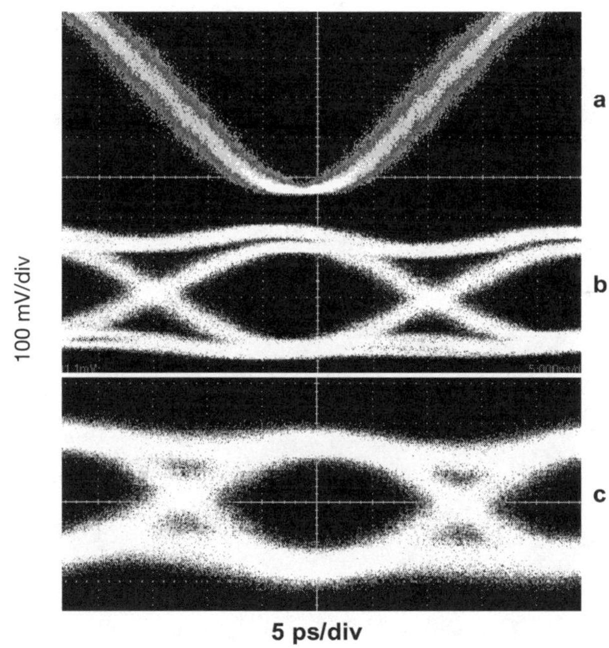

Figure 4.19 Input and output waveforms of the 4:1 multiplexer at 40 Gb/s. (A) 20-GHz clock input; (B) on-wafer measured 40 Gb/s output eye diagram; (C) measured 40-Gb/s output eye diagram of packaged sample.

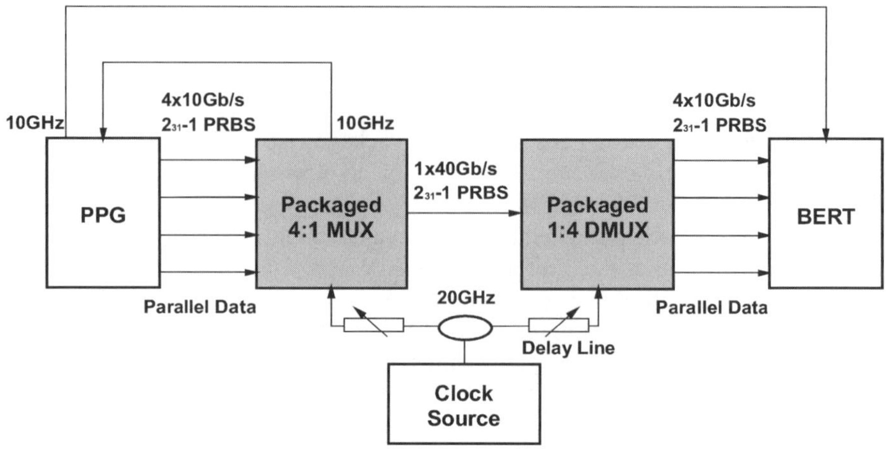

Figure 4.20 Simplified block diagram of the back-to-back multiplexer/demultiplexer test setup.

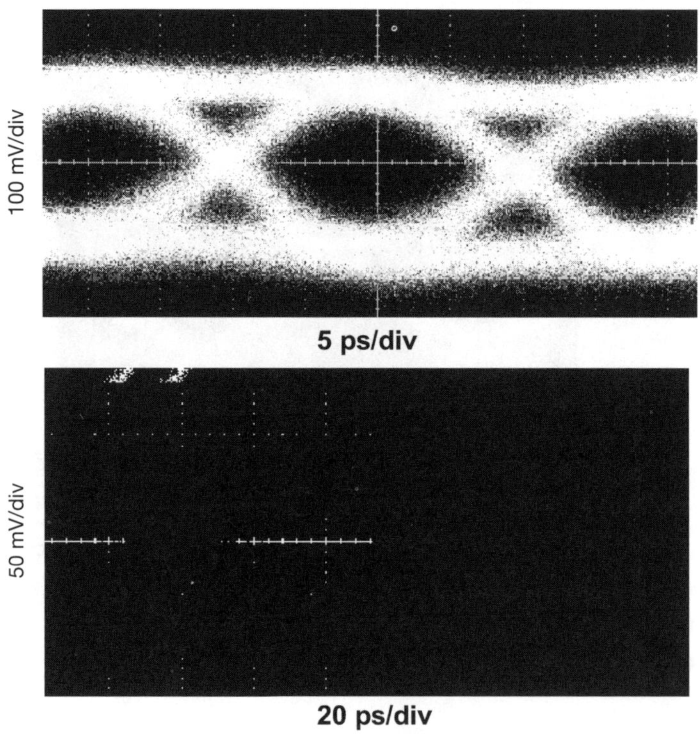

Figure 4.21 Fifty Gb/s output eye diagram of the 4:1 multiplexer (top) and one of the 12.5-Gb/s demultiplexed outputs (bottom).

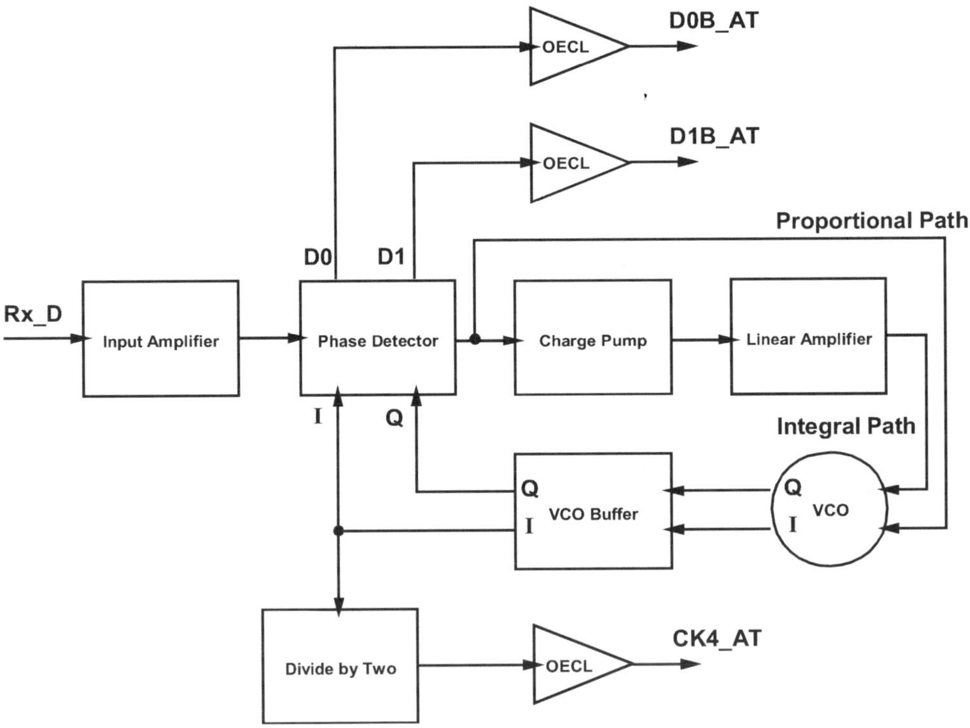

Figure 4.22 CDR block diagram.

Clock and Data Recovery and Clock Multiplier Unit at 40 Gb/s

1. *Overview* In addition to the demultiplexer and multiplexer cores themselves, the CDR and the CMU are required elements of a complete deserializer and serializer, respectively. The function of the clock and data recovery circuit is to extract clock information from the input high-rate serial data stream and then to use that clock to reconstruct the input data. Generally, the signal carrying the data is significantly degraded in the course of its transmission over the high serial data-rate link. The CDR must compensate for the degradation, primarily timing jitter and amplitude reduction, recovering a clean copy of what was transmitted with extremely low error rates. The output data and clock from the CDR is delivered to the demultiplexer. The CMU supports the multiplexing function, taking an input reference clock and multiplying that clock's frequency to generate the set of clocks needed by the multiplexer. In the case of a full-rate architecture, the CMU might also provide a clock that drives a final at-speed latch, enabling very uniform data launch timing at the serializers final output. The clocks generated by the CMU must be very low jitter, and, in the case of a half-rate architecture as was pursued in the 40-Gb/s designs described in the following, the highest-frequency clock must have very uniform duty-cycle characteristics.

In the remainder of this section, the focus will be on the CDR for 40-Gb/s serial data. The general operation of the CMU is exactly as described in the 10-Gb/s work presented earlier, although a half-rate architecture was used.

The challenges faced in executing a 40-Gb/s CDR design are quite similar to those faced in executing the multiplexer and demultiplexer cores. The high data rate pushes the limits of the technology for full-rate architectures, driving the design to half-rate architectures, which unfortunately add complexity while enabling the desirable lower operating frequency.

The 40-Gb/s CDR circuit uses a half-rate clock architecture, which means that the on-chip VCO is running at 20 GHz, in contrast to a full-rate architecture in which the VCO would run at 40 GHz. This architecture was chosen for two main reasons. First, given NPN device performance, it allows sufficient margin to ensure 40-Gb/s operation of the chip with respect to process, temperature, and power-supply variations. Secondly, a full-rate architecture is a more aggressive design that would have required higher power consumption and, most likely, a larger power-supply voltage in order to achieve the required design margin.

The CDR circuit block diagram is shown in Fig. 4.22. In this first-pass design, the loop solely implements a data recovery function. Because of the relatively narrow capture range of a typical CDR, a frequency-acquisition aid loop is usually required to bring the operating frequency of the loop close to that required for the input data rate. Once the loop frequency is close enough, the data loop takes over. A frequency-acquisition aid loop was successfully incorporated and demonstrated operatoinal in a follow-on version of the chip.

The CDR itself extracts a 20-GHz clock from a 40-Gb/s serial input data stream. This clock is then used for data demultiplexing, which, in this design, includes the retiming function. The 2:1 demultiplexing function is actually part of the phase-detector operation. The VCO used in the CDR has two frequency tuning inputs. The first of these is controlled by the output of the charge pump, forming the integral path of the loop (I-filter). The second of these is directly controlled by the output of the phase detector (P-filter), effectively implementing the zero of the loop by direct modulation of the VCO frequency. The I-filter, which has a very large time constant, is responsible for the frequency acquisition of the loop. The P-filter, meanwhile, applies high-frequency modulation pulses to effect phase corrections of the VCO. The amplitude of the P-filter pulses controls the bandwidth of the loop; it is externally adjustable in the implementation described here. Applying phase corrections directly from the output of the PD reduces loop latency and thus jitter generation. The charge pump current of the I-filter controls the loop damping factor and has little effect on the loop bandwidth, assuming that the damping factor is much larger than one (which is the case here because the PD gain is relatively high). This current is also adjustable in the implementation described here The loop filter capacitor is integrated on chip. In addition to the main recovered data output, the chip outputs three additional signals for design evaluation, namely, the two 1:2 demultiplexer data streams and the VCO clock divided by two.

2. *Input Amplifier* The input amplifier schematics is shown in Fig. 4.23. It uses the well-known Cherry-Hooper architecture, which consists of a series chain

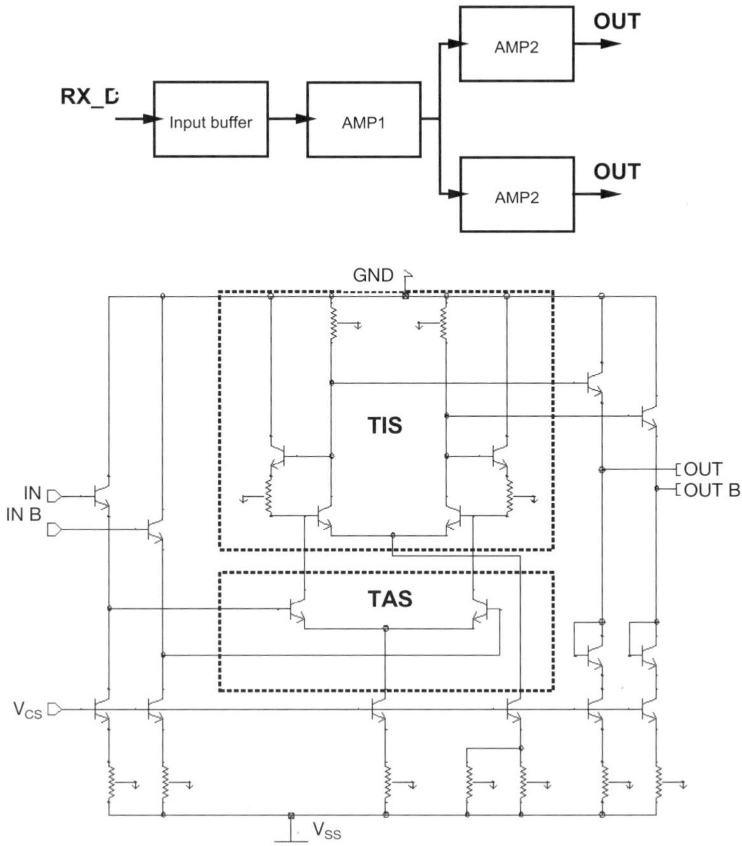

Figure 4.23 Input amplifier.

of an alternating transadmittance stage (TAS) and a transimpedance stage (TIS). The input impedance of the TIS is small compared to the output impedance of the TAS. This results in a mismatch between the stages up to high frequencies, increasing the bandwidth of the amplifier. Because the amplifier fan-out is high (it drives 4 latches of the phase detector), a tree architecture is used where the TIS of a first amplifier is driving the TAS of two second amplifiers. The transimpedance amplifier uses active parallel feedback in order to increase its overall gain-bandwidth product. The input stage of the amplifier consists of a pair of emitter–followers with 50-Ω input termination for impedance matching.

3. *Phase Detector* The phase detector block diagram is shown in Fig. 4.24. This phase detector consists of three double-edge triggered master–slave flip-flops D1, D2, and D3, two latches L1 and L2, one XOR gate X1, and one modified AND gate A1. The flip-flops D1 and D2 sample the incoming data stream on the rising and falling edges of the 20-GHz in-phase (I) and phase quadratics (Q) VCO clock

signals, respectively. Note that the VCO Q signal is 90 degrees out of phase with respect of the VCO I signal. The way that the loop executes data recovery is that when the loop is locked, data transitions will be aligned with the Q signal from the VCO clock, enabling the I clock to be used to sample the data at a favorable time. The way that this is effected is as follows. Depending on the relative phase of the VCO clock with respect to the data, the output of D1 will lead or lag the output of D2. D3 samples the output of D2 on the rising and falling edges of the output of D1 to produce a binary signal indicating whether the VCO clock is leading or lagging the incoming data. In the locked condition, then, the edges of the VCO Q clock signal will be aligned with the transitions of the incoming data. Assuming a 50% duty-cycle VCO clock, and good I and Q matching, the VCO I clock signals then samples the data at the midpoint of each bit interval and therefore acts as a decision circuit with no phase adjustment required over process, temperature, and power-supply variation. The duty cycle of the VCO clock and the I and Q matching are important to ensure effective operation of the phase detector, especially with regard to jitter tolerance, assuming an input data stream for which the optimum sampling point is effectively at the midpoint of each bit.

Because the SONET OC-768 standard allows the run length of random data to be large, the phase detector needs to provide a tristate output when no data transitions are present to avoid excessive VCO frequency drift, which can unlock the loop. The

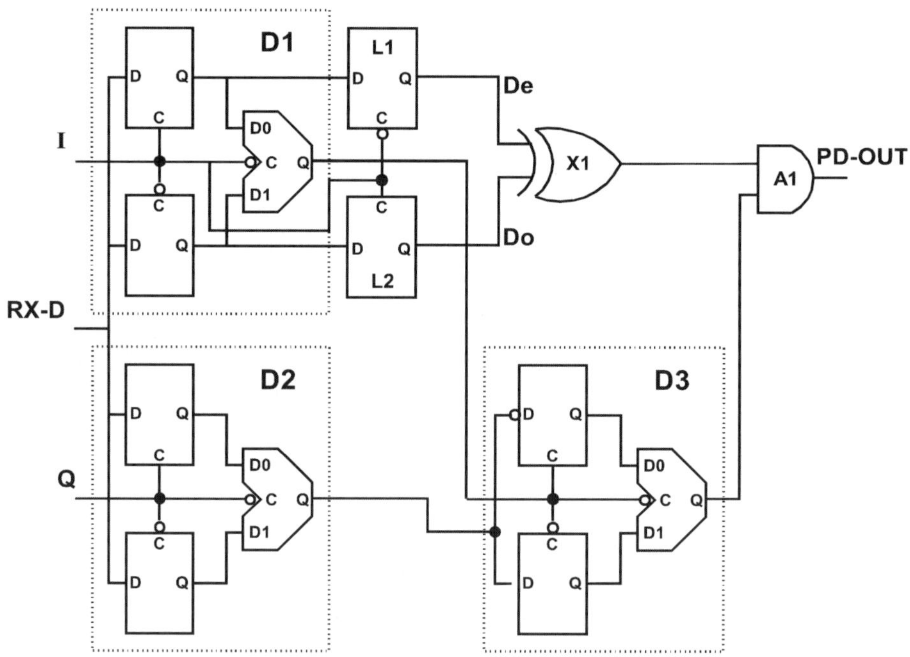

Figure 4.24 Phase detector block diagram.

XOR gate X1 senses the presence of transitions by comparing the outputs of L1 and L2 (these outputs are actually the demultiplexed data, D_Even and D_Odd, which are 90 degrees out of phase). The outputs of X1 and D3 are combined in gate A1 (Fig. 4.25) which generates the tristate output of the phase detector. Gate A1 is a CML AND gate modified so that its output is tristated if its ZB input is asserted.

Two critical constraints that must be satisfied for a successful implementation of this architecture are first the delay matching between A1 and the selector of D3, and second the symmetry of the XOR gate. The former constraint provides good clock-to-data delay matching between L1 and L2 and D3s latches, which is easily achieved, while the latter requires matching between the two inputs to output propagation delay. Reasonable delay mismatch (few picoseconds) will not break the phase detector operation, but instead will slightly modify the dynamic of the loop. Although not catastrophic, this may lead to an increase in jitter generation because of erroneous VCO frequency corrections.

4. *Charge Pump* The charge pump used in the 40-Gb/s CDR uses the same approach as was used in the 10 Gb/s CMU and CDR as described earlier, in which a negative resistance cell is used to cancel filter leakage current. The basic topology for this circuit is shown in Fig. 4.4.

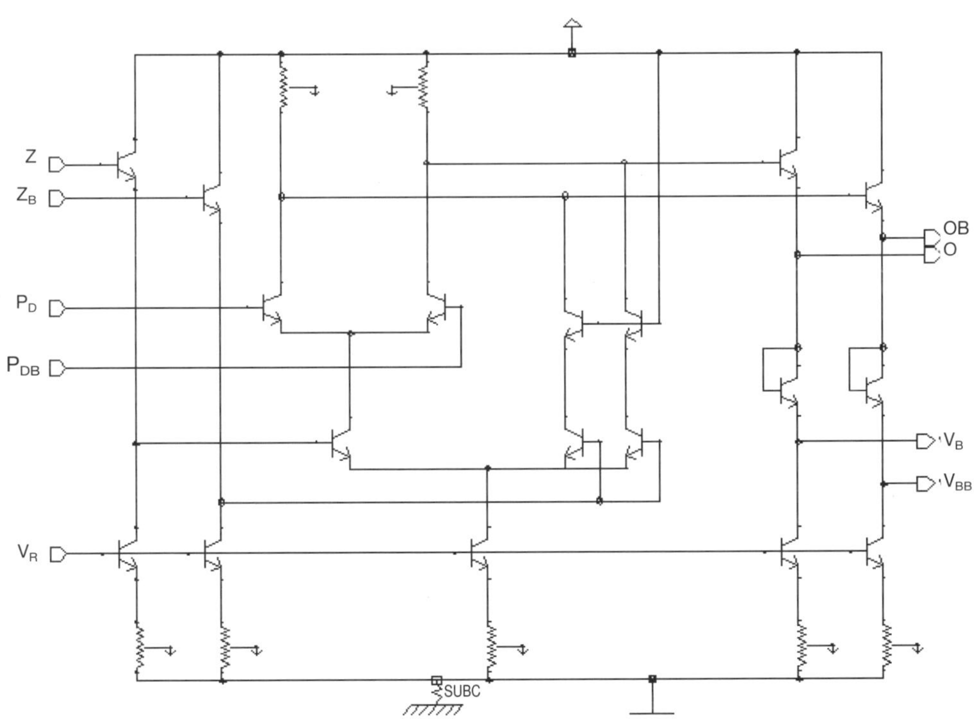

Figure 4.25 Modified AND gate schematic.

5. *20GHz I/Q VCO* Half-rate CDR architectures require, in general, a VCO with precise quadrature phases; these phases are referred to as I and Q. The basic circuit building block in the implementation described here is a CMOS LC oscillator consisting of two back-to-back cross-coupled inverters with an inductor load (Fig. 4.26). The inductor resonates with the varactor capacitance and the gate and drain junction capacitance of the FETs to determine the oscillation frequency. The negative resistance of the cross-coupled FETs acts to overcome the inductor loss as well as the layout wiring loss. In the I/Q oscillator implementation, two such identical oscillator building blocks, labeled A and B in Fig. 32, are coupled to each other by FETs (M5–M8) of the same size as the internal negative resistance FETs (M1–M4), thus creating direct coupling in one direction and cross coupling in the other.

In order to understand the operation of the coupled oscillators, first suppose that the two oscillators synchronize in-phase. In this case, the cross-coupled path from oscillator B to A absorbs the negative resistance current produced by (M1A, M2A) and (M3A, M4A), and the oscillator A stops, which in turns stops oscillator B through the cross-coupling path. The same process applies in reverse if the two oscillators are antiphase. Therefore, oscillations can only coexist in both coupled oscillators when both are synchronized in quadrature. In this mode of operation, the unique phase relationship of the four outputs are 0° at M2A, 180° at M1A, 90° at M2B, and 270° at M1B. The current in the quadrature output VCO is controlled by a large PFET device connected between ground and the top rail of the two oscillators.

6. *Measurement Results* The CDR has been tested on wafer using 40-GHz bandwidth probes and $2^7 - 1$ PRBS input data. The total DC power consumption is

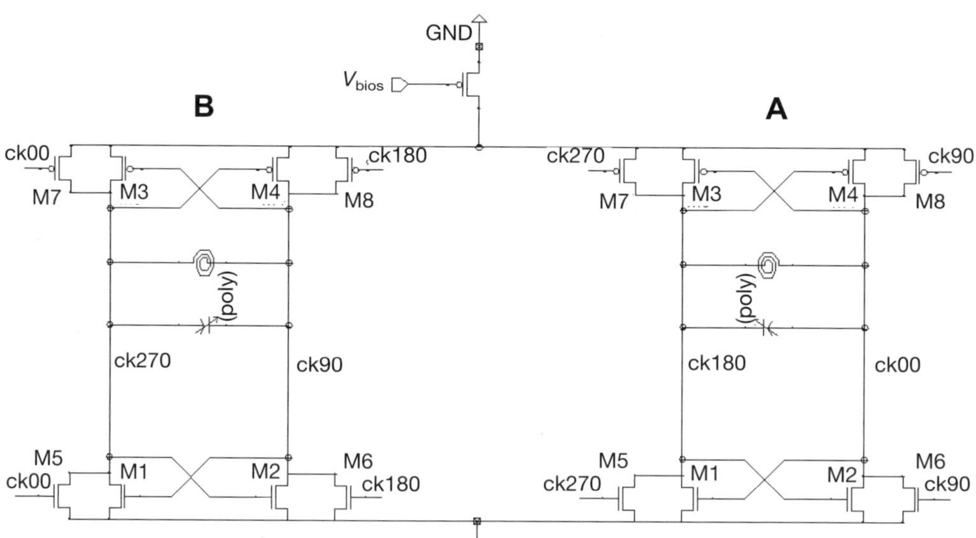

Figure 4.26 Twenty-GHz CMOS VCO with quadrature outputs.

1.3 W at –3.6 V supply voltage. Without the output test buffers, the power consumption is approximately 1.1 W.

The quadrature CMOS VCO oscillation frequency is observed to be 10% higher than expected. It has a phase noise of –97 to –100 dBc/Hz at 1-MHz offset. The VCO consumes about 16 mA from –3.6V. The tuning range is about 600 MHz. This small tuning range ensures a small VCO gain for PLL applications. When locked to PRBS input data, the VCO oscillates between 22.1 and 22.6 GHz, corresponding to a hold range of 2.2%. In a follow-on design, the tuning range of the VCO was extended to 2 Ghz using a band-switching approach like the one used in the 10-Gb/s transceiver work described earlier. At 45-Gb/s PRBS input, the recovered clock phase noise is approximately 0.3 ps or 13 mUI in a 50-kHz–500-MHz bandwidth. Input and output waveforms showing 45-Gb/s CDR operation are shown in Fig. 4.27.

Clock Multiplier Unit As mentioned earlier, the architecture for the clock multiplier unit is the same as that used for the 10-Gb/s clock multiplier unit described earlier in this chapter. The CMU for 40-Gb/s data rate applications, however, was

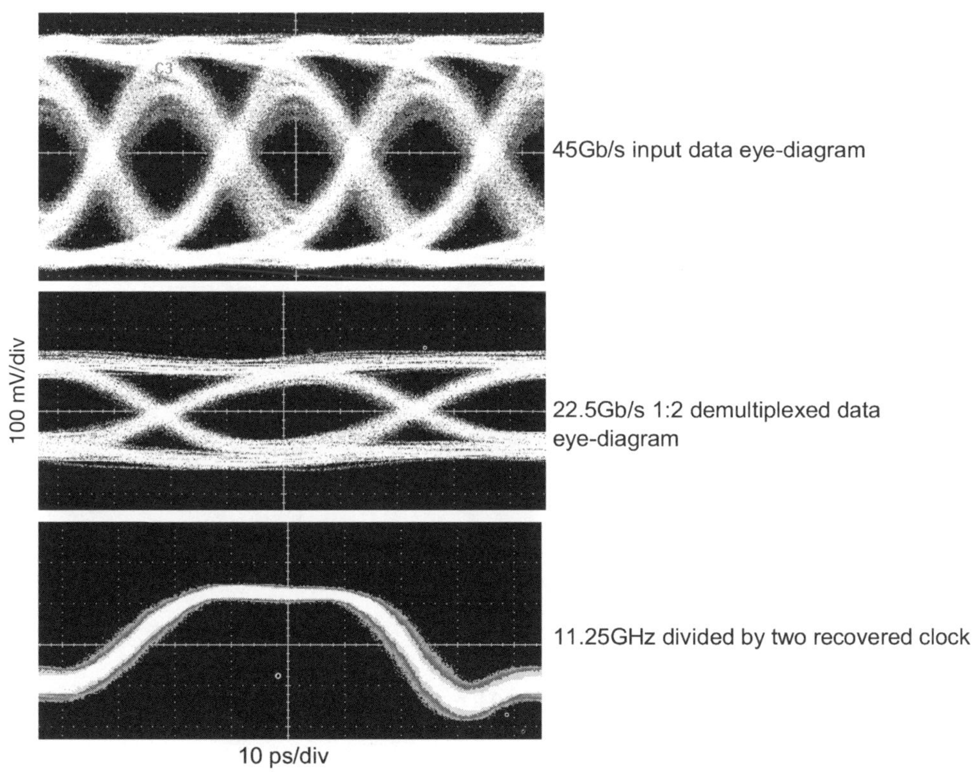

Figure 4.27 Input and output waveforms of the CDR.

designed using a half-rate architecture. The main consequence of this choice is that the VCO runs at half rate. The phase noise performance of this VCO is shown in Fig. 4.28; its architecture is the same as that shown in Fig. 4.10. A system-level consequence of using a half-rate architecture on the transmit path is that careful attention must be paid to duty-cycle symmetry in the 20-GHz output clock; duty cycle variation creates additional deterministic jitter in the output data waveform.

4.1.3 Front-End Circuits at 40 Gb/s

Introduction In addition to the serializer and the deserializer, other key elements of a 40-Gb/s link are the front-end circuits. In this section, details regarding the design and performance of two examples of these circuits implemented in SiGe BiCMOS 7HP will be presented. We will first describe a limiting amplifier, which takes the output of a transimpedance amplifier and generates a signal of sufficient amplitude to satisfy the input sensitivity requirements of the CDR circuit's serial data receiver. We then describe an EAM driver, which takes the output data stream from the serializer and generates sufficient swing and the correct characteristic impedance to satisfy the input requirements of an EAM.

40 Gb/s Limiting Amplifier The 40-Gb/s limiting amplifier described here employs a CML topology to allow DC coupling of the 50-Ω I/O circuits. Input match-

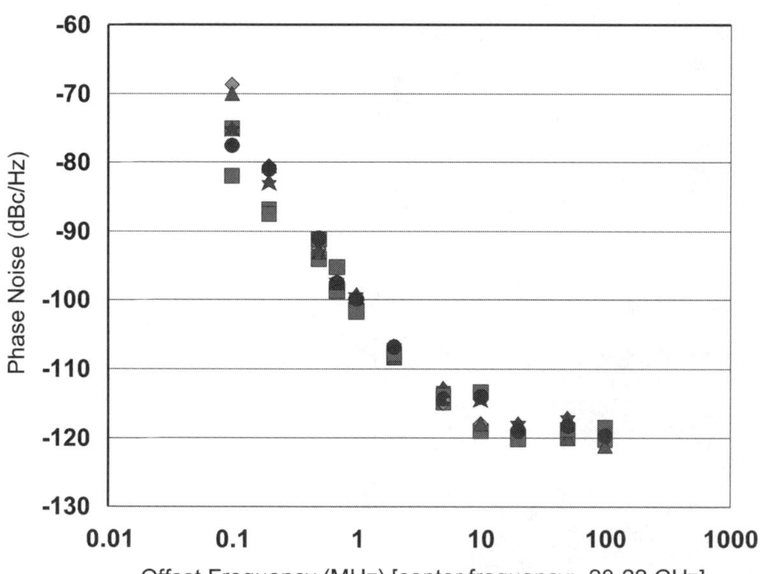

Figure 4.28 VCO phase-noise performance vs. offset frequency for TxPLL VCO operating at 21 GHz under various coarse calibration settings with operating temperatures ranging from 0 to 100°C.

ing is achieved using passive components due to the high impedance of the input stage. Each input signal is terminated by a 50-Ω resistor tied to ground, which also supplies the bias current for the input level-shifter stage. The layout of the input termination network employed a 50-Ω G-S-S-G taper to lump the bond-pad capacitance into the controlled impedance of the taper line, and so avoid a capacitive discontinuity at the bond pad.

The core of the amplifier consists of four 7 dB gain limiting amplifier cells as shown in Fig. 4.29. These amplifiers do not employ feedback to boost gain, since we wished to explore amplifier topologies that would be easier to extend to higher frequencies without fear of phase-margin issues. Each amplifier consists of a differential pair with two diodes as the load followed by an emitter–follower stage to buffer and shift the output to the proper level for the next input stage. Small-signal analysis of this configuration gives a theoretical voltage gain of $g_{m\ load}/g_{m\ diff\ pair} = 2g_m/g_m = 2$, or a gain of 6 dB. However, a large-signal analysis and simulations showed that both the gain and the bandwidth can be increased slightly by employing load diodes slightly smaller than the input differential pair device size. By so doing, the gain per cell in limiting mode is increased to 7 dB.

Although the core of the amplifier is unconditionally stable, employing no feedback, the emitter–follower buffer stage needed careful design to avoid "inductive" behavior causing overshoot in response and increasing amplifier jitter. It should be noted that if feedback amplifiers had been used, they, too, would have required emitter–follower level shifters between stages, and so the same care in design [19].

To support 40-Gb/s I/O, source and load terminations were assumed to minimize potential reflections. The output driver is a cascoded differential amplifier with 50-Ω loads, which provides the termination of a 50-Ω G-S-S-G coplanar taper, identical with the taper used at the input.

Table 4.1 compares simulated and measured performance for the 40-Gb/s limiting amplifier packaged in an experimental ceramic package with V-connector signal launches. The amplifier draws 120 mA from a single –5.2-V supply, dissipating 624 mW while running at 40-Gb/s rates. A 40-Gb/s differential pseudorandom bit

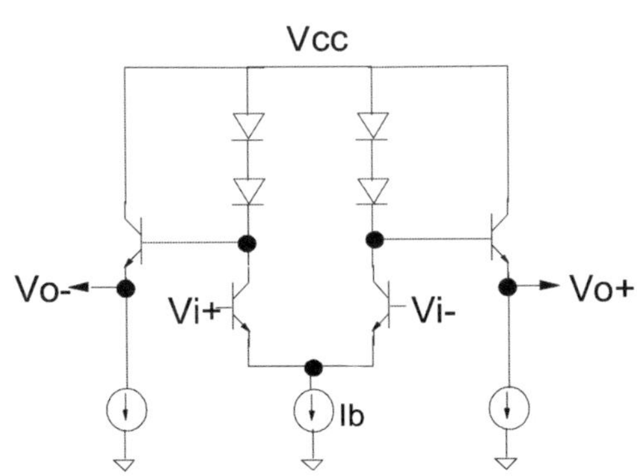

Figure 4.29 Simplified schematic of gain stage with emitter–follower level shifter.

Table 4.1 Comparing Simulated and Measured Performance

Parameter	Extracted Simulation	Measured (de-embedded)
Gain (dB)	26 dB	25 dB
Rise, fall times (ps)	7 ps	8 ps
Jitter, deterministic (ps)	< 1 ps	0.9 ps

stream generated by a super-high-frequency (SHF) pattern generator was deskewed with Anritsu delay lines, attenuated, then fed into the postamp. The 20%–80% rise and fall times of the signal fed to the postamp was 14 ps (12 ps deembedded) after deskewing, attenuation, and cables.

Because the package and connector contribute 2–3 dB in attenuation at these frequencies, some deembedding is necessary to compare the postamp performance to simulations. The measured gain, 25 dB, agrees well with the 26-dB simulated gain for a nominal device at room temperature. Deembedding the scope response yields 7- to 8-ps 20%–80% rise times for the amplifier operating in the limiting mode.

Figure 4.30 shows an example eye diagram of the $2^{31} - 1$ pseudorandom data from the pattern generator (2, top). The amplifier output eye (2, bottom) shows little added jitter and faster edge rates when compared to the input. A clean output eye is maintained for input data from 20 mV to 1 V peak-to-peak differential amplitude or over a 34-dB dynamic range.

Modulator Driver At bit rates much beyond 10 Gb/s, indirect laser modulation is required in order to prevent frequency chirp and concomitant chromatic dispersion penalties in such high-speed links. In this section, we report the results of a single-ended EAM driver designed to deliver over 3 V p-p single-ended drive into 50-Ω termination while consuming 3 W from a single –7.5-V supply. This driver also supplied the adjustable DC bias voltage needed by the EAM device.

As in the postamp, the input and output terminations on this device were on-chip, low parasitic 50-Ω polysilicon resistors used with CPW launches to allow DC coupling of the high-speed signals (Fig. 4.31). Waveform symmetry adjustment was effected by applying an analog voltage to an input offset control pin.

This driver circuit was tested on-wafer using 40-GHz bandwidth coplanar G-S-G probes for the high-speed input and output signals. A representative eye diagram from first-pass hardware at 20 Gb/s is shown in Fig. 4.32. The noise in both eyes is primarily sampling head noise in the scope input, as signals were attenuated with a 20-dB pad to protect the high-speed inputs.

An important aspect of this application is the requirement for large voltage swings—indeed, large enough that there was some skepticism in the III–V semiconductor community that an SiGe-based design would ever succeed in generating the necessary swings with any degree of reliability. This driver is the first SiGe7HP design reported with a single-ended peak-to-peak output swing larger than the collector–emitter breakdown voltage with the base open (BV_{CEO}); swings of 3.2 $V_{p\text{-}p}$, well in excess of BV_{CEO}, are shown in Fig. 4.32. This was possible because of the

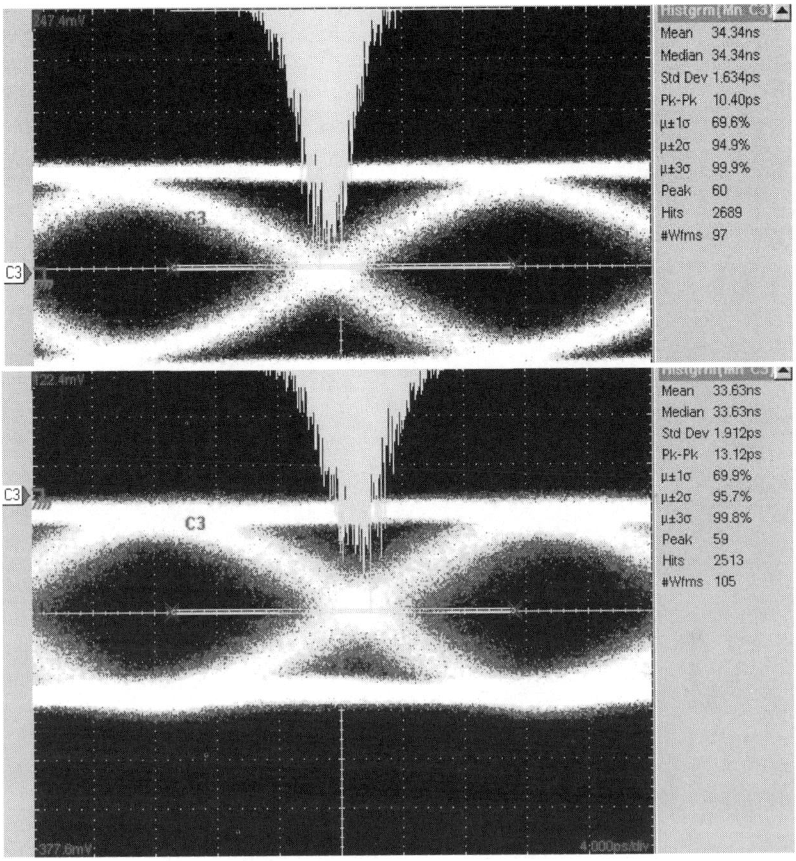

Figure 4.30 Pseudorandom signal from 40-Gb/s pattern generator (top) with 1.6-ps RMS jitter and amplified signal (bottom) with 1.9-ps rms jitter and faster edges.

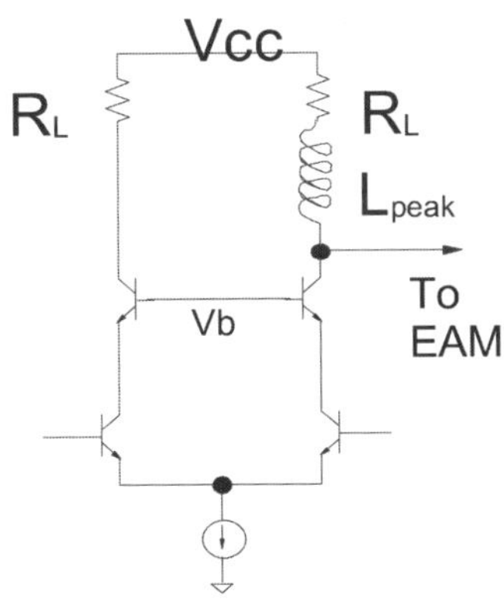

Figure 4.31 Simplified schematic of EAM driver output stage with inductive peaking.

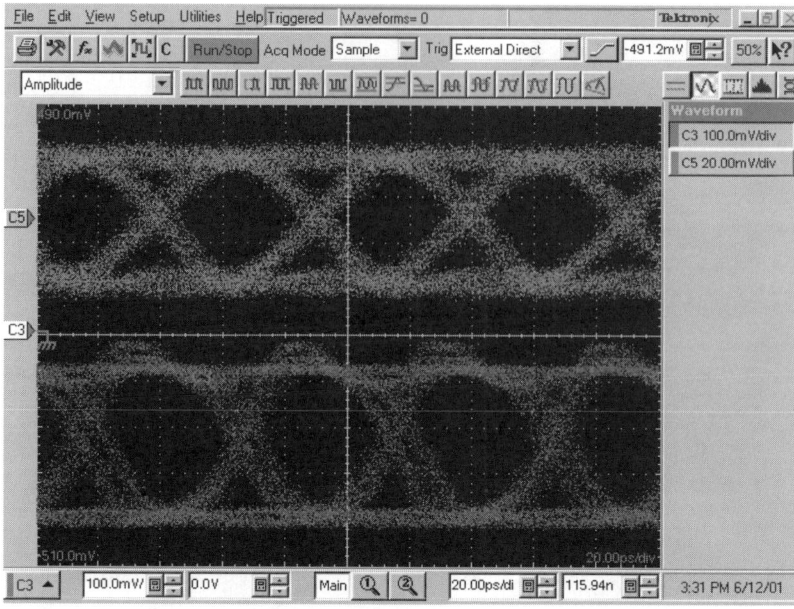

Figure 4.32 Twenty Gb/s signal source (top) and EAM driver output (bottom) showing 3.2-$V_{p\text{-}p}$ single-ended drive voltage.

low-base drive impedance in the design, allowing the devices to operate with V_{CE} of 4 to 5 $V_{p\text{-}p}$—well in excess of the DC-rated BV_{CEO}. The driver was further stressed by increasing the supply voltage to –10 V and running the driver for several hours. Although the waveform was distorted at this high supply voltage, the device exhibited nominal operation with no discernible degradation when the voltage was restored to –7.5 V. Device testing is underway at high dynamic V_{CE} swings to further clarify the safe operating area under low-base drive impedance conditions. These results show that SiGe bipolar technology may be able to address higher voltage drive applications than some had believed.

From Fig. 4.32 it is clear that the output rise/fall times (bottom) are faster than the 20-Gb/s input from the signal source (top). The output signal exhibits deterministic jitter (DJ) in the form of waveform splitting on the falling edge, which was not predicted by simulations. The source of this DJ has not been determined yet, but is most likely due to an interaction between the peaking used to compensate for the additional loading presented by the DC offset devices and reflections in the CPW-to-package interface. The DC offset devices are externally controllable current references that can be used to set the operating point for the EAM. A new design that dispenses with the on-chip bias trim, which was deemed unnecessary, is under way. This redesign should enable operation at higher speeds and reduced DJ, since the peaking will be greatly reduced.

Discussion: Analog Front End Circuit Design for 40 Gb/s Applications

Designing, packaging, and measuring high-speed analog front-end circuits targeting 40 Gb/s data rates is challenging because these efforts push the edge of what is possible while stretching modeling, circuit extraction, and simulation state of the art to the limit. Rules of thumb derived from experience as well as the results of standalone full-wave electromagnetic simulators were used in on-chip interconnect layout, but more care was taken in the design of coplanar taper launches to assure they provided a good match to a 50-Ω environment.

Packages for these circuits were also designed with EM simulators to provide controlled 50-Ω microstrip lines in alumina or organic substrates tapering out to 60-GHz V-connectors. The chips were placed into a cavity in the substrate to make the bonding surface of the die and package coplanar, allowing the shortest possible ribbon bonds to reduce parasitic inductance. At 40 GHz, for example, the magnitude of $j\omega L$ for a 100-pH inductance (approximately a 200-μm-long ribbon bond) is 25 Ω—a huge discontinuity in a 50-Ω environment! We believe that such impedance discontinuities accounted for the performance reduction seen in packaged parts as compared to data collected on-wafer.

Measurements were similarly hampered by passive and active equipment limitations. Electrical cables touting 60-GHz bandwidth and length less than 30 cm actually have 1–2-dB skin-effect losses from DC to 40 GHz that cause intersymbol interference in the signals fed into our high-speed chips. Wafer probing required even longer cables with more loss because bulky test equipment could not be placed on the prober platform. Finally, as seen in the eye diagrams, the pattern generator used to test the circuits had slower rise and fall times than the circuits themselves, and the scope sampling head, rated at 50 GHz, had 7–8-ps intrinsic rise times. For these reasons, de-embedding of measurements was required to estimate actual circuit performance. For example, measured 20%–80% rise/fall times were 10 ps with the intrinsic scope response being 7–8 ps, giving 7–8-ps rise times for de-embedded estimates. As another example, the pattern generator rms jitter was 1.7 ps and the measured postamp output was 1.9-ps rms jitter, yielding an estimate of 0.9-ps rms jitter contributed by the postamp. Since an error detector at 40 Gb/s was not available at the time measurements were taken, we could not use the bathtub curve method to extract jitter information, but even this method likely would have suffered from jitter limitations of the high-speed error detector.

4.1.4 Summary and Conclusions

High-speed data-communication networks of the twenty-first century will continue to require high performance levels from serial links, such as low power and low BER, at an affordable price. This is forcing the IC industry to move to higher levels of integration at lower power levels, which can only be achieved using silicon-based technologies combined with cost-effective packaging and testing approaches. This section has highlighted the challenges of such a serial transmission subsystem design environment from circuit designers' perspective and focused on the integration of key serial link transceiver functions. A wide range of microwave mixed-signal circuit implementations have been described including 10–13-Gb/s serializer/deserializer chips with subpicosecond jitter performance, 56-Gb/s multiplexer/demultiplexer ICs, 40-Gb/s clock-and-data recovery/clock-multiplier phase-locked loops, limiting amplifiers, and EAM drivers. SiGe BiCMOS technologies in production at IBM have been exploited throughout this work. While the 10-Gb/s designs have utilized full-rate architectures, the 40-Gb/s work relied on half-rate clocking schemes. However, thanks to the recent advances in SiGe HBT technology, which provides transistor cutoff frequencies beyond 300 GHz (see the Introduction), it is rapidly becoming possible to implement even the OC-768 functions using a full-rate architecture and keep the system-level power consumption at a very reasonable level.

Based on the leading-edge performance levels described here, it can be foreseen that SiGe-based designs will continue to move into market segments previously believed to be the stronghold of GaAs- and InP-based technologies. With increased emphasis on developing microwave models for active and passive elements as well as the package and board environment, SiGe will be the perfect technology choice for multi-terabit-throughput parallel links of the future. In addition to the broadband communications field, which was the main focus of discussion here, the future microwave test equipment markets and the emerging field of short-range wireless local-area networks that can carry gigabit/s data, will also benefit from these advances in microwave silicon technology and circuit design, which are driven by high-speed communication networks of today.

4.1.5 References

1. M. Soyuer, H. A., Ainspan, M. Meghelli, and J.-O. Plouchart, "Low-Power Multi-GHz and Multi-Gb/s SiGe BiCMOS Circuits," *Proc. IEEE,* vol. 88, no. 10, pp. 1572–1582, October 2000.
2. D. Friedman, M. Meghelli, B. Parker, H. Ainspan, and M. Soyuer, "Sub-Picosecond Jitter SiGe BiCMOS Transmit and Receive PLLs for 12.5 Gbaud Serial Data Communication," in *IEEE Symposium of VLSI Circuits Digest of Technical Papers,* pp. 132–135, June 2000.
3. J. F. Ewen, A. X. Widmer, M. Soyuer, K. R. Wrenner, B. Parker, and H. A. Ainspan, "Single-Chip 1062-Mbaud CMOS Transceiver for Serial Data Communications," in *IEEE ISSCC Digest of Technical Papers,* pp. 32–33, February 1995.
4. T. H. Lee, and J. F. Bulzacchelli, "A 155-MHz Clock Recovery Delay-and Phase-Locked Loop," *IEEE J. Solid-State Circuits,* vol. SC–27, pp. 1736–1746, December 1992.
5. M. Meghelli, B. Parker, H. Ainspan, and M. Soyuer, "SiGe BiCMOS 3.3V Clock and Data Recovery Circuits for 10Gb/s Serial Transmission Systems," *IEEE J. Solid-State Circuits,* vol. 35, pp. 1992–1995, December 2000.
6. J. F. Ewen, M. Soyuer, A. X. Widmer, K. R. Wrenner, B. D. Parker, and H. A. Ainspan, "CMOS Circuits for Gb/s Serial Data Communication," *IBM J. Res. Developing,* vol. 39, pp. 73–81, January/March 1995.
7. A. X. Widmer and P. A. Franaszek, "A DC-Balanced, Partitioned-Block, 8B/10B Transmission Code," *IBM J. Res. Develop.,* pp. 440–451, September 1983.

8. D. Friedman, M. Meghelli, B. Parker, J. Yang, H. Ainspan, and M. Soyuer, "A Single-Chip 12.5 Gbaud Transceiver for Serial Data Communication," in *IEEE Symposium of VLSI Circuits Digest of Technical Papers,* pp. 145–148, June 2001.
9. S. Ueno, K. Watanabe, T. Kato, T. Shinohara, K. Mikami, T. Hashimoto, A. Takai, K. Washio, R. Takeyari, and T. Harada, "A Single-Bit 10 Gb/s Transceiver LSI Using SiGe SOI BiCMOS," in *IEEE ISSCC Digest of Technical Papers,* pp. 82–83, February 2001.
10. Y. Greshishchev, P. Schvan, J. L. Showell, Mu-Lian Xu, J. J. Ojha, and J. E. Rogers, "A Fully Integrated SiGe Receiver IC for 10-Gb/s Data Rate," *IEEE J. Solid-State Circuits,* vol. 35, pp. 1949–1957, December 2000.
11. J. Savoj and B. Razavi, "A 10 Gb/s CMOS Clock and Data Recovery Circuit with Frequency Detection," in *IEEE ISSCC Digest of Technical Papers,* pp. 78–79, February 2001.
12. M. Green, A. Momtaz, K. Vakilian, Xin Wang, Keh-Chee Jen, D. Chung, Jun Cao, M. Caresosa, A. Hairapetian, I. Fujimori, and Yijun Cai, "OC-192 Transmitter in Standard 0.18 μm CMOS," in *IEEE ISSCC Digest of Technical Papers,* pp. 248–249, February 2002.
13. J. Cao, A. Momtaz, K. Vakilian, M. Green, D. Chung, Keh-Chee Jen, M. Caresosa, B. Tan, I. Fujimori, and A. Hairapetian, "OC-192 Receiver in Standard 0.18 μm CMOS," in *IEEE ISSCC Digest of Technical Papers,* pp. 250–251, February 2002.
14. M. Meghelli, M. Bouche, and A. Konczykowska, "Very High Speed Integrated Circuits: Design Methodology and Applications for Optical Communications," in *ECCTD Proceedings, Budapest, Hungary,* pp. 1366–1370, September 1997.
15. M. Reinhold, C. Dorschky, E. Rose, R. Pullela, P. Mayer, F. Kunz, Y. Baeyens, T. Link, and J.-P. Mattia, "A Fully-Integrated 40 Gb/s Clock and Data Recovery/1:4 DEMUX IC in SiGe Technology," *IEEE J. Solid-State Circuits,* vol. 36, pp. 1937–1945, December 2001.
16. W. Pöhlmann, "A Silicon-Bipolar Amplifier for 10 Gb/s with 45 dB Gain," *IEEE J. Solid-State Circuits,* vol. 29, pp. 551–556, May 1994.
17. M. Meghelli, A. V. Rylyakov, and Lei Shan, "50 Gb/s SiGe BiCMOS 4:1 Multiplexer and 1:4 Demultiplexer for Serial Communication Systems," in *IEEE ISSCC Digest of Technical Papers,* pp. 260–261, February 2002.
18. L. Shan, M. Meghelli, Joong-Ho, Kim, J. Trewhella, M. Taubenblatt, and M. Oprysko, "Millimeter Wave Package Design: A Comparison of Simulation and Measurement Results," in *IEEE 10th Topical Meeting on Electrical Performance of Electrical Packaging,* pp. 29–34, 2001.
19. Y. Greshishchev and P. Schvan, " A 60-dB Gain, 55-dB Dynamic Range, 10-Gb/s Broad-band SiGe HBT Limiting Amplifier," *IEEE J. Solid-State Circuits,* vol. 34, pp. 1914–1920 , December 1999.

4.2 WIRELESS DESIGN: A DIRECT CONVERSION RECEIVER IC FOR WCDMA MOBILE SYSTEMS

As a first step toward addressing the radio hardware needs of a universal mobile telecommunications system (UMTS) mobile transceiver for the European cellular phone market, a single-mode 3-V wideband CDMA (WCDMA) direct-conversion receiver IC has been designed and fabricated using IBM SiGe 6HP BiCMOS technology. The receiver design has been targeted to address the industry needs of high integration, low power, and low cost, while meeting all WCDMA RF system performance requirements [1] with adequate design margin to accommodate expected operating voltage, temperature, and fabrication process variations. This section presents an overview of the receiver system design, including key system performance requirements derived from the WCDMA specification, followed by detailed descriptions of the design and measured performance of the receiver circuits. The system performance of the IC is characterized in a receiver test bed that uses a software baseband processor to compute link BER and estimated code channel SNR. RF packaging issues and the chosen packaging technique for test and evaluation of the receiver IC are also discussed.

4.2.1 WCDMA Receiver System Design

A high level block diagram of the direct conversion architecture chosen for the receiver design is shown in Fig. 4.33. Direct-conversion works by mixing a received signal against in-phase (I) and quadrature (Q) local oscillators tuned to the center frequency of the desired radio channel. The mixer produces products at twice the local oscillator frequency and at baseband. Low-pass filtering following the downconverter mixers removes the high-frequency mix product and any interference signals near the received channel so the baseband I/Q signals can be independently amplified and converted to digital representation with low precision (6 to 8 bit) A/D converters.

The advantages of the direct-conversion architecture over a classic "superheterodyne" (or multiple-stage downmix) approach include elimination of a second frequency synthesizer, elimination of an intermediate-frequency (IF) filter, reduction of spurious mixer products, and the potential to efficiently accommodate multiple radio standards with different channel bandwidths over a wide frequency band. Although these are attractive advantages, several difficulties arise in the design of a

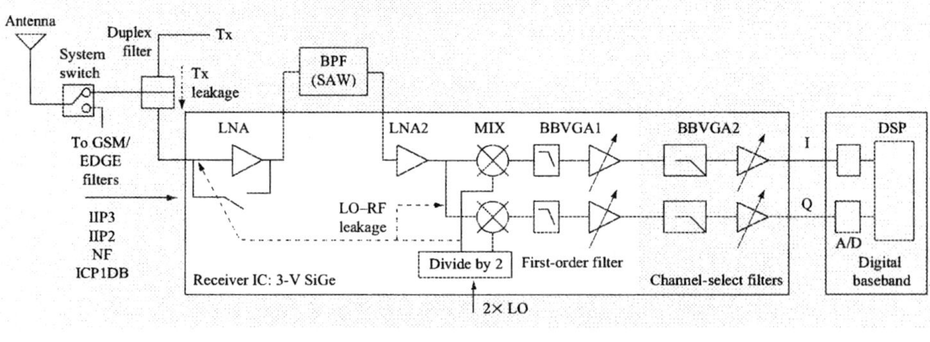

Figure 4.33 Direct-conversion WCDMA FDD receiver system design.

practical direct-conversion receiver. Key among the disadvantages are static DC and dynamic low-frequency distortion terms that fall on top of the desired signal at baseband. These distortion terms arise from coupling between the local-oscillator and RF signal paths (LO-RF) and second-order intermodulation in the quadrature mixers and subsequent baseband amplifiers and filters that translate any signal with amplitude variations to the baseband frequency range where the desired signal lies. These problems are addressed through a combination of circuit-design techniques that achieve high second-order linearity and selection of a receiver architecture, which mitigates the LO-RF coupling problem.

The receiver architecture shown in Fig. 4.33 minimizes LO-RF leakage by driving the LO port of the IC at twice the desired channel frequency, according to the technique described by Lie and Kennedy [2]. This eliminates a leakage path arising from coupling of the LO input to the LNA input through the IC bondwires. The 2X LO is divided on-chip to generate quadrature LO signals for the downmixers. Attenuation of any LO-RF leakage in the mixer circuitry is also provided by the LNA2 reverse isolation. This design approach allows the LNA to be powered down and bypassed while keeping the LO power at the antenna well below WCDMA specification requirements.

A first-order pole on the output of the quadrature mixers is used to attenuate high-frequency distortion terms such as the transmitter leakage signal and both in-band and out-of-band interference signals. Because the desired signal can still be at a relatively small level at the mixer output, a low-noise variable gain amplifier (BB-VGA1) must be used to increase the signal level sufficiently so that the input noise of the active-channel select filter does not degrade system sensitivity. The channel-select filter and following variable-gain amplifiers have not been integrated onto the IC described in this section, but another prototype design that incorporates on-chip channel-select filters has been fabricated already, and the variable-gain amplifiers are planned to be added to a future fully integrated receiver IC system-on-chip.

Receiver System Performance Requirements Key system performance requirements for the direct-conversion receiver include LNA-referred NF, second- and third-order linearity (IIP2, IIP3), input 1-dB compression point (ICP1dB), quadrature accuracy, I/Q balance, in-channel phase distortion, and LO phase noise. Because the front-end switch/RF filter characteristics can vary significantly, depending on vendor and design, the LNA referred receiver performance requirements such as NF, IIP2, IIP3, and ICP1dB cannot be directly derived from the WCDMA specification. Some important operating-condition parameters needed to complete the performance requirements include maximum transmitter leakage power at the LNA input port, insertion loss and blocking characteristics of the antenna switch/duplexer filter, and performance of the baseband digital-signal processor (DSP) demodulator. Assumptions made for each of these parameters are covered in the following sections.

Transmitter Leakage at LNA Input The receiver IC is designed for use in a WCDMA frequency-domain duplex (FDD) mobile device. A FDD transceiver achieves full duplex functionality by using separate transmit and receive frequencies that are simultaneously active. As shown in Fig. 4.33, the radio must use a 3-port "duplexer" filter to isolate the relatively high-power transmit signal from a small received signal at the antenna. Because the isolation of the duplexer is limited, some amount of residual transmitter leakage appears at the receiver LNA input. For the system design considered here, a maximum operating transmitter leakage of –22 dBm at the LNA input is assumed. This level arises under the assumption of +28 dBm at the transmit PA output (or duplexer transmit port), and 50-dB isolation from the duplexer transmit output to receive input port. This leakage term is one of the key parameters that drives the compression and out-of-band linearity requirements of the receiver.

Duplexer Filter Frequency Response Out-of-band third-order linearity requirements arise from transmitter leakage at the LNA mixing with blocker signals below the receiver band that produce third-order intermodulation distortion (IM3D) in the receiver channel passband. The WCDMA specification [1] defines three blocking bands with corresponding blocker power levels that must be tolerated by the receiver. A summary of the frequency bands relevant to the transmit/blocker IM3 distortion problem is given in Table 4.2. In each band, an estimate of the minimum amount of insertion loss provided by the duplexer is given. Cascading the blocker power levels at the antenna with minimum switch/duplexer attenuation allows determination of maximum blocker power level at the LNA input so the band-specific LNA-referred third-order linearity and input compression requirements for the receiver IC can be estimated.

Required Es/No and Receiver Implementation Loss The WCDMA receiver RF tests use coded BER as the performance metric to assess specification compliance. To pass the receiver tests, a coded BER of <0.1% is required under a range of test conditions designed to stress sensitivity, linearity, and selectivity of the receiver. From reference simulation results, the 12.2-kpbs modulation used for the RF tests requires approximately 0.9 dB Es/No (channel-symbol energy-to-noise power spec-

Table 4.2 Blocking Levels at LNA Input

Band	Frequency Range (in MHz)	Power Level at Antenna (dBm)	Min Attenuation (dB)	Max Attenuation	Power Level at LNA (dBm)
Transmit band	1920–1980	+27	50	—	–22
Blocking band 1	2050–2075	–44	2	—	–46
Blocking band 2	2025–2050	–30	10	—	–40
Blocking band 3 above Tx band	2015–2025	–15	25	—	–40
Blocking band 3 below Tx band	1679–1840	–15	44	—	–59

tral density ratio), or de-spread channel symbol SNR, to achieve 0.1% BER. To allow for practical receiver degradations such as imperfect channel estimation, filter amplitude/phase distortion, and quadrature inaccuracy in the downconverter, the 0.9-dB Es/No is increased by a receiver "implementation loss" factor. Although a high-performance receiver/DSP combination may be able to achieve an implementation loss of less than 1 dB, a conservative implementation loss of 2 dB (as in [3]) is chosen to build margin into the system spec requirements. With this loss, a de-spread channel symbol SNR of 2.9 dB must be maintained to pass the BER requirement for the RF tests.

The required channel E_s/N_0 can be related to E_b/N_0 (information-bit energy-to-noise power spectral density ratio) for comparing the relative energy efficiency of different communication system designs. Since WCDMA channel symbols are transmitted at a symbol rate of 30 K symbols per second for a bit rate of 12.2 Kb/s, the bits per symbol ratio (E_b/E_s) is 30/12.2, or approximately 4 dB. Given a 0.9-dB E_s/N_0 requirement for 0.1% BER, and a 2-dB implementation loss, the required E_b/N_0 is therefore ~7 dB.

Receiver Performance Requirements The receiver system performance specifications are intended to assure that a receiver design can pass the WCDMA receiver RF tests with the assumed RF front-end filtering and baseband DSP parameters. An overview of some of the key specifications is given in the following sections. Where noted, some of the specifications are unique/critical to the direct-conversion receiver architecture. Details of computations necessary to derive most noise and linearity performance requirements for the receiver from assumed operating conditions are covered in [3] and are not repeated here. Some performance requirements cited here may differ substantially from those presented in [3] due to different assumptions on transmit leakage power at the LNA and duplexer blocking characteristics.

Receiver Noise Noise performance of the receiver is characterized by its input-referred noise figure. The noise figure is measured across the received signal bandwidth (3.84 MHz) using a matched filter instead of at one "spot noise" frequency, so that it is relevant to achievable system performance. Since the receiver pass band may have significant gain ripple in the cascaded LNA → SAW → LNA2 → MIXER transfer function, the noise figure must be characterized at each of the 12 center frequencies across the UMTS terrestrial radio-access (UTRA) receive band (2110 to 2170 MHz). The noise figure requirement of the receiver is driven by the received power level specified in the WCDMA sensitivity test (–117-dBm code channel power at the antenna) and the largest expected insertion loss from the antenna to the receiver LNA. A LNA-referred noise figure of less than 5 dB is targeted to provide design margin with 4-dB loss in the front-end RF switch/filter [3].

Third Order Linearity Third-order linearity performance of a receiver is characterized by its input-referred IIP3. Third-order nonlinearity in a receiver causes a distortion product to appear in-channel from two interferers offset from the desired channel in frequency by Δf and $2\Delta f$. The interference signals occur both within the receiver pass band (in-band IM3 interference) and outside the receiver pass band (out-of-band IM3 interference). The in-band IM3 requirement is determined with an explicit receiver IM test in the WCDMA specs [1] (10 MHz offset CW tone, 20 MHz offset WCDMA interferer both at a power level of –46 dBm at antenna), but the out-of-band IM3 requirement is only implicitly defined by the need of the FDD receiver to pass the out-of-band blocking desense tests in the presence of a large transmit leakage signal at the antenna [3]. From the WCDMA specifications, there are four blocking regions (summarized in Table 4.2) that can produce in-channel IM3 products when mixed with the transmitter leakage signal. Targeted requirements for the in-band and four out-of-band IM3 frequency regions are summarized later in the performance results section in Table 4.5.

Second-Order Linearity Second-order linearity performance of a receiver is characterized by its input-referred IIP2 for both in-band and out-of-band interference signals. Second-order nonlinearity in an amplifier results in distortion products at baseband that can add time-varying DC offsets and in-channel distortion to the downconverted signal. The DC offset problem can be easily addressed in the WCDMA application by using a high-pass filter to remove the lower 1 to 5 kHz of the signal bandwidth [4, p. 1895]. However, higher frequency in-channel distortion cannot be filtered away, so a constraint must be placed on the maximum allowable second-order nonlinearity as a function of blocker frequency.

The in-band IIP2 requirement arises from the maximum in-band blocker that must be tolerated by the receiver. The WCDMA RF specs define this blocker level at a 15-MHz offset from the desired received channel with a power level of –44 dBm at the antenna. The out-of-band IIP2 requirement is dominated by the large transmit leakage signal (–22 dBm) at the LNA input. The large IIP2 requirement at the transmit offset frequency can be met without imposing unrealizable constraints on the IIP2 of the downconverter by assuring that the interstage LNA/LNA2 RF filter has sufficient Tx/Rx isolation (a minimum level of approximately 25 dB is desired for the system design described here). A summary of the targeted IIP2 values for the in-band and transmit-band blockers is given later in the performance results section in Table 4.5.

Input 1-dB Compression Input 1-dB compression points are specified for both in-band (high and low LNA RF gain) and transmit band (high LNA RF gain only) frequency signals. The high-gain in-band compression requirement is derived from the in-band IM test condition, which produces a composite –43-dBm signal level at the antenna input with the receiver in high-gain mode. To provide enough headroom for the "crest factor" (the peak power level above average power level) of the received WCDMA interferer signal, 10-dB margin is added to the average input power level to result in a desired antenna-referred in-band compression point of –33 dBm. A minimum switch/duplexer loss of 2 dB results in a targeted high-gain in-band 1-dB compression point of –35 dBm or better referred to the LNA input of the receiver IC. A low-gain in-band compression point of –17 dBm is chosen to handle the WCDMA maximum-input level test, which places a –25-dBm signal at the LNA input. The low-gain compression specification does not place a more stringent requirement on the LNA2/mixer circuitry than the high-gain requirement since the LNA is switched into low-gain state in this mode, which lowers the LNA to LNA2 voltage gain by about 18 dB.

Transmit band compression is derived from the maximum assumed transmitter leakage power at the LNA input. A leakage of –22 dBm with 8-dB margin added to accommodate the WCDMA mobile unit crest factor (the mobile transmit crest factor is less than the base transmit, or received signal, crest factor due to fewer simultaneous code channels) of the handset transmit signal results in a desired minimum ICP1dB of –14 dBm at the LNA input for transmit band (1920 to 1980 MHz) signals. Depending on the characteristics of the RF interstage band-pass filter, this ICP1dB may be limited by either the LNA or the LNA2/mixer. If this critical ICP1dB is not met, the receiver may be desensed by the transmitter leakage under conditions of small received power and high transmit power due to small-signal gain compression and/or noise figure degradation in the receiver amplifiers/mixers.

Quadrature Accuracy If the quadrature downconverter has other than a 90 degree phase shift between the I and Q downmixers, power from the I channel projects onto the Q channel and vice-versa. A quadrature accuracy of ±5 degrees or better between the I and Q channels for the downmixer is chosen to limit system performance degradation to acceptable levels in the WCDMA application. This accuracy may need to be improved and/or compensated with digital signal processing for future multimode designs to accommodate higher-order modulation schemes that are more susceptible to I/Q crosstalk distortion.

I/Q Amplitude Match, Phase Match, and Group-Delay Distortion The I/Q signals at baseband propagate through separate filters and amplifiers, which can have different amounts of gain and phase versus frequency and add group-delay distortion to the signal. Targeted I/Q amplitude match (i.e., the ratio of average I channel signal level to Q channel signal level) for the receiver IC at the BBVGA1 output is ±0.25 dB. Phase match and group-delay distortion need not be characterized for the current version of the IC, since it does not have an integrated channel-select filter, and the low-pass filters following the mixers cut off well above the channel bandwidth, so phase distortion arising from in-band differential group delay or signal distortion from phase mismatch between the I and Q channels is negligible at the BBVGA1 I/Q output ports. Similar to the I/Q quadrature error problem, there is also potential to compensate these baseband distortion effects using digital signal processing and relax requirements on the analog circuit designs if desired.

LO-RF Leakage A direct conversion receiver is susceptible to desense by a third-order nonlinear distortion product resulting from cross modulation [5] of the transmitter leakage signal onto the LO-RF leakage signal at the LNA input. Mathematically, this distortion product arises from convolution of the transmitter leakage signal with itself (producing a frequency spectrum product at baseband) and the

CW LO leakage term at the antenna (thereby centering the baseband spectrum from the transmitter directly over the desired received signal). Although a large LO-RF leakage can theoretically be handled by improving the input third-order linearity, this is not a practical solution because a high linearity requirement would result in unacceptably large power draw. Therefore, a constraint is placed on the LO-RF leakage level so that it does not increase the needed out-of-band third-order linearity any more than the level needed from the worst-case Tx/blocking band linearity requirement. A maximum high-LNA-gain mode LO-RF leakage power of –80 dBm at the LNA input has been chosen as a design target to reduce the Tx-LO cross-modulation distortion to a level small enough not to degrade sensitivity performance of the receiver while keeping the LNA IIP3 requirement within bounds of the Tx/blocking band linearity requirements. A low-LNA-gain mode LO-RF leakage power at the antenna of –60 dBm must also be maintained to comply with WCDMA standards.

4.2.2 Receiver Circuit Design

The receiver chip has the basic architecture described by the block diagram of Fig. 4.34. The signal from the antenna passes through a duplexer and into the 50-Ω unbalanced input of LNA1. This switched-gain LNA provides either 14 dB of gain or

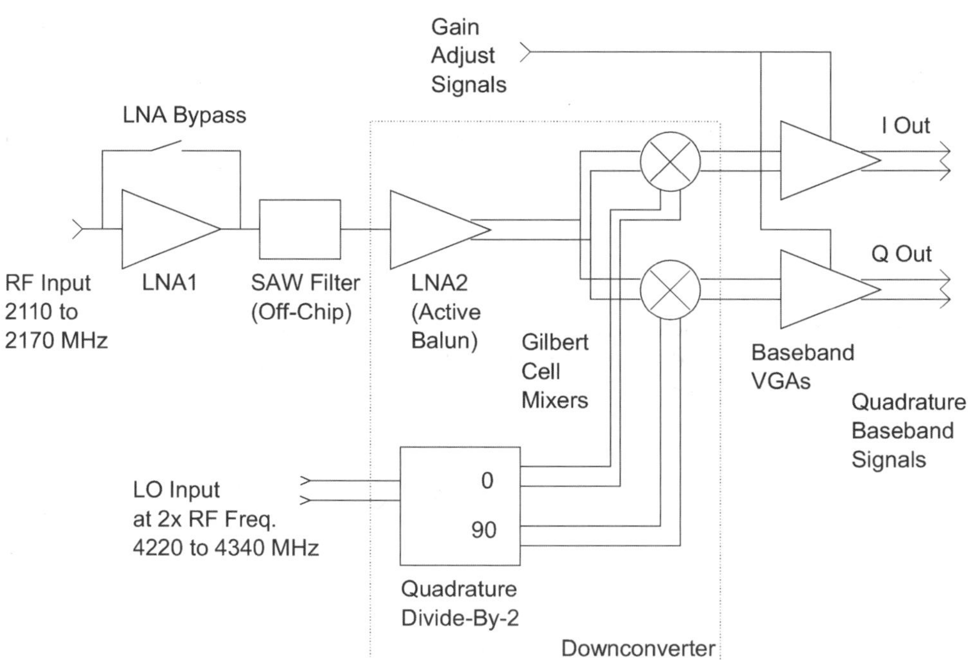

Figure 4.34 A block diagram of the WCDMA receiver chip.

4 dB of loss, depending on the strength of the input signal level. Following LNA1, the 50-Ω signal goes off-chip to a band-select surface acoustic wave (SAW) filter, which attenuates RF signals outside the 2110- to 2170-MHz WCDMA band, easing linearity requirements further downstream. In particular, the SAW attenuates the handset's own transmit signal in the 1920- to 1980-MHz band, which appears at the LNA1 input due to finite isolation in the duplexer. The output of the SAW filter comes back on-chip to the 50?Ω input of LNA2, which has 12 dB of gain and acts as an active balun to provide differential signals to the two mixers. To minimize LO-to-RF leakage, the LO signal comes on chip at twice the frequency of the desired LO signal; hence, a divide-by-two circuit is used on-chip to produce two quadrature, differential LO signals. The mixer outputs go through a first-order low-pass filter at 4 MHz to attenuate adjacent-channel blocking signals, then to the baseband variable-gain amplifier (BBVGA). This BBVGA has five selectable gain states of +16, +10, +4, –2, and –8 dB, respectively.

The chip is fabricated in IBM SiGe BiCMOS 6HP technology. This technology has standard NPN bipolar transistors with a peak f_T of 47 GHz, high-breakdown NPN transistors with a peak f_T of 27 GHz, 2.5-V CMOS transistors with 0.24-μm drawn channel lengths, and 3.3-V CMOS transistors with 0.4-μm drawn channel lengths. In addition, it features polysilicon resistors, metal-to-metal capacitors, and high-Q inductors using a thick final aluminum layer.

The receive chip uses no external components except for the SAW filter shown in Fig. 4.34 and a DC blocking capacitor at the LNA1 input. LNA1 uses bondwire inductances for degeneration and impedance matching, as well as on-chip inductors. Total current consumption is 14.5 mA with LNA1 on and 10.5 mA with LNA1 off.

Low-Noise Amplifier Design A simplified schematic of LNA1 is shown in Fig. 4.35. This LNA has two gain states. In high-gain mode (BYP=0), the LNA is biased at 4 mA and has a gain of about 14 dB; in bypass mode (BYP=1), the bias current for LNA1 is switched off and the signal is routed around the gain stage through a MOSFET switch. The LNA consumes no power in bypass mode, allowing for reduced system power consumption for high signal-level conditions. In both modes, the LNA is matched to 50 Ω at the input and output, targeted for a specification of $S_{11} < -12$ dB and $S_{22} < -15$ dB. A PTAT bias circuit is included on-chip, setting Q1's transconductance (I_c/V_T) approximately constant over temperature. This PTAT is derived from an on-chip bandgap reference included in the downconverter.

In the high-gain mode, amplification is provided by a common-emitter (Q1) with inductive degeneration (L_E). In sizing the bipolar device, the collector current density is chosen to achieve the lowest minimum noise figure ($\mathrm{NF}_{\min}$), while the emitter length is chosen such that the optimal source impedance, $R_{S,\mathrm{opt}}$, has a real part of 50 Ω [6,7].* Inductive degeneration, which affects neither $\mathrm{NF}_{\min}$ nor $R_{S,\mathrm{opt}}$, is then

*As described, this design procedure sets the total current (current density times emitter area); however, since $\mathrm{NF}_{\min}$ versus current density has a wide minimum, the current can readily be optimized together with $\mathrm{NF}_{\min}$ and $R_{S,\mathrm{opt}}$.

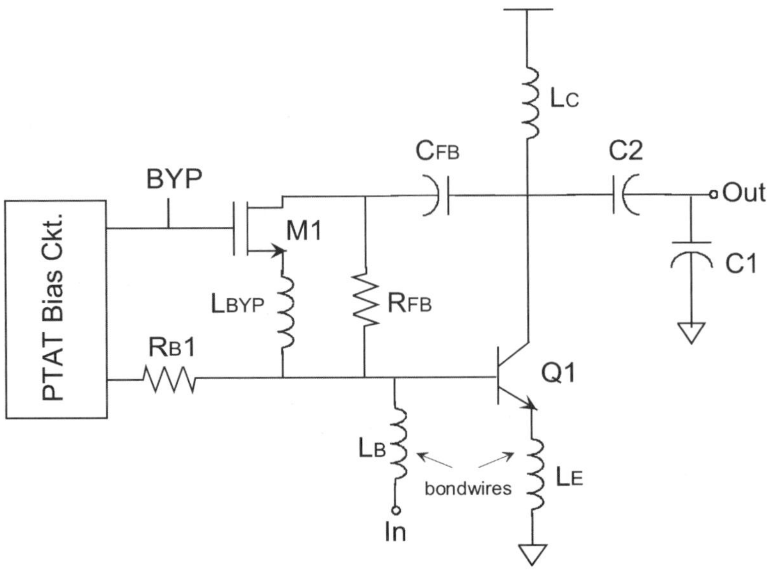

Figure 4.35 A simplified schematic of LNA1.

used to provide a real, 50-Ω term to the input impedance. Thus, optimum noise matching and power matching are obtained simultaneously. A small amount of feedback (i.e., the impedance of R_{FB} in series with C_{FB} is large) is used from the base to the collector to ease the input match, as well as facilitate matching in the bypass mode. As a result, the input bondwire parasitic inductance is enough to complete the input 50-Ω match. This weak feedback, since it includes a resistor, will slightly increase NF_{min} and slightly modify Z_{opt}. Finally, the output match is implemented as a pi-network, with a shunt inductor to the supply and a capacitive transformer.

In the bypass mode, Q1 is powered down, and M1 is switched on. This routes the signal from the input matching network (L_B) through L_{byp}, M1, and C_{FB}, to the output matching network. Since the LNA is now passive in this mode, its linearity is very high. Due to the on-resistance of switch M1 and the loss in the matching elements, there is approximately 4 dB of loss in the bypass mode, meaning that the NF in this mode is also about 4 dB. Matching the input and output in the bypass mode can be challenging, since the input and output matching networks for the high-gain mode are still present in the signal path. To realize a 50-Ω input and output match in either high-gain or bypass mode, the impedance seen looking into the base or collector node has to be approximately the same for both modes. This requires an additional inductor (L_{byp}) to be added in series with M1.

The targeted in-band IIP3 for this LNA in the high-gain mode is +4 dBm at 4 mA for a 3-V supply. The linearity efficiency, defined as IIP3 in milliwatts divided by the DC power consumption in milliwatts, is then 0.2—an aggressive target. The

Table 4.3 A Summary of Measured and Simulated Performance Data for LNA1

Frequency = 2140 MHz	Measured High-Gain Mode	Simulated High-Gain Mode	Measured Bypass Mode	Simulated Bypass Mode
Current	3.9 mA	4.2 mA	0 mA	0 mA
Supply voltage	2.85 V	2.85 V	2.85 V	2.85 V
Gain (S_{21})	13.2 dB	13.2 dB	–4.0 dB	–3.0 dB
Noise figure	1.8 dB	1.4 dB	3.6 dB	3.1 dB
S_{11}	–10.7 dB	–10.9 dB	–12.2 dB	–19.8 dB
S_{22}	–12.0 dB	–13.2 dB	–13 dB	–29.2 dB
IIP3 (10 MHz)	+5.6 dBm	+6.0 dBm	+20.2 dBm	+22.9 dBm
IIP3 (70 MHz)	+3.9 dBm	+6.4 dBm	+23.5 dBm	+23.0 dBm
ICP1dB	–11.5 dBm	–8.5 dBm	NA	NA

easiest way to increase IIP3 is to increase the bias current. However, when power consumption is constrained, this is not viable. A second method to improve IIP3, used in this design, is to use feedback (e.g., inductive degeneration). A third method is to minimize the impedance looking into the bias circuit from the base of Q1 at the tone frequency (f_1-f_2) [8]. The impedance should, however, still be large at RF frequency to prevent NF and gain degradation. Thus, a moderate value for R_{B1} is chosen to achieve a compromise between NF and IIP3. Finally, to improve IIP3, the voltage gain from the RF input to the base should be kept to a minimum; thus, the input Q is kept low.

The large-signal linearity for LNA1 was targeted for an input-referred 1-dB compression point (ICP1dB) of better than –12 dBm. This specification is targeted at meeting receiver sensitivity with a –22-dBm transmit leakage signal present at the LNA1 input. Not only should gain degradation be taken into account, but NF degradation should as well. Under large-signal conditions, the NF will increase,† hence, one can define a 1-dB NF *expansion* point. Simulations performed using SpectreRF show that this expansion point occurs when the input power level is one-fourth the ICP1dB (in milliwatts), or 4 dB below ICP1dB. These simulations also show that the NF increase (in dB) is approximately 2.5 * P_{block}/ICP1dB, where P_{block} is the input blocker level, in milliwatts, and ICP1dB is in milliwatts. Thus, the NF expands by 2.5 dB at ICP1dB. For a –22-dBm transmit leakage signal and a –12-dBm ICP1dB, the LNA gain degradation is under 0.1 dB and NF degradation is under 0.25 dB.

The measured characteristics of LNA1 are summarized in Table 4.3. These measurements were performed on a receiver IC that was directly bonded to the board (chip-on-board), with the output SAW filter removed. The gain and NF results listed have the loss due to the input and output coplanar waveguides de-embeded. In high-gain mode, the gain (S_{21}) is 13.2 dB and the NF is 1.8 dB. Input and output return loss are –10.7 and –12 dB, respectively. The reverse isolation in high-gain

†For example, shot noise is proportional to current.

mode is 22 dB. The in-band IIP3 (10-MHz tone spacing) is +5.6 dBm, the out-of-band IIP3 (70-MHz tone spacing) is +3.9 dBm, and ICP1dB is –11.5 dBm. The LNA consumes 3.9 mA from a 2.85-V supply. In bypass mode, the gain and NF are –4.0 and 3.6 dB, respectively. Again, the LNA is matched input and output, with S_{11} and S_{22} better than –12 dB. Finally, IIP3 is +20 dBm in bypass mode.

Downconverter Design The downconverter is that portion of the RF IC enclosed within the dashed box of Fig. 4.34, including LNA2, the mixers, and the quadrature divider. Before settling on the architecture shown in Fig. 4.34, several other options were considered, as detailed in Fig. 4.36. The options differ on whether a single-ended or differential output is taken from the band-select SAW filter, and whether or not a second stage of amplification is used prior to the mixers.

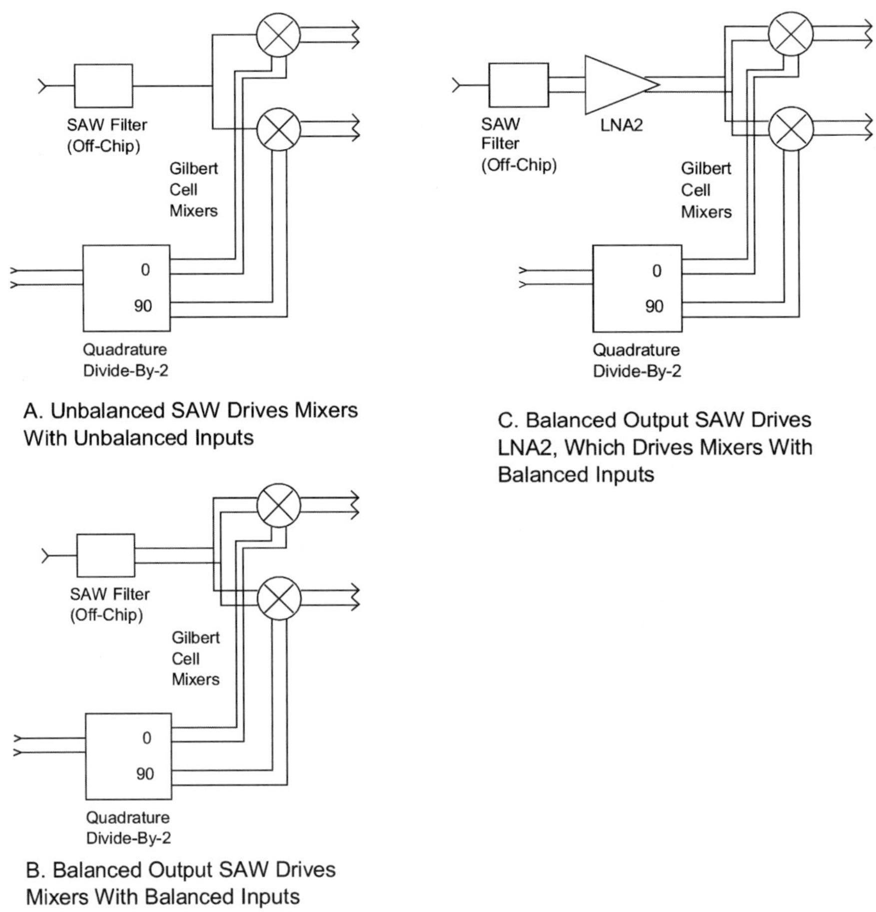

Figure 4.36 Block diagrams of alternative downconverter architectures.

Probably any of these architectures could be made to meet WCDMA system requirements. But in Fig. 4.36A, it is difficult to achieve very high IIP2, because the mixer input is unbalanced and second-order distortion products may not completely cancel in the mixer stage. Also, isolation between the LO signal and the RF input is relatively poor in Fig. 4.36 unless a common-base input stage is used for the mixer.

The use of a differential output filter in Fig. 4.36B improves the mixer balance and IIP2, and it also improves the LO-RF isolation because a certain portion of the LO leakage is out-of-phase and cancels. If the mixer circuit in Fig. 4.36B is a common-emitter circuit and has high input impedance, it can be difficult to achieve a good impedance match to the 200-Ω filter output using only on-chip components.

The use of both a differential output filter and a second stage of amplification (LNA2) in Fig. 4.36C, as described by [2], provides good IIP2 and LO-RF isolation. However, the linearity requirements of the mixers are increased by the gain of LNA2. Therefore, the gain of LNA2 must be kept low enough that the mixers have reasonable IIP3 requirements.

The architecture that we followed in Fig. 4.34 is the same as that shown in Fig. 4.36C, except that an unbalanced output is taken from the filter. The use of an unbalanced SAW filter does not impair IIP2 to the degree that it does in Fig. 4.36A, because second-order distortion products generated in LNA2 can be removed by filtering prior to the mixers. The gain of LNA2 must again be kept low in order to most easily achieve good IIP3, but in Fig. 4.34 it is not necessary to accommodate the effective incremental 6-dB voltage gain of the filter as it is in Fig. 4.36C if a standard 50-Ω single-ended to 200-Ω differential output filter is used. The choice of an unbalanced input to LNA2 was made in part to accommodate the SAW filter that we wished to use.

A simplified schematic of LNA2 is shown in Fig. 4.37. It employs inductive degeneration (L_e1 and L_e2) to increase the linear range of a standard differential pair (Q1 and Q2). A tuned *RLC* load is used so that the gain peaks in the 2110- to 2170-MHz WCDMA band, but the circuit Q is kept low by resistors R1 and R2, so that the circuit gain and frequency response are not sensitive to process variations. Load capacitance of the mixers and interconnect can be absorbed into C1 and C2. Shunt feedback is applied through $R_{fb}1$ and $C_{fb}1$ to reduce distortion and make input matching easier, and $R_{fb}2$ and $C_{fb}2$ are used to balance the output level of the inverting and noninverting outputs. Lmatch and Cmatch make up an impedance matching network to match the unbalanced input of LNA2 to 50 Ω.

The output of LNA2 is AC coupled to the mixers. The combination of the load inductors L1 and L2 and the AC coupling capacitors (not shown) forms a second-order high-pass filter that removes even-order distortion products generated in LNA2 and prevents them from unbalancing the mixers. All five inductors in LNA2 are fabricated on the chip. LNA2 is biased at 3.2 mA.

A simplified schematic of the mixers is shown in Fig. 4.38. Devices Q8–Q11 form a conventional doubled-balanced Gilbert cell mixer. The transconductor portion of the mixer (devices Q4–Q7) uses a Schmook (or multitanh) cell to expand the linear input range [9]. This cell gives a better trade-off between noise and linearity than a conventional resistively degenerated differential pair. Achieving high IIP3

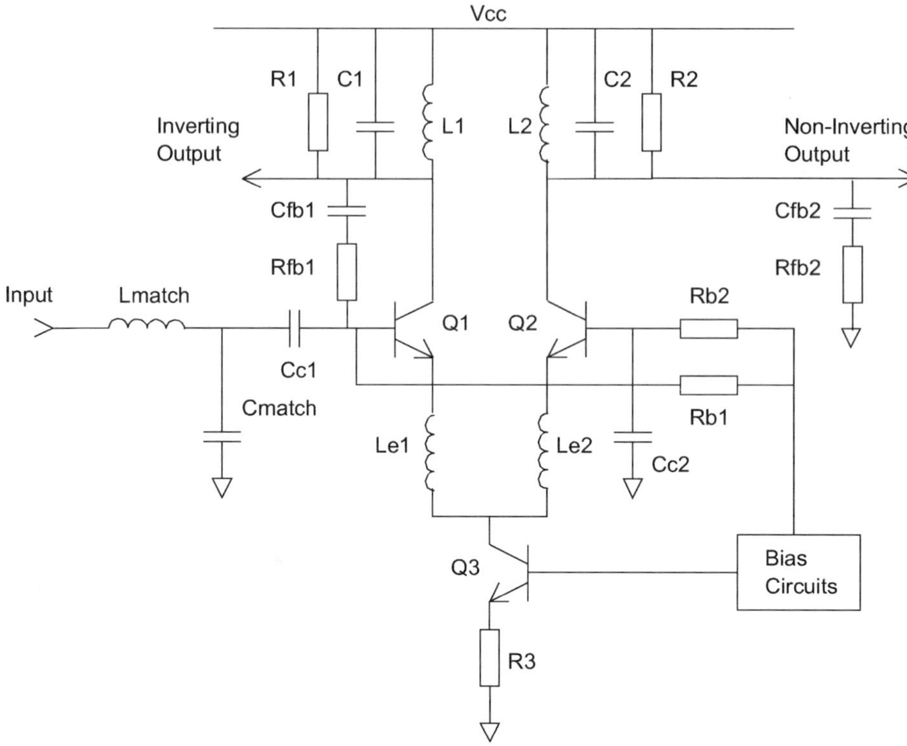

Figure 4.37 A simplified schematic of LNA2.

with this cell depends on careful selection of the emitter area ratio between the Q4–Q5 and Q6–Q7 device pairs and the voltage drop across resistors R6 and R7. Note that biasing circuitry is not shown in Fig. 4.37. The mixers are biased at 1.2 mA each.

The quadrature divider is a conventional ECL D-flip-flop configured as a divide-by-two. The double-frequency LO signal comes on-chip differentially at approximately –10 dBm and is capacitor coupled into the clock input of the ECL flip-flop.The flip-flop clock input is matched to 100 Ω at 4280 MHz. The flip-flop quadrature outputs are buffered and applied to the mixer LO inputs. The quadrature divider consumes 2.05 mA.

The BBVGA simplified schematic is shown in Fig. 4.39. There are five transconductance cells (Gm-cells) sharing a common input buffer, a common output buffer, and common load resistors. Only one of the Gm-cells is biased at a time, and the transconductance of that cell (together with the load resistors) determines the BBVGA gain. An RC filter at the input (components $R_{in}1$, $R_{in}2$, and C_{in}) has a cutoff frequency of 4 MHz. The two (I and Q) BBVGA amplifiers consume 2 mA.

Separate measurement results are available for the downconverter and BBVGA. These measurements were made by wafer probing of a breakout site including only

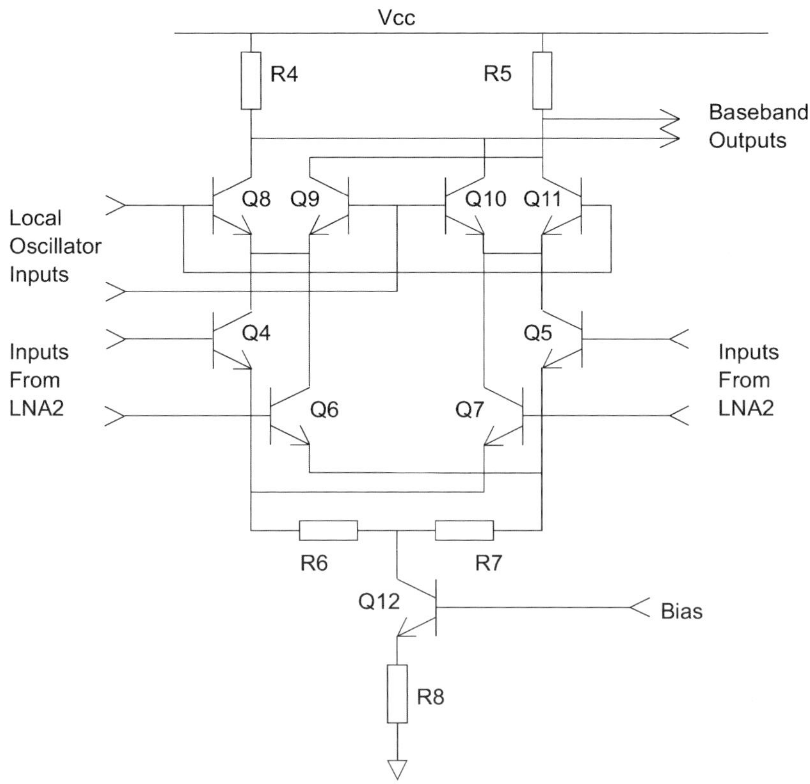

Figure 4.38 A simplified schematic of the mixers. Biasing arrangements are not shown.

the downconverter and BBVGA, without LNA1. Table 4.4 summarizes some of these measurement results and compares them to nominal simulated performance. The agreement between measurement and simulation is quite good. Gains, noise figures, and linearity measurements are all within 2 dB of nominal simulations. The measured IIP3 of the downconverter was in the range of –5.7 dBm to –6.3 dBm for all 12 chips tested (using 10-MHz and 19.5-MHz downconverted tones). This compares to a nominal simulated IIP3 of –4.1 dBm and a worst-case simulated IIP3 of –7.7 dBm at –20±C, with resistor values 2.2 sigma high and bipolar transconductance 2.2 sigma low. Based on these downconverter IIP3 measurements, nominal parts will meet WCDMA cascaded IIP3 requirements with several dB of margin. However, there may not be quite enough margin to meet cascaded IIP3 requirements for extreme variations of process and temperature. The amplitude balance of the two quadrature channels was better than 0.1 dB, and the quadrature error was less than 1.5 degrees, for all 12 samples tested. Ten of the 12 samples had quadrature error of less than 0.8 degrees. A die photograph of the fabricated chip is shown in Fig. 4.40.

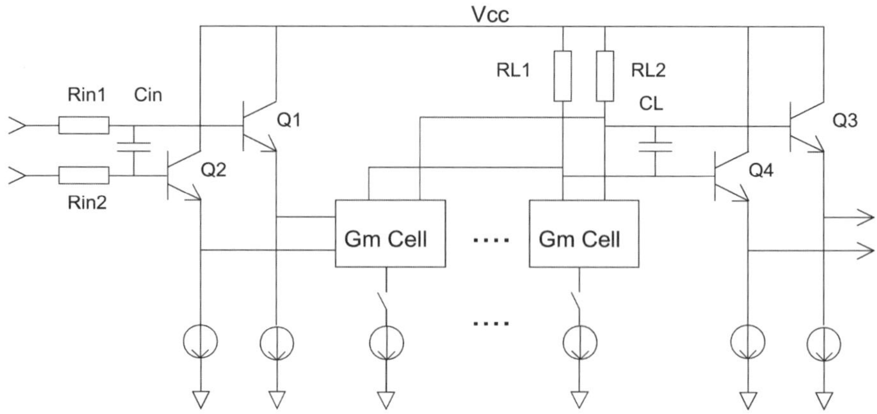

Figure 4.39 A simplified schematic of the BBVGA.

4.2.3 RF Packaging and Evaluation Board Design

The chip package is an essential and integrated part of an RF semiconductor product. Certain limitations on the package performance exist, which have to be accounted for during the design process. Some important package parameters that affect system performance of an RF IC design include isolation (crosstalk, signal leakage), insertion-loss (dielectric- and resistive loss), and impedance mismatch.

Table 4.4 A Summary of Measured and Simulated Performance Data for the Downconverter[a]

	Nominal Simulation	Measured Data
Downconverter gain	18.6 dB	17.4 dB
Downconverter NF	10.9 dB	10.5 dB
Downconverter IIP3	–4.1 dBm	–6.0 dBm
Downconverter IIP2	> +39 dBm	> +39 dBm
Downconverter ICP1dB	–16.9 dBm	–16.2 dBm
LNA2 input return loss (S_{11})	< –20 dB	< –18.6 dB
2140-MHz LO @ LNA2 input	< –82 dBm	–96 dBm
I channel vs. Q channel amplitude imbalance		< 0.1 dB
I channel vs. Q channel quadrature error		< 1.5 degrees
Downconverter current	9 mA	8.1 mA
Downconverter + BBVGA current	11 mA	10.5 mA

[a]The downconverter current includes the bandgap reference and biasing circuitry. All figures stated in dBm refer to the 50-Ω unbalanced input of LNA2.

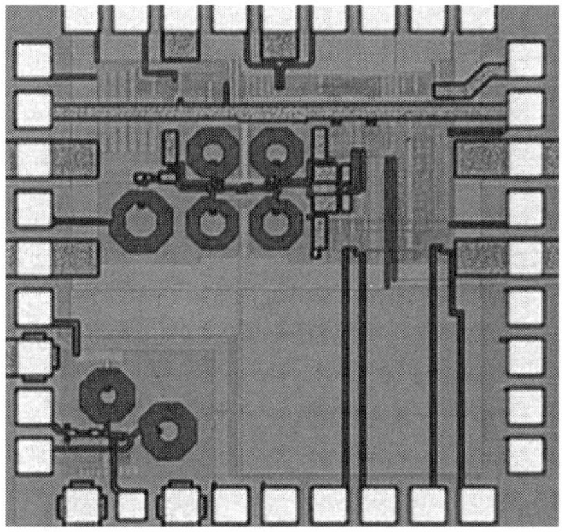

Figure 4.40 A die photograph of the chip. The die size (including the inner set of pads) is 2.07 mm × 2.07 mm.

The WCDMA receiver architecture, described earlier, imposes certain requirements on the signal isolation between package pins. The off-chip SAW filter, for instance, has to attain at minimum 25-dB, and ideally 30- to 40-dB or more, attenuation in the transmitter band. This can only be achieved if crosstalk between package pins is minimized. Another important system parameter is the insertion loss, which is affected by the dielectric loss of both the board and package material. To minimize this problem in measuring performance of the receiver IC, a low-loss dielectric board material with a loss tangent of less than 0.002 at about 2 GHz has been used for the evaluation-board design. Finally, impedance matching is critical to maintain desired receiver noise figure and gain performance. In particular, optimization of the LNA circuit design requires that the impedance of the package pins is taken into account. Variations of the bondwire length may lead to a large impedance mismatch, which in turn results in a high return loss for the LNA. Design goals for S_{11} and S_{22} (input and output return loss) of the LNA are better than −12 dB and −15 dB, respectively.

Evaluation Board Design Figure 4.41 shows a picture of the evaluation board that was used to measure the performance of the receiver chip. The board was fabricated from a low-dielectric and low-loss Teflon-based laminate on top of an FR-4 carrier. The board loss is 1/10 dB/in., and the size of the board is 1.1 in. × 2.2 in. The picture shows the connectors for LNA input for the quadrature baseband signals (I/Q) and for the local oscillator. The receiver IC is located close to the SAW filter and is shown without glob-top encapsulation.

4.2.4 System Performance Tests

In order to measure the system performance parameters of the RF chip, a receiver test-bed, including a software baseband processor, has been developed. A block di-

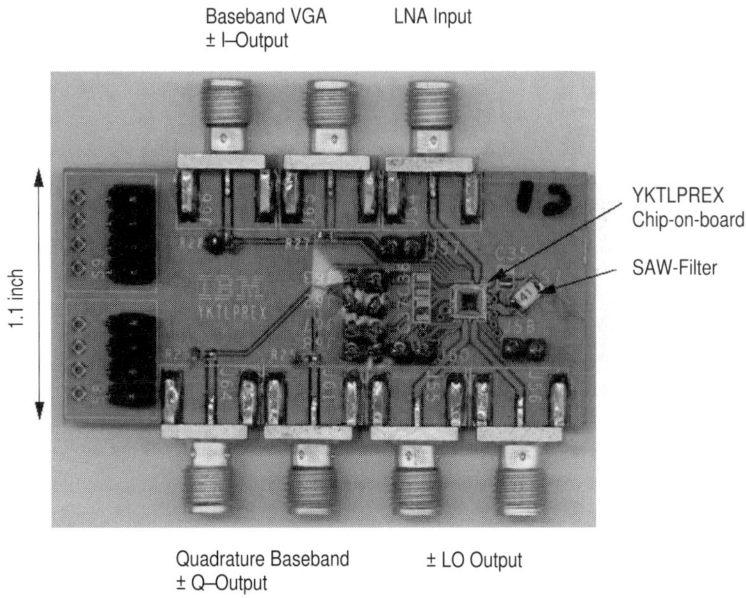

Figure 4.41 A photograph of the receiver chip evaluation board.

agram of the test-bed setup is shown in Fig. 4.42. The test bed provides controlled signals to the RF IC, which allows determination of its noise, linearity, compression characteristics, wideband transfer function, and BER sensitivity. A detailed description of the test-bed design follows.

Receiver Test-Bed By using two arbitrary waveform generators together with two vector signal generators, a desired WCDMA signal, and if necessary a WCDMA blocker, can be injected at the RF input of the device under test (DUT) (RF IC). A synthesized signal generator is used as LO. The RF chip is followed by a "golden" (i.e., negligible noise/linearity) analog baseband (ABB) consisting of differential to single-ended converters, some filtering (does not affect the signal bandwidth), and variable gain stages. The performance of the golden ABB (<15 nV/rtHz input noise, high-input impedance, IIP3 with 10-MHz tone spacing: >16 dBV, IIP2 with 15-MHz tone spacing: >46 dBV) enables the exact measurement of noise and linearity parameters of the RF IC without any added degradation from the baseband circuitry. The ABB is followed by a personal computer (PC) with a peripheral component interconnect (PCI) A/D converter card that captures the I and Q signals with 12 bits at a rate of 15.36 Ms/s. The high sampling rate allows accurate WCDMA matched channel filtering (no analog channel-select filter is yet included in the RxIC) to be performed in the PC. The software on this PC determines the de-spread channel symbol SNR as well as the BER (described more in detail in the next sec-

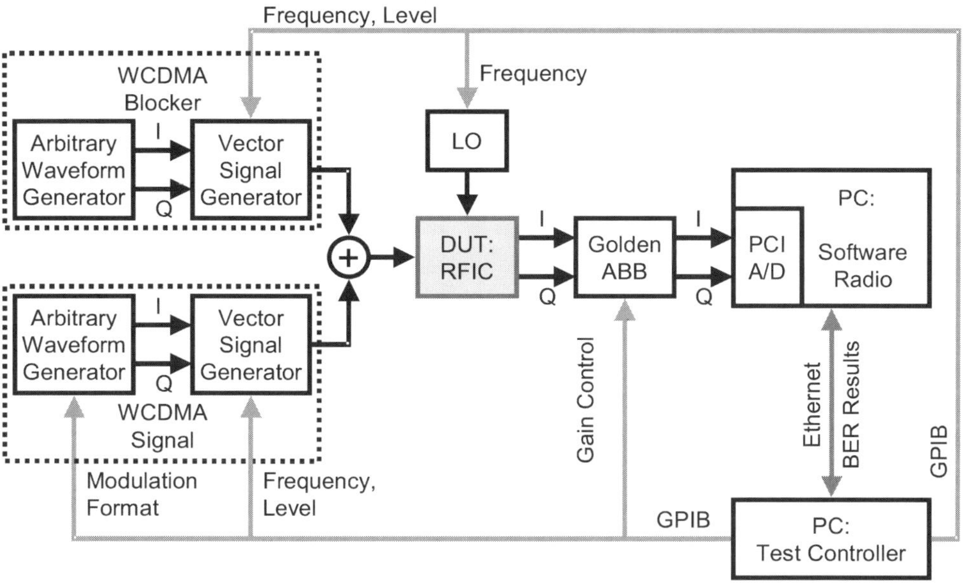

Figure 4.42 Test-bed measurement setup.

tion). All instruments are controlled from a second PC over a general-purpose interface bus (GPIB) Ethernet bus to enable automatic measurements.

By replacing the vector signal generators with ordinary CW sources and using a spectrum analyzer at the baseband output, all the analog parameters like gain, IIP2, IIP3, ICP1dB can be captured automatically over the interesting frequency range. The NF is measured by connecting a noise source at the RF input and finding the noise power in the WCDMA channel bandwidth at the RFIC baseband I/Q outputs using the software baseband processor. In this way an accurate WCDMA filter function (matched filter) is applied, resulting in a noise figure relevant to achievable system performance. Measuring two rms power values by switching the noise source on and off yields the NF. Important for all the measurements is a proper gain calibration in the 2-GHz, as well as in the baseband, frequency range.

Software Baseband Processor A PC with a 2-channel PCI A/D card is used to emulate the functionality of a WCDMA baseband processor to determine system BER and block-error rate (BLER) performance. A simplified diagram of the major functional blocks making up the software baseband processor is given in Fig. 4.43. After capturing a buffer of samples, the software first performs matched filtering and time-frequency synchronization to detect the presence of the signal, digitally remove any large frequency offset, and align the demodulator to the correct IQ sample, channel symbol, and frame boundaries. Once time-frequency synchronization is completed, the demodulator de-spreads the channel symbols and applies a chan-

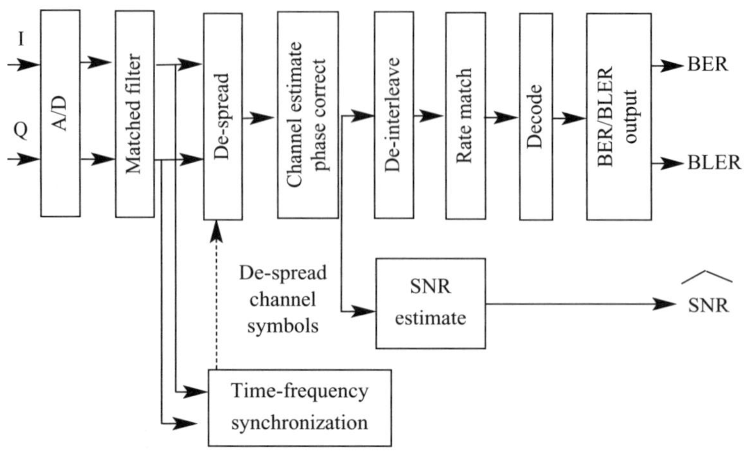

Figure 4.43 Software baseband processor.

nel-estimation algorithm to produce a phase-coherent estimate of the transmitted channel symbols. The channel-estimation algorithm used has an implementation loss of approximately 0.2 dB, which is nearly ideal. It is possible to achieve this very low implementation loss because the channel is static (constant gain/frequency offset) for the RFIC tests considered here. The recovered channel symbols are de-interleaved, rate-matched to account for any code-symbol puncturing or repetition used at the transmitter, and decoded using a convolutional decoder for 12.2-Kb/s data rate or turbo decoder for 64-Kb/s through 384-Kb/s data rates. The information bits output from the decoder are used for BER/BLER counting. The software also implements an algorithm that estimates the SNR of the de-spread channel symbols. In a static Gaussian channel, the SNR output can be used instead of BER to evaluate system performance because it converges to a stable value faster than bit-error rate. Fading channels were not implemented in the test setup because noise and linearity, the most important performance characteristics of the RFIC, can be found using only static channels. The software demodulator also implements an algorithm that estimates the SNR of the recovered de-spread channel symbols. In a static Gaussian channel, the SNR output can be used instead of BER to evaluate system performance since it converges to a stable value faster than BER. The channel symbol recovery algorithm used has an implementation loss of approximately 1 dB. This implementation loss can be reduced, but was left at 1 dB to conservatively estimate the achievable system BER performance.

The software baseband processor decodes any of the four formats used for WCDMA specification compliance testing (12.2 kbps for RF noise, linearity, and blocking tests, with 64 kbps, 144 kbps, and 384 kbps rates added for system performance tests in both static and fading channels [1]). In operation, eight consecutive slots of modulation are repeatedly transmitted by the arbitrary waveform generator at the transmitter. The data payload for each slot of data is created using a pseudo-random linear-feedback shift-register (LFSR) generator, which the baseband processor can automatically synchronize to for BER counting. The PCI A/D card used has a memory buffer that permits approximately 14 consecutive 20-ms 12.2-kbps slots of data to be stored in memory. The software baseband processor decodes these slots in nonreal time and accumulates averaged BER/BLER/SNR statistics that are sent to the test system controller PC upon completion of a measurement interval (typically 5000 blocks of data for a single BER point).

System Test Results Here the system test results of the RFIC on the evaluation board, which can be seen in Fig. 4.41, are presented. With a voltage supply of 3 V, the chip needs 14.5 mA or 10.5 mA with the LNA in high-gain or bypass mode, respectively. In Fig. 4.44 the measured frequency response of the RxIC in high gain mode (both LNA and BBVGA) is shown. Most likely due to a board layout or bondwire coupling problem, the Tx suppression (minimum: 24 dB at 1980 MHz) is not as good as expected from the SAW filter specifications. As can be seen later, the RxIC is still within 1 dB of meeting all specifications for out-of-band linearity and compression. The measured NF at all 12 WCDMA receive channels is plotted together with the frequency response in Fig. 4.45. The antiproportional behavior of both curves (i.e., higher noise figure when the gain is lower) can be seen very well as expected. Even in the worst-case channel for noise figure (channel with lowest

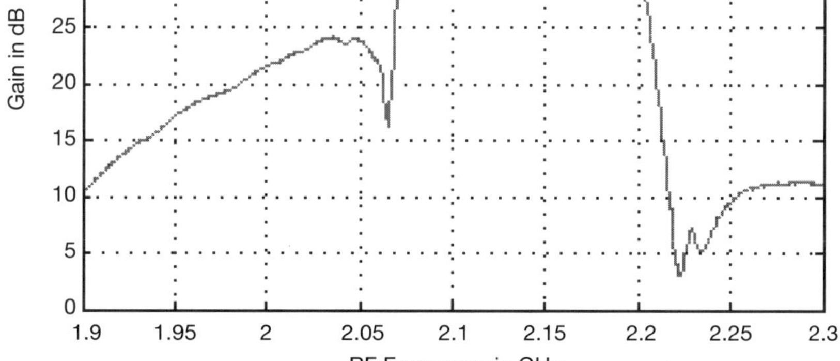

Figure 4.44 Measured frequency response (LNA: high gain; BBVGA: high gain).

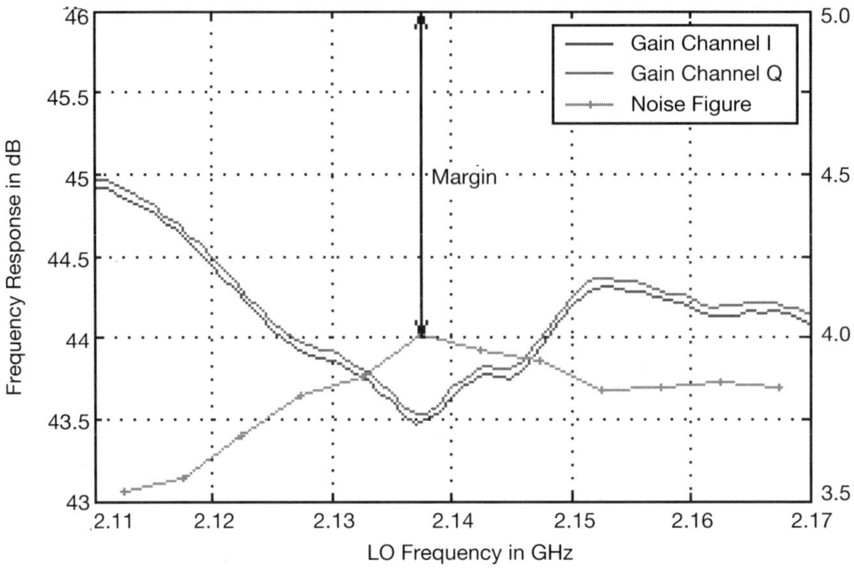

Figure 4.45 Measured frequency response and noise figure in the receive band (LNA: high gain; BBVGA: high gain).

gain), a margin of 1 dB is maintained below the required 5 dB. The maximum gain ripple in the receive band (2110 MHz–2170 MHz) is ±0.75 dB. Additionally, a constant offset of 0.06 dB between I and Q can be seen that matches exactly the on-wafer measurements of the downconverter given in Table 4.4.

Figure 4.46 shows the measured IIP3 in the receive band for 10-MHz tone spacing again for all 12 channels. The four curves represent the two cases of placing the two tones below or above the LO frequency for both baseband channel outputs (I and Q). This illustrates the importance of having an automated measurement setup that measures the system performance parameters at all possible settings. The worst of all obtained results is then compared with the specification. For in-band IIP3 there is still a 1-dB margin in the worst case. The same kind of exhaustive measurements have been performed for all the other important test parameters. For out-of-band IIP3, for example, all possible combinations of LO frequencies and CW tones anywhere in the Tx band and the blocker band (1, 2, 3 above and below Tx) at output I and Q have been used to determine the worst case value for each of the blocking bands. The worst results out of all measurements are given in Table 4.5 which shows that under nominal conditions (temperature and voltage) with the assumed operating conditions the presented chip is expected to either meet or exceed all system specifications except for two out-of-band IIP3 cases, which miss the design goal by about 1 dB due to the poor Tx suppression in the external SAW filter as discussed earlier. Finally, the extremely low LO leakage at the LNA input (over 20-dB margin compared to the system design goal) is pointed out because this is one of the major concerns of direct

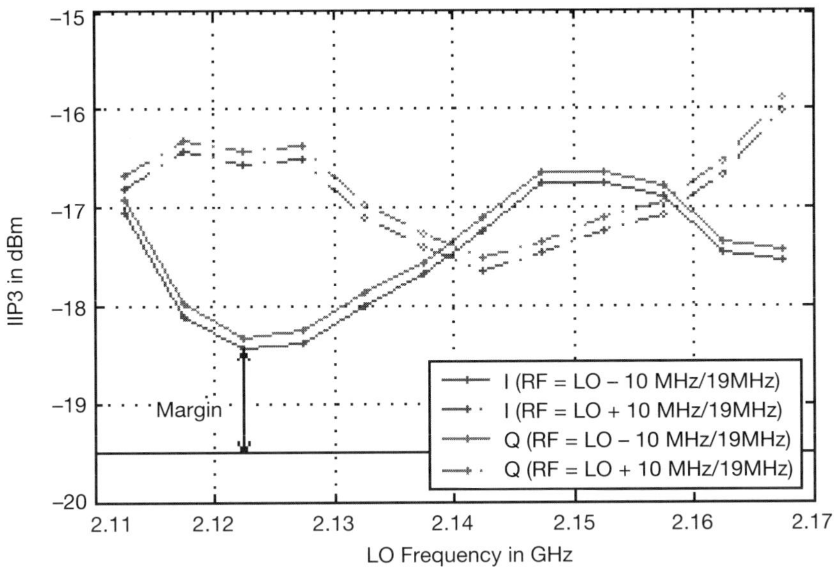

Figure 4.46 Measured IIP3 in the receive band (LNA: high gain; BBVGA: high gain).

Table 4.5 System Performance Results (measured worst-case figures)

Parameter	Unit	Goal	Measured	Description
Noise figure	dB	<5	4	Largest NF across 12 WCDMA channels
IIP3 (in-band)	dBm	>–19.5	–18.5	In-band IM3, 10-MHz tone spacing
IIP3 (band 1, Tx)	dBm	>–4.7	0	Band 1/Tx IM3
IIP3 (band 2, Tx)	dBm	>+1.3	0	Band 2/Tx IM3
IIP3 (band 3, Tx)	dBm	>+1.3	0.8	Band 3 above Tx/Tx IM3
IIP3 (Tx, band 3)	dBm	>+0.8	1.3	Tx/Band 3 below Tx IM3
IIP2 (in-band)	dBm	>+10	30	In-band IM2, tested at ±15-MHz offset, 1-MHz tone spacing
IIP2 (Tx band)	dBm	>+72	77	Tx band IM2, 1-MHz tone spacing
ICP1dB_Hi (in-band)	dBm	>–35	–33.2	In-band 1-dB compression, high LNA gain
ICP1dB_Lo (in-band)	dBm	>–17	–10	In-band 1-dB compression, low LNA gain. Low BBVGA gain
ICP1dB (Tx)	dBm	>–14	–15	Tx band compression, high LNA gain
LO-RF_Hi leakage	dBm	<–80	–101	LO leakage power at LNA input, high LNA gain
LO-RF_Lo leakage	dBm	<–60	–90	LO leakage power at LNA, low LNA gain
I/Q amplitude match	dB	<0.25	0.06	I/Q match at BBVGA out

conversion receiver designs. Further device characterization is required to determine performance over temperature extremes and process variations.

In order to prove the validity of the system considerations and simulations that lead to the specifications in Table 4.5, measurements with modulated WCDMA signals have to be performed. In Fig. 4.47 results of the sensitivity test according to [1] are given for the worst NF (see Fig. 4.45) at channel 6 (LO = 2137.5 MHz). For each measurement point 5000 blocks have been sent and received by the software baseband processor that was described earlier. A BER of 0.1% is reached at a de-spread channel symbol SNR of approximately 1 dB, which verifies that the algorithm in the software radio works properly. Compared to the required sensitivity of −121 dBm there is a margin of over 2 dB. Figure 4.48 shows the de-spread channel symbol SNR at the same LO frequency with a fixed signal power of −121 dBm at LNA input (−117-dBm sensitivity at antenna with 4-dB duplexer loss) dependent on the power of a Tx blocker at 1977.5 MHz. The Tx blocker has been placed at the upper end of the Tx band, because there it will experience the lowest attenuation (see Fig. 4.44), which is the worst case for this test. It can be seen that the receiver will be desensed by approximately 0.2 dB at sensitivity with a Tx blocker level of −22 dBm at the LNA input.

4.2.5 Summary and Conclusions

A direct-conversion receiver system design optimized for application in low-power WCDMA mobile systems has been described. Key features of the system design in-

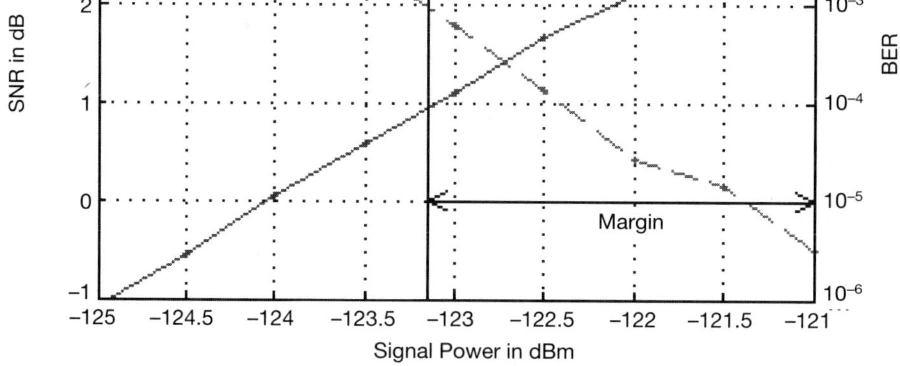

Figure 4.47 Measured sensitivity for LO = 2137.5 MHz (worst case: highest NF).

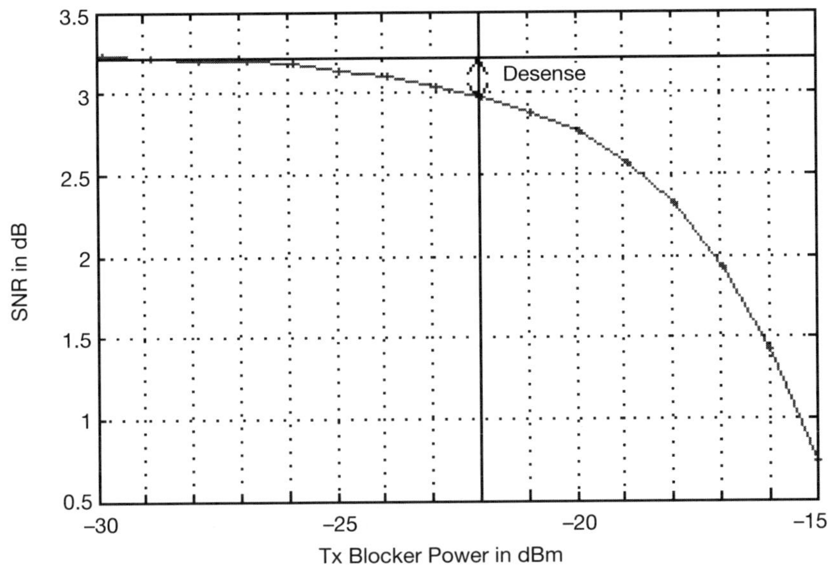

Figure 4.48 Measured Tx blocker desensing for a –121 dBm signal at LO = 2137.5 MHz and a WCDMA blocker at 1977.5 MHz (worst case: highest NF; lowest 1-dB compression point in transmit band).

clude a bypassable LNA, which enables system current savings of 4 mA in low-gain mode while maintaining LO leakage well below specification requirements and a single-ended input active downconverter (LNA2/Mixer/2× LO Quadrature Generator) which meets the noise/linearity requirements for the WCDMA application under expected nominal operating conditions while drawing only approximately 8.1 mA from a 3-V supply. To optimize the receiver design for the WCDMA application, assumptions are required for the maximum transmitter leakage signal that must be tolerated at the receiver LNA input and blocking characteristics of the receive band-pass filtering prior to the LNA. Improved isolation of the transmit leakage and out-of-band blocker signals relaxes linearity requirements on the receiver design, which can lead to lower-power designs. Measured performance results of the receiver IC show that it is expected to easily meet the second-order linearity and LO-RF isolation requirements, which are critical to the direct-conversion architecture. In-band third-order linearity margins may not be large enough to pass the needed system requirement under all operating conditions and will be improved in future design iterations. Out-of-band third-order linearity performance misses the desired worst-case targets by about 1 dB in some cases, but this is primarily due to poor transmitter leakage attenuation through the interstage band-pass RF filter on the evaluation board. The discussed receiver design is expected to meet all needed WCDMA RF performance requirements under nominal operating conditions when this external filter problem is resolved. A measured LNA-referred BER sensitivity of approximately –123 dBm received code channel power with a simultaneous transmitter interference signal of –22 dBm at the LNA input indicates the design is capable of working well in the demanding FDD WCDMA application.

4.2.6 References

1. Third-Generation Partnership Project (3GPP), "UE Radio Transmission and Reception (FDD)," *Technical Specification 25.101,* vol. 3.0.1, April 2000.
2. D. Y. C. Lie, J. Kennedy, D. Livezey, B. Yang, T. Robinson, N. Sornin, T. Beukema, L. E. Larson, A. Senior, J. Blonski, N. Swanberg, P. Pawlowski, D. Gonya, X. Yuan, J. Mecke, and H. Zamat, "A Direct-Conversion W-CDMA Front-End Receiver Chip with LO leakage of –105dBm in a 0.25um SiGe BiCMOS Technology," *Proceedings of the IEEE 2002 Radio Frequency Integrated Circuit Symposium, Digest of Papers,* Seattle, Washington, USA, June 2–4, 2002, pp. 31–34.
3. O. K. Jensen, T. E. Kolding, C. R. Iversen, S. Laursen, R. V. Reynisson, J. H. Mikkelser, E. Pederson, M. B. Jenner, T. Larsen, "RF Receiver Requirements for 3G WCDMA Mobile Equipment," *Microwave J.,* vol. 43, no. 2, pp. 22–46, February 2000.
4. A. Parssinen, J. Jussila, J. Ryynanen, L. Sumanen, K. A. I. Haloren, "A 2-GHz Wide-Band Direct Conversion Receiver for WCDMA Applications," *IEEE J. Solid-State Circuits,* vol. 34, no. 12, pp. 1893–1903, December 1999.
5. L. W. Couch, *Digital and Analog Communication Systems,* p. 149, Macmillan, 1983.
6. H. Fukui, "The Noise Performance of Microwave Transistors," *IEEE Trans. Electron Devices,* vol. ED-13, pp. 329–341, March 1966.
7. S. P. Voinigescu et al., "A Scalable High-Frequency Noise Model for Bipolar Transistors with Application to Optimal Transistor Sizing for Low-Noise Amplifier Design," *IEEE. J. Solid-State Circuits,* vol. 32, pp. 1430–1439, September 1997.
8. K. L. Fong and R. G. Meyer, "High-Frequency Nonlinearity Analysis of Common-Emitter and Differential-Pair Transconductance Stages," *IEEE J. Solid-State Circuits,* vol. 33, pp. 548–555, April 1998.
9. B. Gilbert, "The Multi-tanh Principle: A Tutorial Overview," *IEEE J. Solid State Circuits,* vol. 33, no. 1, p. 2, January 1998.

4.3 WIRELESS DESIGN: ERICSSON POWER-AMPLIFIER DESIGN

Power amplifiers (PA) are a core component in the high-growth wireless communications industry, rapidly evolving in both architecture and communication protocols [1] to satisfy ever-increasing bandwidth requirements. Bipolar transistors are the critical PA building block due to their power-handling capabilities at high frequencies. GaAs HBTs have dominated such applications; however, bandgap-engineered SiGe HBTs are an emerging alternative due to their ability to provide high integration and to reduce cost [2]. Further, SiGe has several advantages over GaAs for PA applications: (i) A heat-conductive substrate that drives down the chip area and contributes to chip robustness;(ii) temperature-insensitive current-gain behavior providing immunity to thermal runaway; and (iii) high current density.

The challenge faced by SiGe-based PA technologies is providing sufficient high-voltage immunity without compromising PA performance. PA devices must withstand high-voltage excursions and large current densities to survive in wireless environments. Involved/complex circuit solutions to this problem add cost and degrade performance. Thus, for SiGe to be a compelling alternative for wireless PAs, SiGe HBT ruggedness needs to be enhanced without degrading RF performance.

Noting that technology is a key factor to successful PA design, in this chapter we demonstrate for the first time a fully manufacturable 0.5-μm SiGe BiCMOS technology where a very nonuniform collector design has provided a significant improvement in HBT ruggedness while maintaining a performance suitable for PA development across several different wireless standards.

4.3.1 0.5-μm SiGe BiCMOS PA Technology

The 0.5-μm 3.3-V BiCMOS technology (which was described in Section 1.1) contains two SiGe HBTs, a high-performance HBT with f_T/f_{max} of 50/90 GHz and a high breakdown HBT with f_T/f_{max} of 25/80 GHz. Several kinds of resistors complement a high-Q varactor, a Schottky barrier diode, and silicon and metal-to-metal capacitors.

For circuit applications requiring both ruggedness and performance, the high-breakdown SiGe HBT has been designed to achieve a BV_{CBO} greater than 20 V and minimum BV_{CEO} of 6.5 V. This technology follows a "base-equals-gate" integration scheme [3], with a buried subcollector and deep and shallow trench isolation. AC performance degradation resulting from the device modifications to increase breakdown voltage was avoided by the addition of a precisely placed collector implant. This implant acts to suppress the Kirk effect without contributing significantly to the carrier heating that leads to device breakdown. Figure 4.49 shows the experi-

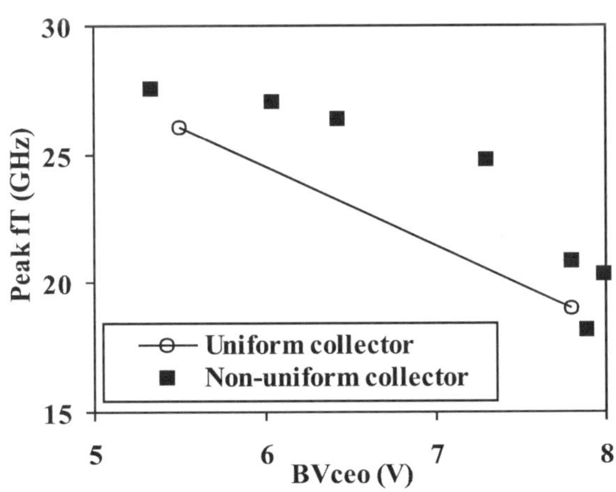

Figure 4.49 Leverage of nonuniform collector profiles for SiGe HBT BV_{CEO}-f_T trade-off, as observed from single transistors.

mentally observed performance/ruggedness trade-off provided by this collector implant approach. Under optimal implant conditions, f_T for the high-breakdown device peaks at a collector current density of 0.4 mA/μm^2, enabling significant device scaling. As an open-base situation will rarely be witnessed in wireless PA applications [1], the BV_{CER} measurements shown in Fig. 4.50 are expected to reflect the critical ruggedness characteristics of the high-breakdown HBT.

Figure 4.51 compares the characteristics of this technology's high-breakdown HBT to those of other published devices [1,4,5]. The ruggedness/speed characteristics of this device are seen to depart favorably from industry trends for both SiGe and Si transistors and approach those of GaAs. Figure 4.52 plots the f_T/f_{max} versus I_c characteristics for the high-breakdown transistor, indicating the large RF-performance range spanned by this technology.

4.3.2 HBT Safe Operating Area

Apart from straightforward junction breakdown, other phenomena play a role in limiting the region of operation (ROO) of an HBT as a PA, such as second breakdown, HBT self-heating and electromigration. We have explored comprehensively the ROO of the high-breakdown NPN device by recording the emitter current as a function of base–emitter forward bias for different values of V_{CB} and looking for avalanche conditions. We define ROO limits for the device by the V_{CB} and J_E at which avalanche commences. The condition at which J_E becomes independent of V_{BE}, denoted in Fig. 4.53 as "J_E at breakover," is the device operating limit due to HBT runaway. Operation of the device up to these limits is nondestructive and completely reversible. As shown in Fig. 4.53, the preferred operating point at peak f_T for the HBT falls well within the device ROO limits, and so PA applications will not be limited by the ROO constraints.

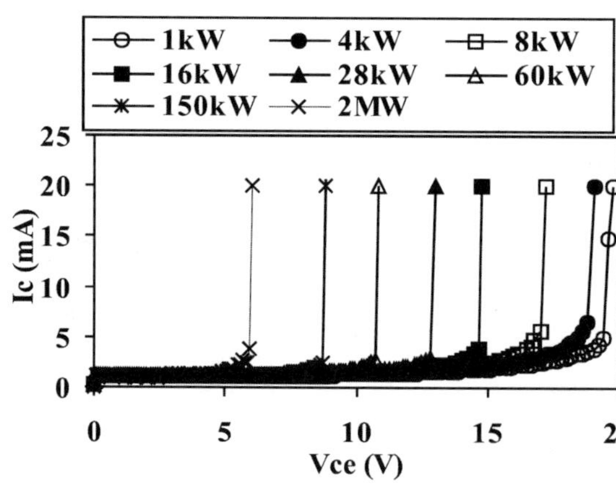

Figure 4.50 Effect of external base resistance on high-breakdown HBT breakdown. These characteristics indicate that actual device ruggedness is significantly greater than suggested by BV_{CEO}. Device size is 2 × 0.5 μm × 20 μm.

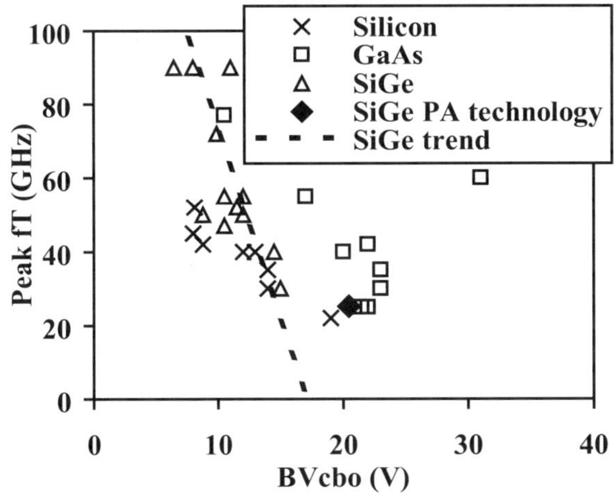

Figure 4.51 Comparison of published technology $f_T - BV_{\mathrm{CBO}}$ trade-off. Characteristics of SiGe technology discussed in this section appear to be significantly more favorable than expected from industry trends. (Based on data from Jos [1], Deixler et al. [4], and Schuegraf et al. [5]). Omitting GaAs.

4.3.3 Circuit Design and Performance

The current PA design approaches are targeting the Global System for Mobile Communication (GSM) and General Packet Radio Service (GPRS) standards, both using the Gaussian minimum shift-keying (GMSK) modulation scheme and the WCDMA standard using hybrid phase-shift keying (HPSK) modulation. For

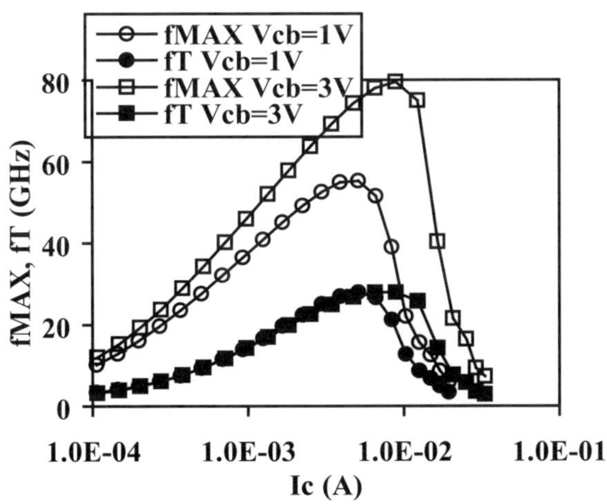

Figure 4.52 f_T and f_{MAX} characteristics of high-breakdown SiGe HBT. Device size is 2 × 0.5 μm × 20 μm. f_{MAX} defined by U extrapolation with –20 dB/dec slope.

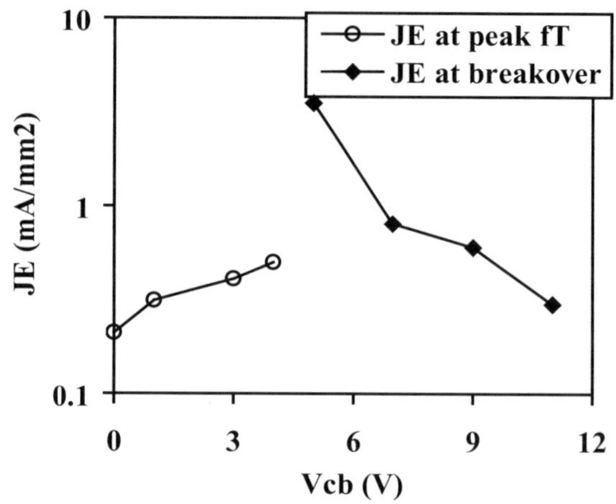

Figure 4.53 Region of Operation (ROO) measurements for the 0.5 μm × 20 μm high-breakdown HBT. "J_E at breakover," is the device operating limit due to HBT runaway. The preferred operating point for the device, at or below peak f_T, is observed to be well within the ROO limit.

GMSK standards, the signal envelope is constant, and as a consequence linearity of the PA is not an issue.

GMSK Modulation Type PA Design The quad-band GMSK PA is serving the GSM bands at 850 MHz, 900 MHz, and 1800 MHz together with the American 1900-MHz Personal Communication Service (PCS) GSM band. The main figures of merits are output power, efficiency, and robustness. The output-power requirements for GSM and GPRS are 33 dBm from the antenna. In order to provide sufficient margin for losses in antenna near components, the output from the PA should exceed 34.5 dBm. Such high output-power imposes severe demands on both the voltage immunity of the output device and the thermal stability of the entire design. For the GPRS standard, multiple time-slot operation further increases the requirement on the thermal performance and robustness of the power amplifier. The current PA design has been demonstrated to operate at a 4-time-slot time-division scheme without performance degradation.

The circuit topology chosen is a single-ended two- or three-stage solution with built-in biasing circuitry and power control. Interstage matching is performed both on- and off-chip. The off-chip components are either designed directly into a low-temperature cured ceramic (LTCC) substrate or mounted on top of the substrate surface. The die is mounted onto the substrate surface by means of C4 flip-chip technology, thereby providing extremely low parasitic inductance, good thermal conductivity, together with a small footprint. No external components outside of the module are needed for matching or decoupling. The output power and efficiency characteristics of the PA for 900 MHz are shown in Figure 4.54. Note the high PAE

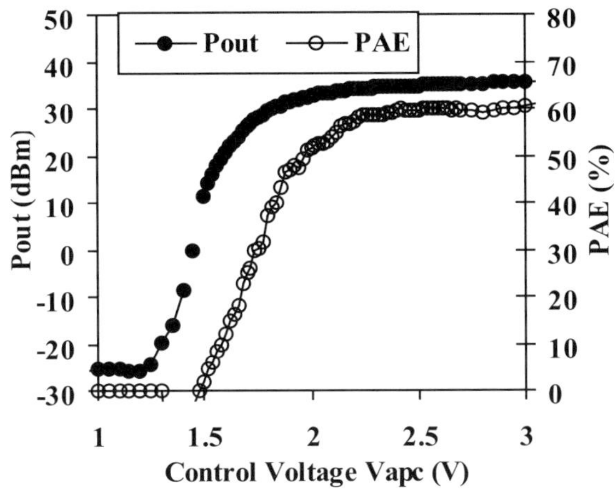

Figure 4.54 GMSK PA characteristics showing output power and power-added efficiency (PAE) at 900 MHz and P_{in} = 8.5 dBm.

at higher output levels. The slope of the power curve as a function of control voltage, as can be seen in Fig. 4.54, determines the controllability of the PA. The slope is close to constant, thus providing a transfer function making the power control easier to implement. The constant current gain as a function of temperature, together with the high current density of the PA SiGe HBT technology, makes it possible to make use of smaller output devices, thereby minimizing parasitic effects. The low parasitics increase the inherent gain of the devices, which together with an efficient biasing scheme can be utilized for achieving high efficiency.

Concerns have been raised on using Si technologies in PA applications for wireless standards requiring high outputs at high efficiency and the changing load characteristic of the GSM/GPRS standards. The high-breakdown voltage of the SiGe PA technology, $BV_{CBO} \geqq = 20$ V, constant current gain as a function of temperature and an Si substrate thermal conductivity significantly exceeding that of GaAs, makes this technology extremely well suited for applications requiring both ruggedness and high output power, such as the multiple-slot-operation GPRS standard. A fully functional test was performed into a load presenting a voltage standing-wave ratio (VSWR) of 50:1 at the SiGe PA output through all phase angles at a 5-V supply voltage. The DC current was kept at a level that would give 35-dBm output power in a 50-Ω load at 5-V supply voltage.

HPSK Modulation Type PA Design The HPSK modulation scheme, together with the filtering used in the WCDMA standard, puts specific requirements on the PA performance. The varying signal envelope makes the linearity of the transceiver chain and the PA essential. The linearity requirement is formulated as adjacent-channel leakage ratio (ACLR1) describing how much power is leaking into the next channel as a result of nonlinearities in the transmitter chain. The power leaking into the next nearer-channel is termed ACLR2. The 3GPP standard (25.100.v5.3.0 June 2002) requires the linearity to be ACLR1 < –33 dBc and ACLR2 < –43 dBc.

The circuit topology is a two-stage solution with built-in biasing circuitry and linearization. Interstage matching is performed on chip. The off-chip components are either designed directly into a LTCC substrate or mounted on top of the substrate surface. The die is mounted onto the substrate surface by means of C4 flip-chip technology, providing extremely low parasitic inductance and good thermal conductivity, together with a small footprint. No external components outside of the module are needed for matching or decoupling. Simulated output power and PAE characteristics of the PA at 1.95 GHz are shown in Fig. 4.55. Measured WCDMA output power and linearity data from an early prototype module are shown in Fig. 4.56. Note that the adjacent-channel power rejection (ACPR) characteristics exceed the WCDMA requirement all the way up to the 1 dB compression point.

The SiGe PA technology demonstrates high early voltages even at lower collector–emitter voltages, and pushes the onset of avalanche breakdown toward the higher voltage regime, thus increasing the linear region of operation for the output transistor. Increasing the linear range of operation can be explored for the WCDMA PA application.

4.3.4 Summary

A PA design has been demonstrated for GSM, GPRS, and WCDMA wireless standards using the SiGe 0.5-μm BiCMOS technology specifically tailored to meet the divergent requirements of high VSWR robustness at high output power and high linearity. The GMSK PA shown here exceeds the 10:1 VSWR requirement at 35 dBm with a nominal efficiency of 57%. For the WCDMA PA, an ACLR1 < –33

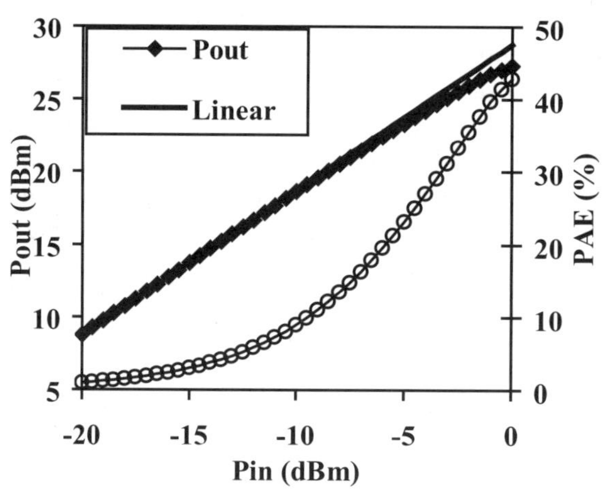

Figure 4.55 Simulated WCDMA PA characteristic showing output power and power added efficiency (PAE) at 1.95 GHz.

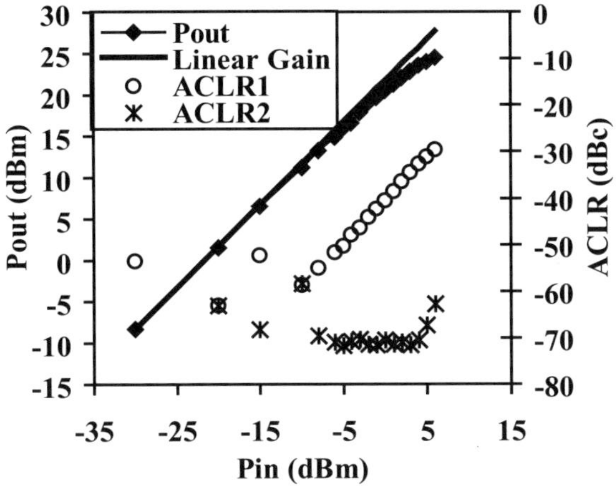

Figure 4.56 Measured output power and linearity data from WCDMA prototype hardware. ACLR1 < –32 dBc and ACLR2 < –48 dBc at 1-dB compression point.

dBc and ACLR2 < –43 dBc is achieved up to and even beyond the 1-dB compression point. The SiGe HBT was optimized to simultaneously provide ruggedness and speed and is compatible with the base 0.5-μm CMOS. The favorable thermal properties, lower cost of wafer processing, and the higher integration capabilities demonstrated by these SiGe PAs make them a compelling choice for wireless applications.

4.3.5 References

1. R. Jos, "Technology Developments Driving an Evolution of Cellular Phone Power Amplifiers to Integrated RF Front-End Modules," *IEEE J. Solid-State Cir.,* vol. 36, pp. 1382–1389, September 2001.
2. D. L. Harame, D.C. Ahlgren, D. D. Coolbaugh, J. S. Dunn, G. G. Freeman, J. D. Gillis, R. A. Groves, G. N. Henderson, R. A. Johnson, A. J. Joseph, S. Subbanna, A. M. Victor, K. M. Watson, C. S. Webster, and P. J. Zampardi, "Current Status and Future Trends of SiGe BiCMOS Technology," *IEEE Trans. Electron Devices,* vol. 48, pp. 2575–2594, November 2001.
3. D. L. Harame, J. H. Comfort, J. D. Cressler, E. F. Crabbe, J. Y.-C. Sun, B. S. Meyerson, and T. Tice, "Si/SiGe Epitaxial Base Transistors II: Process Integration and Analog Applications," *IEEE Trans. Electron Devices,* vol. 42, pp. 469–482, March 1995.
4. P. Deixler, H. G. A. Huizing, J. J. T. M. Donkers, J. H. Klootwijk, D. Hartskeerl, W. B. De Boer, R. J. Havens, R. Van der Toorn, J. C. J. Paasschens, W. J. Kloosterman, J. G. M. Van Berkum, D. Terpstra, and J. W. Slotboom, "Explorations for High Performance SiGe-Heterojunction Bipolar Transistor Integration," *Proceedings of IEEE Bipolar/BiCMOS Circuits and Technology Meeting 2001,* pp. 30–33, 2001.
5. K. Schuegraf, M. Racanelli, A. Kalburge, B. Shen, C. Hu, D. Chapek, D. Howard, D. Quon, D. Feiler, D. Dornisch, G. U'Ren, H. Abdul-Ridha, J. Zheng, J. Zhang, and K.

Bell, "0.18 μm SiGe BiCMOS Technology for Wireless and 40 Gb/s Communication Products," *Proceedings of IEEE Bipolar/BiCMOS Circuits and Technology Meeting 2001,* pp. 147–150, 2001.

4.4 MEMORY DESIGN: A 32-WORD BY 32-BIT THREE-PORT BIPOLAR REGISTER FILE

The purpose of this work (performed by researchers at Rensselaer Polytechnic Institute (RPI)) was to design a memory circuit that could operate in a high-speed digital system such as a microprocessor. The IBM SiGe 5HP technology was chosen because it offered a bipolar device with significantly faster switching speeds (an f_T of 48 GHz and an f_{MAX} of 69 GHz) than were available with CMOS and Si bipolar technologies [1], when this work started. In fact, more recently, an SiGe HBT technology has been developed with an f_T of 120 GHz and an f_{MAX} of 100 GHz for the HBT [2], discussed in Chapter 1. The device yields for these technologies are still sufficient to design complex digital and mixed-signal systems, however. This is especially true given the availability of CMOS devices within these technologies. Multiple ports were considered essential for the register file to read multiple data words and write data simultaneously. Two read ports and one write port were chosen, since even the simplest pipelined microprocessor must perform an arithmetic logic unit (ALU) operation on two data words and write the results of a previous operation within a clock cycle.

4.4.1 Register File Overview

The register file has a size of 32 words by 32 bits that can be accessed simultaneously by two read ports and a single write port. A block diagram for the register file, shown in Fig. 4.57, illustrates that the register file contains seven distinct types of functional blocks. These are the memory-cell array, the read address decoders and word line drivers, the write address decoder and word line drivers, the bit line drivers, the sense amplifiers, the output latches, and the comparators. The memory-cell array is responsible for storing the 1024 bits of data. It is arranged in a grid of 32 rows by 32 columns of memory cells. Each read address decoder is responsible for decoding a 5-bit address to determine which of the 32 rows is selected for each read operation. The 64 read word line drivers drive the read word lines accordingly. The write-enable circuit determines whether or not a write operation is to take place. If so, the 5-bit write address is decoded to select the memory cell row in which new data will be written. The 32 write word line drivers are responsible for driving the write word lines accordingly. The 32-bit line drivers provide input data to the memory cells during write operations using the write bit lines. The 64 sense amplifiers detect the values on the two sets of read bit lines and translate them into differential voltages corresponding to two output data words. The comparators are used to detect whether or not a write operation is occurring for a row that is being read by either read port. The 64 output latches are responsible for selecting and stor-

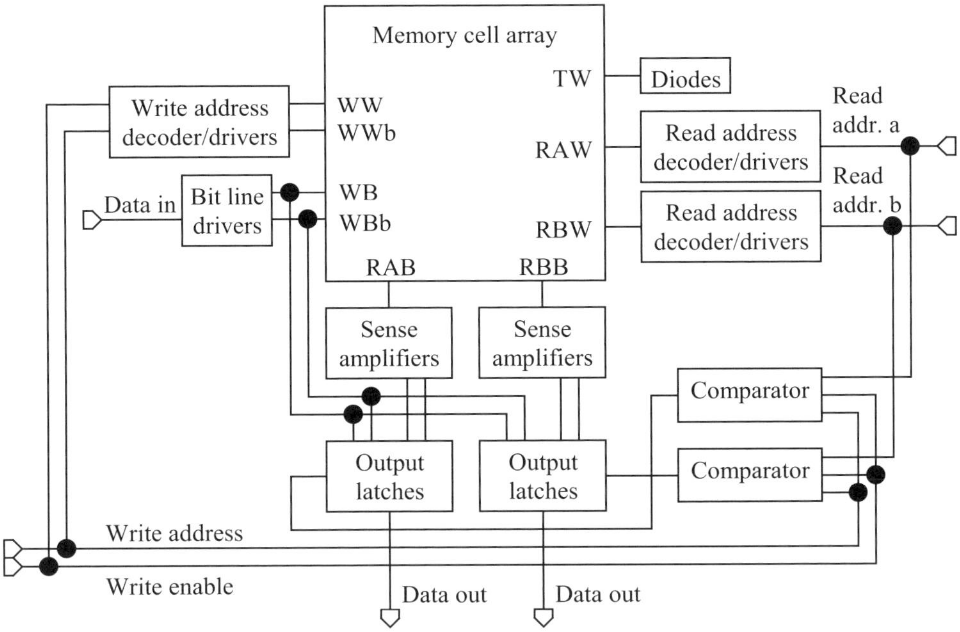

Figure 4.57 Register-file block diagram.

ing the appropriate data from either the sense amplifiers or the bit line drivers based on the results of the comparators.

4.4.2 Memory-Cell Design

The memory cell, shown in Fig. 4.58, consists of four current switches and two collector resistors [3]. During a write operation, 17 mA of current flows through WW, while no current flows through WWb. About 0.54 mA of the current flowing through WW is directed through either Q_D or Q_{Db}, depending on the differential voltage applied between WB and WBb by the corresponding bit line driver. Each bit line driver is simply an ECL buffer that drives nodes WB and WBb for a particular column of memory cells, producing a differential voltage across these lines that corresponds to the value to be written into one of the memory cells in the column. The current through Q_D or Q_{Db} produces a differential voltage between MC and MCb. The magnitude of this voltage is determined primarily by the size of the collector resistors. At the end of the write operation, the current flowing through WW is redirected through WWb. About 0.54 mA of current flows through either Q_F or Q_{Fb}, depending on whether MCb or MC is at a higher potential as a result of the write operation. The positive feedback configuration of the devices Q_F or Q_{Fb} maintains the differential voltage between MC and MCb for as long as current is flowing through WWb. In this way, the memory cell stores data.

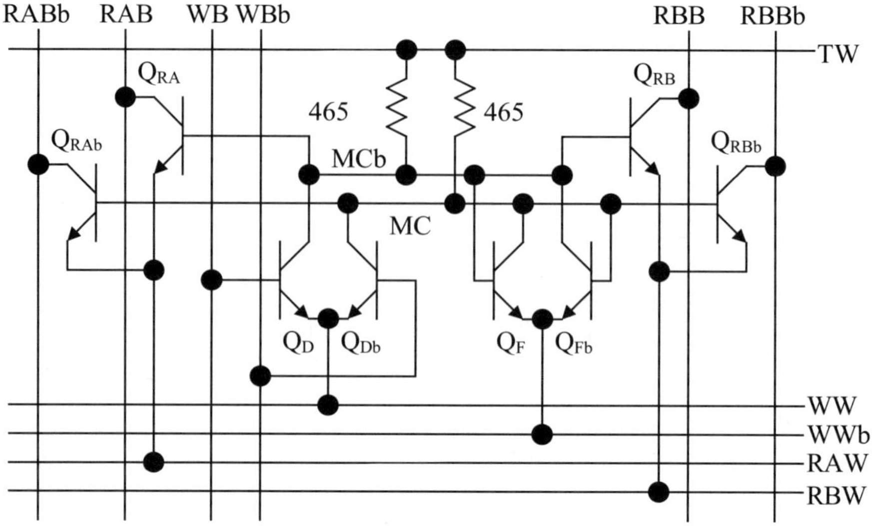

Figure 4.58 Three-port memory-cell schematics.

When a memory-cell row is selected for a read operation through read port A, 17 mA of current flows through RAW. For a particular memory cell, about 0.54 mA of this current either flows through Q_{RA} or Q_{RAb}, depending on whether MCb or MC is at a higher potential. This in turn causes most of this current to flow through either RAB or RABb. The value stored in the memory cell is determined by the sense amplifier connected to RAB and RABb, based on the current flowing through these two bit lines. The value stored in the memory cell can be read simultaneously through read port B using similar methods.

4.4.3 Read Address-Decoder Design

The read address decoder, similar to the design used in a BiCMOS register file [4], is composed of two stages, as shown in Fig. 4.59. In the first stage, wired-OR techniques are used to separately decode the lower two bits and the upper three bits of the address. The 2-bit decoder uses two CML buffers to drive four wired-OR lines. The CML buffers drive the wired-OR lines for the 2-bit decoder in a manner such that two wired-OR lines receive an inverted version and two wired-OR lines receive a noninverted version of each address bit, as Fig. 4.59 illustrates. Since exactly one wired-OR line receives a low voltage on both input terminals, it is the only one of the four wired-OR lines to produce a low output voltage. The wired-OR line that produces a low output voltage indicates the decoded value of the 2-bit decoder.

The 3-bit decoder decodes the upper three bits of an address in a manner similar to that of the stage-one 2-bit decoder. However, in this case, eight wired-OR lines are required. For each wired-OR line, three CML buffers drive the three emitter-coupled input devices. Since each CML buffer drives four wired-OR lines from both the inverting and noninverting output nodes, the pattern is such that for a given 3-bit address pattern, the wired-OR line corresponding to the encoded pattern is driven low, while all seven other wired-OR lines are driven high.

The second stage of each read address decoder consists of 32 single-ended 2-input ECL NOR gates. This well-known NOR implementation is shown in Fig. 4.59 [5–7]. The emitter–follower output stage of this gate contains an additional diode to produce an output signal that properly biases the corresponding read word line driver. One input of each NOR gate is connected to one of the set of four wired-OR lines from the stage-one 2-bit decoder, while the other input is connected to one of the set of eight wired-OR lines from the stage one 3-bit decoder, as illustrated in Fig. 4.59. There are exactly 32 ways to choose unique pairs from the two sets of wired-OR lines, providing a unique pair for each NOR gate. Since only one of the wired-OR lines in each set is low for any given address, only one ECL NOR gate will receive two low signals as input. This NOR gate produces a high output signal, indicating that the corresponding row is selected, while all the other NOR gates produce low output signals.

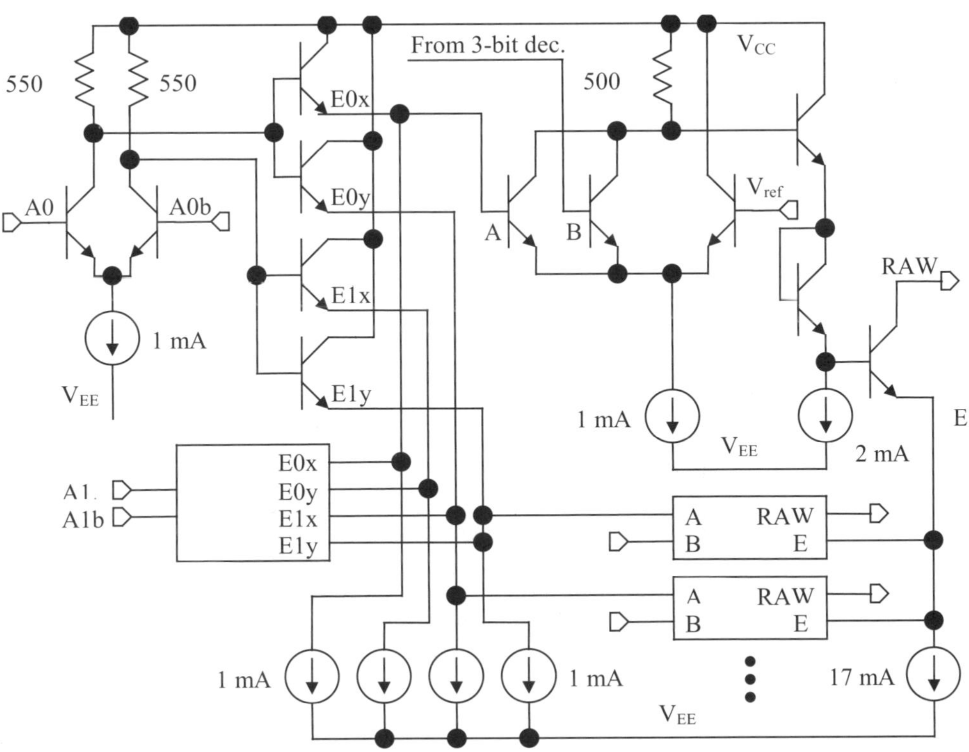

Figure 4.59 A portion of the read address-decoder schematics. Included is the stage-one 2-bit decoder, a portion of the stage two decoder, and some of the read word line drivers.

4.4.4 Read Word Line-Driver Design

Each read word line driver consists of a single large device capable of handling the current necessary to drive a read port in every memory cell for a single row. The base of each device is driven by the corresponding NOR gate from the second stage of the decoder. The emitters of all 32 read word line driver devices for a given port are connected to a single current source, as illustrated in Fig. 4.59. Since only one of the NOR gates is producing a high output signal, V_{BE} of the read word driver device connected to this gate is significantly larger than V_{BE} of the other read word line-driver devices. Therefore, nearly all the current sunk by the current source flows through the selected read word line-driver device, while the other read word line-driver devices are cut off. Since the collector of each read word line driver device is connected to the appropriate read word line of the corresponding row of memory cells, current only flows through the selected read word line of each read port.

4.4.5 Write Address-Decoder and Word Line-Driver Design

The write address decoder, illustrated in Fig. 4.60, operates in a manner similar to that of the read address decoder with two notable exceptions. The first is the addition of a write-enable circuit, which prevents the decoder from selecting any of the rows of memory cells for a write operation unless the write-enable signal is asserted. This circuit, shown in Fig. 4.60, consists of a CML buffer with differential inputs and a single-ended output that drives four additional input devices to the set of four wired-OR lines used in the stage one 2-bit decoder. Because the CML buffer in the write-enable circuit drives the same value onto all four wired-OR lines, when the buffer input is high, indicating a write operation is enabled, a low voltage is placed on one of the inputs to each wired-OR line. The low voltages do not alter the states of the four wired-OR lines, allowing the first stage of the decoder to operate in the same manner as in the read address decoder. However, when the CML buffer input is low, indicating no write operation is to take place, the write-enable circuit places a high voltage on one of the inputs to each of the four wired-OR lines. In this case, all four wired-OR lines are forced high, which ensures that each of the NOR gates in the second stage of the decoder will receive a high voltage on at least one of its input terminals, and hence, produce a low output signal. Therefore, when no write operation is enabled, each row of memory cells is driven in a manner that preserves the values currently stored in the memory cells.

The second way in which the write address decoder differs from the read address decoder is in the modification to the ECL NOR gates to accommodate the write word line drivers. Each ECL NOR gate for the write address decoder provides a differential output, which is necessary to drive a write word line driver. Schematics for the write address-decoder ECL NOR gate are shown in Fig. 4.60. The write word line driver, also shown in Fig. 4.60, consists of a pair of emitter-coupled devices with emitters attached to a current source. When the NOR gate output is low, no write operation is occurring. Therefore, in this case most of the current flows through a device connected to WWb. During a write operation for a particular row,

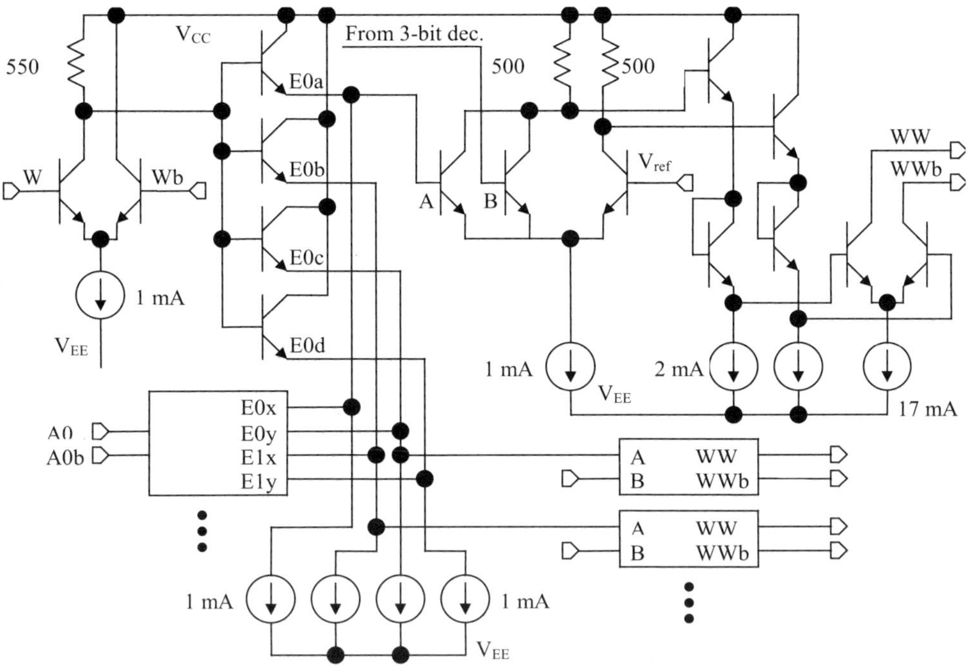

Figure 4.60 A portion of the write address-decoder schematics. Included is the stage-one 2-bit decoder, a portion of the stage two decoder, and some of the write word line drivers.

the NOR gate output becomes high, causing most of the current to flow through the device connected to WW until the NOR gate output becomes low again.

4.4.6 Sense-Amplifier Design

An active sense amplifier is used to convert the differential current on a pair of read bit lines into an output differential voltage. Of course, this could have been accomplished simply by using a pair of pull-up resistors on the bit lines. However, since the bit lines must traverse the entire memory cell array and are connected to a number of devices, they have a large parasitic capacitance. This means that, using the pull-up resistors, a significant lag in the change in differential voltage across the bit lines would be observed as a result of a change in the differential bit-line currents. For this reason, a common-base sense amplifier, shown in Fig. 4.61, is often used [8]. In this circuit, Q_{RB} and Q_{RBb} provide the majority of the current that flows through the bit lines. Most of this current also flows through the collector resistors, producing a differential voltage across O and Ob that is proportional to the difference in current flowing on the two read bit lines. In this way, the bit lines are isolated from the differential output voltage. Therefore, the capacitance on the sense-amplifier output node is significantly reduced with respect to the capacitance on the

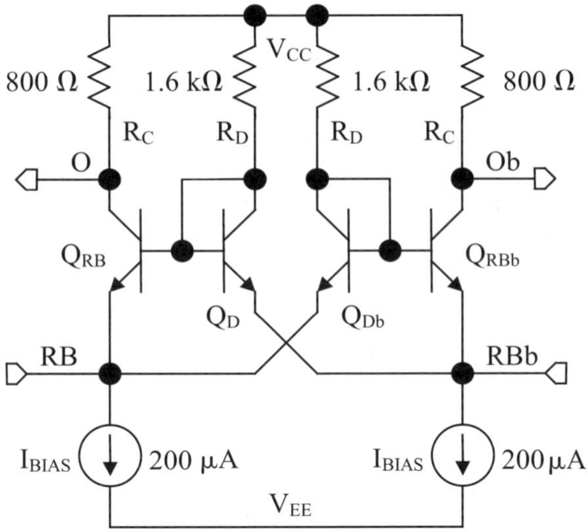

Figure 4.61 Sense-amplifier schematics.

read bit lines. The sense amplifier current sources provide a small amount of current to allow Q_{RB} and Q_{RBb} to remain forward biased regardless of which bit line is conducting current. Therefore, the V_{BE} of the device conducting current from one of the bit lines is not much larger than the V_{BE} of the device that is only conducting current from one of the bias current sources. This means that only a small voltage shift on the bit lines is required to significantly shift the operating points of Q_{RB} and Q_{RBb}. These devices, in turn, can rapidly produce a change in the output differential voltage due to the high voltage gain of the sense amplifier and lower capacitance on nodes O and Ob with respect to the capacitance on the bit lines.

The sense amplifier uses a cross-coupled bias network to provide reference voltages for the base nodes of Q_{RB} and Q_{RBb}. This provides better rejection for common-mode noise on the bit lines. If a fixed reference voltage is supplied to the base nodes of Q_{RB} and Q_{RBb} instead, a shift in the common-mode voltage on the bit lines results in a change in the differential output voltage. The shift may be enough to mask the normal bit line current, corrupting the sense-amplifier output. With the cross-coupled bias scheme, however, the reference voltages at the base nodes of Q_{RB} and Q_{RBb} shift in response to the common-mode voltage noise on the bit lines, leaving V_{BE} for Q_{RB} and Q_{RBb}, relatively unchanged. This allows the differential output voltage to remain, for the most part, unchanged.

Since the voltage of every read bit line is biased about one V_{BE} drop below V_{CC} by the sense amplifiers, it is necessary to ensure that the memory-cell current switches that direct current between the read bit lines are biased appropriately to avoid saturation. To accomplish this, a single diode was connected between V_{CC} and the shaped wire compact conductor (TW) for each row of memory cells to shift the level of MC and MCb down by one V_{BE} drop. This solution requires 32 diodes capable of handling the current drawn by the write word line drivers. This solution consumes much less power and die area than placing emitter followers in each memory cell to shift the voltage of MC and MCb down by one V_{BE} drop.

SPICE simulations were used to estimate the propagation delay through the memory cell and sense amplifier analytically. This propagation delay, along with the read bit-line voltage swing, as a function of the sense-amplifier bias currents according to simulations, are shown in Fig. 4.62. From these results, one finds a diminishing return in the decrease in propagation delay as the sense-amplifier bias currents are increased. This directly correlates with the nonlinear decrease in the read bit-line voltage swing as the sense amplifier bias currents are increased, as Fig. 4.62 illustrates. This is not unexpected given that the voltage swing on the read bit lines is a key factor in the propagation delay through the memory cell and sense amplifier, due to the large read bit line capacitance.

4.4.7 Output Latch Design

Each read port contains a set of output latches to capture the register-file output. Normally, this output comes from the sense amplifiers. However, when a write operation is occurring in a row that is also being read by one of the read ports, the normal read access is delayed while new data is written into the memory cells. This worst-case scenario will limit the read access time of the register file under practical

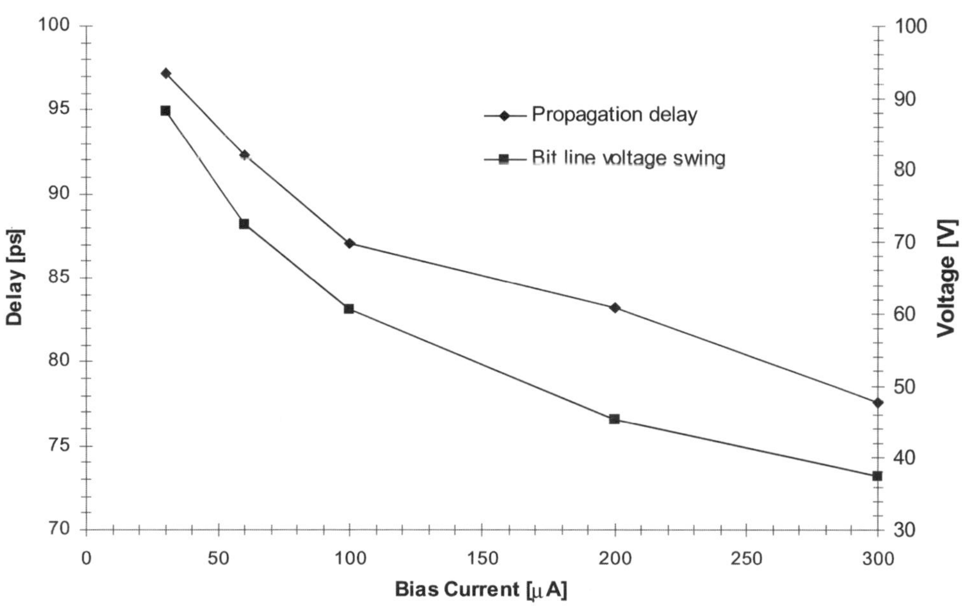

Figure 4.62 Simulated read-bit line voltage swing and propagation delay through the memory cell and sense amplifier as a function of the sense-amplifier bias currents.

circumstances, since the required read access time for most designs must be met under all circumstances. To prevent the read access time from being limited by this special case, it is possible to bypass the memory cells under these circumstances and send the data on the write bit lines directly to the output latches. To implement this scheme, a 2-to–1 multiplexer is required for each output latch to select whether the data to be stored in the latch are coming from a sense amplifier or a bit line driver. This multiplexer can be integrated into the output latch current tree [8,9]. When a value is being written into the latch, the differential voltage between the multiplexer select lines determines whether the bit line-driver data or the sense amplifier data is written to the output latch.

4.4.8 Comparator

To determine whether the data written into the output latches of a read port should come from the sense amplifiers of that port or from the bit line drivers, it is necessary to compare the read address of that port with the write address to determine whether or not they match. It is also necessary to check the write-enable signal to determine whether or not a write operation is occurring. The circuit can be constructed using five exclusive-NOR (XNOR) gates to compare each pair of bits from the two addresses. Three AND gates are also used to determine whether or not all five XNOR gate output values are high (indicating the two addresses match), and whether or not a write operation is enabled. The circuit output is high only in the case where the two addresses match and a write operation is enabled. All the logic gates in the comparator are differential ECL or CML circuits.

4.4.9 Test Chip Design

A test chip, similar to one designed to test a 2-kb SRAM [10], was designed to test both the functionality and performance of the register file. A block diagram of the test chip is shown in Fig. 4.63. During normal operation, three 5-bit counters supply the register file with repeating patterns of sequential addresses for read and write operations. A LFSR supplies the register file columns with a repeating pseudorandom 63-bit pattern that is used as input data for write operations [11]. Using these circuits, data can be written to the register file and then read using either read port to determine whether or not the register file is functioning properly. A scan mode also exists in which the counters and LFSR are linked in a scan chain that allows data to be serially loaded into the counters and LFSR. This is useful for offsetting the addresses of the counters by a set amount. Either an on-chip VCO or an externally generated clock can be used to clock the test chip.

A number of multiplexers are used to select between the signals on the test chip that can be viewed externally. In this way, half the register file columns using each read port can be viewed. Also, the read port A comparator match signal, the top bit of each counter, one of the LFSR bits, and the on-chip VCO can be viewed by setting the multiplexer select lines appropriately. A set of sampling latches is provided on the test chip to work with the register-file output latches to produce master–slave

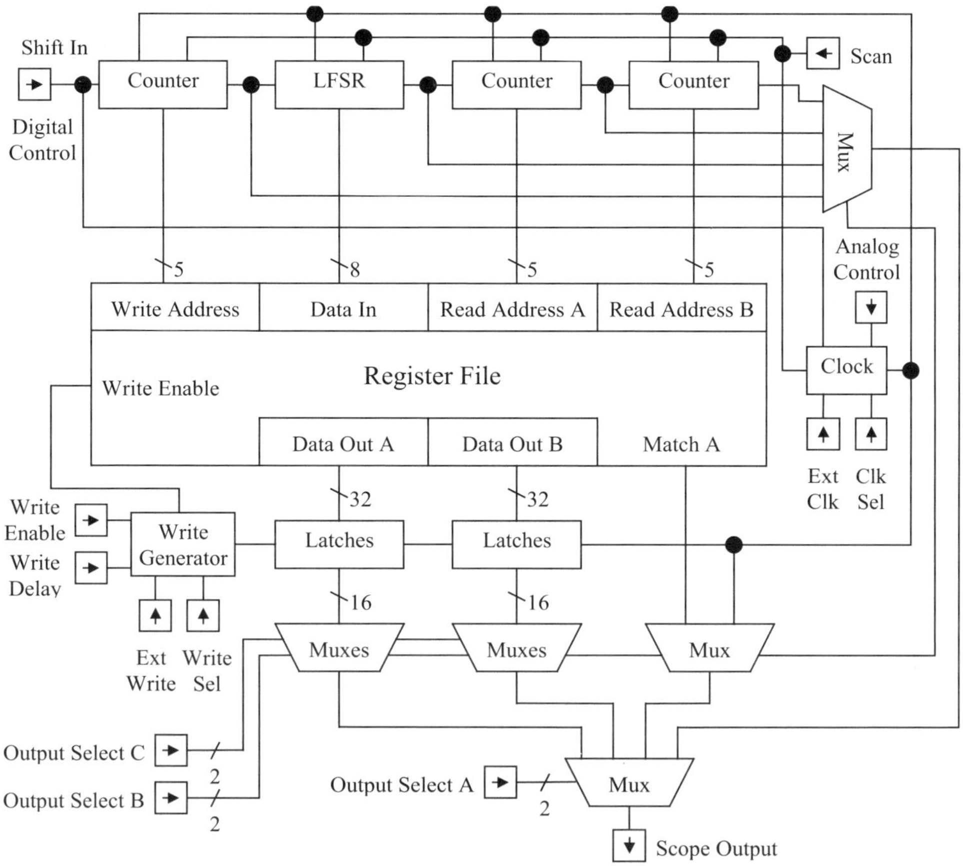

Figure 4.63 Register-file test-chip block diagram.

latches that provide edge-triggering capability. These master–slave latches sample the register-file output data on the falling edge of the clock. Therefore, given the address counters increment addresses on rising clock edges, the on-chip register-file read access time is at most half the clock period, assuming the correct data are latched. Therefore, by increasing the clock frequency until a failure in the expected pattern occurs, the on-chip read access time of the register file in this pipelined system can be determined. Similarly, the on-chip write access time can be estimated by writing in known data at various clock frequencies, and then reading out the data at a slow rate to determine if the data were written correctly.

4.4.10 Physical Characteristics

The dimensions of the register file memory cell are 20 μm by 37 μm. The large aspect ratio of the memory cell is due to the fact that the word lines must conduct much larger currents than the bit lines. Therefore, the word lines must be made much wider than the bit lines to accommodate the larger current. The register file has overall dimensions of 1.0 mm by 1.8 mm. It is composed of 11,365 HBTs and 3769 resistors. A die micrograph of the register file test chip is shown in Fig. 4.64. It has dimensions of 2.6 mm by 2.2 mm. An additional 1949 HBTs and 898 additional resistors were used in the test chip. Most of the additional area in the test chip beyond that of the register file is used for I/O pads and power routing.

4.4.11 Measured Results

The register file test chip die were tested as bare die on a Tektronix probe station using two GGB Picoprobe Multi-Contact Wedge Probes. Each probe contained six signal pins and two sets of power and ground pins. A clock signal was provided to the test chip by a Weinschel Engineering 430A Sweep Oscillator through a 50-Ω cable. Test-chip output signals were measured using a Tektronix 11801A Digital Sampling Oscilloscope with an SD-24 TDR/Sampling Head and an SD-51 Trigger Head, also through 50-Ω cables. Since the access times of interest are those occurring within the internal test-chip pipeline, the exclusion of packaging parasitics as a

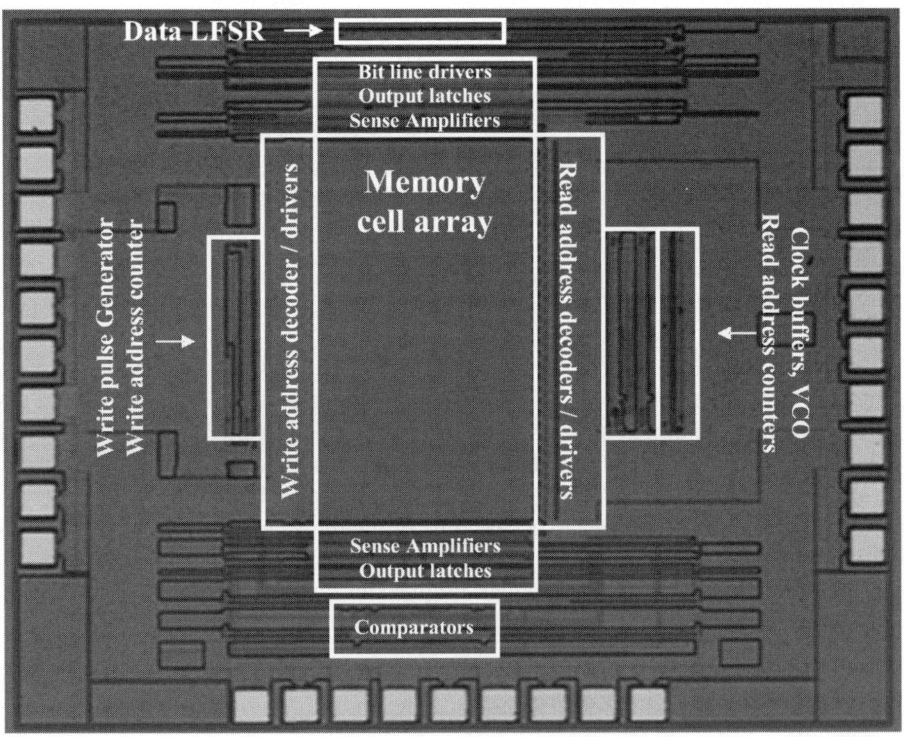

Figure 4.64 Register-file test-chip die micrograph.

result of testing bare die is not an important factor in the measurements. Only one wafer was available for testing.

The read address counters for port A and port B on the register-file test chip operate with maximum clock frequencies of 5.8 GHz and 5.0 GHz, respectively, on average. The write address counter operates with a maximum clock frequency of only 4.0 GHz on average, however, due to the distance of the write address counter from the external clock pad on the test chip. The data LFSR operates with a maximum clock frequency of 5.5 GHz. These clock rates are sufficient for the test chip to test the performance of the register file. The power dissipation of the register file test chip is 6.6 W on average, using a 4.5-V supply.

The best measured on-chip column read access time for the register file was 290 ps. This is illustrated by the measured waveforms in Fig. 4.65, which are generated at the "scope output" node on the test chip, shown in Fig. 4.63. The waveform on the left contains a 32-bit sequence that is a portion of the 63-bit LFSR pattern. During a series of read-after-write operations, however, the entire 63-bit LFSR pattern is observed, as illustrated in Fig. 4.66, along with the corresponding match signal. The on-chip read-after-write access time for this particular column was 340 ps. One notes that the voltage is not constant in the test-chip output waveforms over multiple clock cycles when the digital value remains constant, as Fig. 4.65 and Fig. 4.66 illustrate. This appears to be a superposition of the clock waveform on top of the output waveforms. Simulations show that this behavior is a result of the influence of the clock input on master–slave latch output signals. This distortion of the latched signals does not, however, introduce errors in the digital logic at normal operating frequencies.

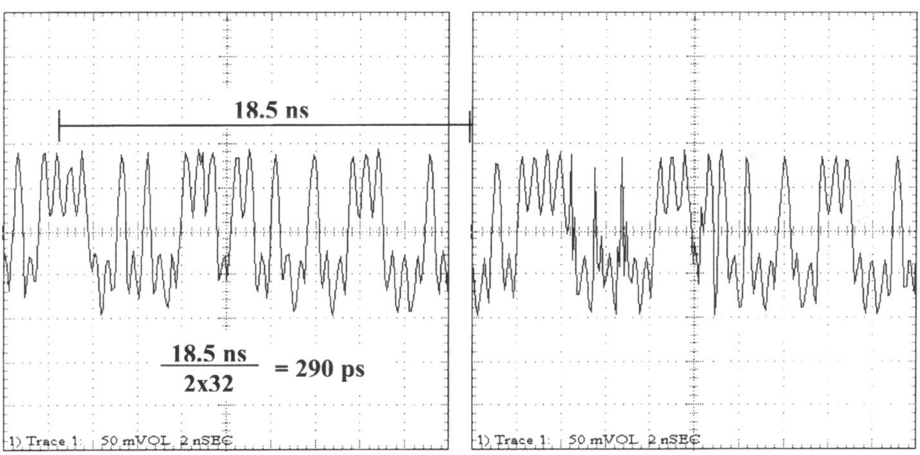

Figure 4.65 Measured test-chip output signal with register file read port A selected for viewing. The left-hand scope plot indicates a correctly read 32-bit pattern from one of the columns. The shortest pulse widths are 580 ps, indicating a 290-ps read access time for the selected column. The right-hand scope plot demonstrates errors in reading out the 32-bit pattern from the same column using a slightly higher clock frequency.

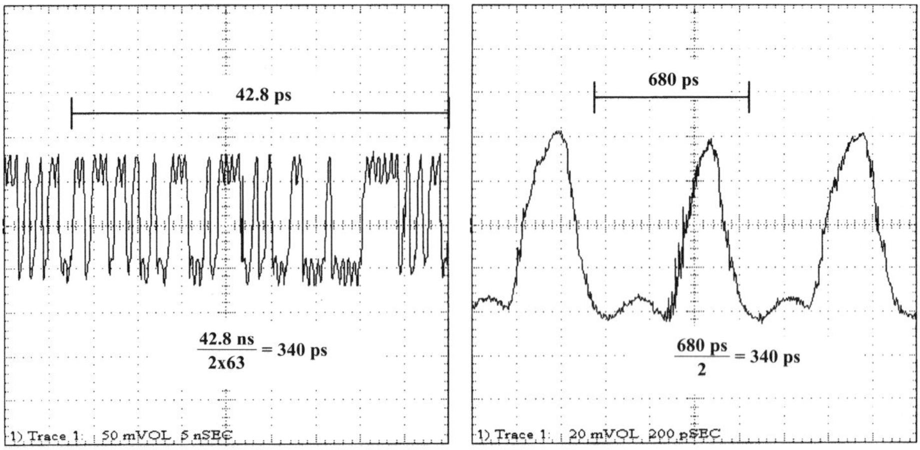

Figure 4.66 Measured test-chip output signals during a read-after-write operation. The left-hand scope plot is a 63-bit pattern read out of one of the columns through read port A while in write mode. The right-hand figure is the match signal for port A, indicating that a read-after-write operation is occurring. The shortest pulse widths for the left-hand waveform are 680 ps, indicating a 340-ps read after write access time for the selected column.

It is the worst-case column access time for a particular die that limits the overall register-file performance. Therefore, even the best measured die has an overall read access time longer than 290 ps. The best measured register-file die has a read access time of 350 ps for port A, while the port B read access time is 360 ps, based on four measured columns for each port. This die has a read-after-write access time of 320 ps for port A, and an on-chip read-after-write access time of 350 ps for port B. The write access time for this register file is 340 ps, with a write-enable pulse width of 170 ps, while the estimated power dissipation is 4.7 W using a 4.5-V supply. The difference in access times between the two read ports is due to layout parasitics, since in other respects the two ports are identical.

Summarizing the performance of the poorest measured register file, the read access time for port A is 330 ps, while the read access time for port B is 380 ps, again based on four measured columns for each port. This die has an read-after-write access time of 340 ps for port A, which limits the read access time for this port, and a read-after-write access time of 370 ps for port B. The write access time for this register file is 250 ps, with a write-enable pulse width of 130 ps, while the estimated power dissipation is also 4.7 W using a 4.5-V supply.

4.4.12 Summary

In this section, we have described a novel system-level design for a 32-word by 32-bit three port bipolar register file designed using the IBM SiGe 5HP technology. It has dimensions of 1.0 μm by 1.8 μm. The on-chip read access time for this register file is between 340 and 350 ps using read port A, while the on-chip read access time using read port B is between 360 and 380 ps. The on-chip write access time is between 250 and 340 ps, using a write-enable pulse with a width between 130 and 170 ps. The estimated power dissipation is 4.7 W using a 4.5-V supply. The short access times are made possible by the fast switching speeds of the SiGe HBTs, combined with the ECL-style circuits used in the design.

4.4.13 References

1. D. C. Ahlgren, G. Freeman, S. Subbanna, R. Groves, D. Greenberg, T. Malinowski, D. Nguyeme Ngoe, S. J. Jeng, K. Ktein, K. Schonenberg, D. Kiesling, B. Martin, S. Wu, D. Marame, B. Meyerson, "A SiGe HBT BiCMOS Technology for Mixed Signal RF Applications," in *Proceedings of 1997 Bipolar/BiCMOS Circuits and Technology Meeting, Minneapolis, MN,* pp. 195–197, September 1997.
2. S. Subbanna, L. Larson, G. Freeman, D. Ahlgren, K. Stein, C. Dieking, J. Merke, A. Rincon, P. Bacon, P. Groves, M. Seymor, D. Harame, J. Dunn, D. Rowe, W. Chon, D. Herman, B. Meyerson, "Silicon-Germanium BiCMOS Technology and a CAD Environment for 2–40 GHz VLSI Mixed-Signal ICs," in *Proceedings of IEEE 2001 Custom Integrated Circuits Conference, San Diego, CA,* pp. 559–566, May 2001.
3. D. Chang and R. Hook, "A Sub-Five Nanosecond ECL 128 × 18 Three Port Register File," in *Proceedings of 1987 Bipolar Circuits and Technology Meeting, Minneapolis, MN,* pp. 98–100, September1987,.
4. C. C. Chao and B. A. Wooley, "A 1.3-ns 32-Word × 32-Bit Three-Port BiCMOS Register File," *IEEE J. Solid-State Circuits,* vol. 31, no. 6, pp. 758–766, June 1996.
5. H. Haznedar, *Digital Microelectronics,* Benjamin/Cummings Publishing, Redwood City, CA, 1991.
6. D. A. Hodges and H. C. Jackson, *Analysis and Design of Digital Integrated Circuits,* 2nd edition, McGraw-Hill, New York, 1988.
7. J. F. Wakerly, *Digital Design Principles and Practices,* Prentice Hall, Englewood Cliffs, NJ, 1990.
8. M. Horowitz, M. Slamowitz, B. Rose, and M. Johnston, "A 3.5 ns, 1 Watt, ECL Register File," in *IEEE ISSCC Digest of Technical Papers,* vol. 33, pp. 68–69, 267, February 1990.
9. H. J. Greub, J. F. McDonald, T. Creedon, and T. Yamaguchi, "High Performance Standard Cell Library and Modeling Techniques for Differential Advanced Bipolar Current Tree Logic," *IEEE J. Solid-State Circuits,* vol. 26, no. 5, pp. 749–762, May 1991.
10. C. A. Maier, H. Grant, B. Philhower, S. Steidl, A. Gary, M. Ernest, S. Carlough, P. Campbell, J. F. McDonald, "Embedded At-Speed Testing Schemes with Low Overhead for High Speed Digital Circuits on Multi-Chip Modules," in *Proceedings of IEEE International Conference on Innovative Systems in Silicon,* Austin, TX, pp. 210–216, October 1996.
11. M. Abramovici, M. A. Breuer, and A. D. Friedman, *Digital Systems Testing and Testable Design,* Computer Science Press (imprint of W. H. Freeman and Company), New York, 1990.

APPENDIX

SUMMARY OF IBM FOUNDRY OFFERINGS

IBM has a diverse offering of RF/Analog and Mixed Signal Technologies across several generations. In production, these technologies span the 0.5-, 0.4-, 0.25-, and 0.18-μm lithography tool sets and both SiGe BiCMOS and RF-CMOS. In development and early production are the 0.13- and 0.10-μm lithographic technologies, which are not included here. These technologies cover a wide range of applications from high-performance wireline to cost-sensitive wireless and storage applications. Tables A.1 and A.2 give a broad overview of the 0.5–0.18 μm SiGe BiCMOS technologies, pointing out key features and the passive offerings. At each lithographic node there is the original technology, a high performance at the time of the development, and a series of "derivative" offerings for special-purpose applications. These derivatives roughly fall into three groups:

1. *Updating* the older technology with improved passives and additional devices
2. *Low complexity* lower-cost (and lower-performance) technologies for commodity markets like wireless
3. The development of *high-breakdown* technologies for specialized applications

A detailed description of seven SiGe BiCMOS technologies is included below.

Silicon Germanium: Technology, Modeling, and Design. By Singh, Harame, and Oprysko
ISBN 0-471-44653-X

Table A.1 Active Device Offering, with Key Technology Parameters

Feature	5HP/AM/DM	5PA	5HPE	6HP	7HP	7WL
Isolation	STI/DT	STI/DT	STI	STI/DT	STI/DT	STI/TI
HP f_T/f_{MAX} (GHz)	51.5/65	51.5/65	43/45	47/65	120/100	60/70
HB f_T/f_{MAX} (GHz)	29/50	23/50	19/35	27/50	30/50	30/50
HP BV_{CEO} (V)	3.3	3.3	3.3	3.3	1.8	3.3
HB BV_{CEO} (V)	5.5	7	9.6	5.7	4.2	6
Min. W_E Drawn (μm)	0.5	0.5	0.40	0.25	0.18	0.18
CMOS Generation	5S0	5S0	5SF	6SF	7SF	7SF
CMOS L_G drawn (μm)	0.5	0.5	0.4	0.25	0.18	0.18
CMOS supply (V)	3.3	3.3	3.3, 5.0	2.5, 3.3	1.8,3.3	1.8,3.3
BEOL metal type	Al	Al	Al	Al	Cu + Al	Cu + Al
	1X	1X	1.16X	0.89X	1.24X	1.24X

CMOS technology can also be used for many RF/analog and mixed-signal applications. From an RF/analog point of view, we can break CMOS into three groups:

1. Digital CMOS process and infrastructure
2. Digital CMOS process with RF/analog and mixed-signal infrastructure
3. Augmented CMOS with modified CMOS process, additional devices, and an RF/analog infrastructure

Table A.2 Passives Offering

	5HP 5AM	5DM	5PA	5HPE	6HP	6DM	7HP	7WL
Poly (ohm) resistors	220	220	220	220 3000	210 3600	225 3500	270 1600	270 1600
BEOL resistor	No	130	No	No	No	50	142	50
MOSCAP 5 V (fF/μm²)	1.5	1.5	1.5	1.2	3.1	3.4	2.5	No
Poly-poly cap (fF/μm²)	No	No	No	1.6	No	No	No	No
Stack finite end-of-the-line (FEOL) caps	No	No	No	2.8	No	No	No	No
MIMCAP (fF/μm²)	0.7	1.35	0.3, 0.7, or 1.35	1.35	1.35	1.35	1.0	2
Stack MIMs	No	Dual	No	Yes	Yes	Yes	No	Dual
Varactor options	CB Jct MOS	CB Jct MOS HA	CB Jct	CB Jct MOS	CB Jct MOS HA	CB Jct MOS HA	MOS Jct	CB Jct MOS HA
Thick last metal	LM or AM	Dual metal	AM	AM	AM	Dual metal	LY + AM	ML or AM or DM

The term RF-CMOS Technology will be applied only to the third case. A detailed description of IBM's 0.25- and 0.18-μm RF-CMOS technologies is given below.

DETAILED OFFERING OUTLINE

SiGe BiCMOS

1. Overview of 0.5-μm SiGe BiCMOS offering (5HP/AM)
2. Overview of 0.5-μm dual-metal SiGe BiCMOS offering
3. Overview of 0.35-μm SiGe BiCMOS (5HPE)
4. Overview of 0.25-μm SiGe BiCMOS (6HP)
5. Overview of 0.25-μm dual-metal SiGe BiCMOS (6DM)
6. Overview of 0.18-μm SiGe BiCMOS (7WL)
7. Overview of 0.18-μm high-performance SiGe BiCMOS (7HP)

RF-CMOS

1. Overview of 0.25-μm RF-CMOS offering (6RF)
2. Overview of 0.18-μm RF-CMOS offering (7RF)

SiGe BiCMOS TECHNOLOGY EXAMPLES

1. Overview of 0.5-μm SiGe BiCMOS Offering (5HP/5AM/5DM)

0.5 μm BiCMOS technology for RF and wired communications
SiGe NPN HBT (50 GHz f_T)
Standard and high-breakdown HBTs available
CMOS 5S0 FETs
0.36-μm L_{eff}, 7.8-nm gate oxide, 3.3 V
ASIC library compatibility
Technology features
Deep- and shallow-trench isolation
Aluminum BEOL
Tungsten local interconnect
Thick last-level dielectric and metal optimized for high-Q inductors
Suite of elements for analog designers
MOS caps, MIM capacitor, inductors,
Polysilicon and single crystal resistors
ESD protection
Base–collector diode varactor, Schottky-barrier diode

NPN	
High-performance SiGe HBT	$f_T = 50$ GHz, $BV_{CEO} = 3.35$ V, $BV_{CBO} = 10.5$ V
High-breakdown SiGe HBT	$f_T = 27$ GHz, $BV_{CEO} = 5.5$ V, $BV_{CBO} = 14$ V
FET	
NFET	$L_{eff} = 0.39$ μm, $I_{DSAT} = 470$ μA/μm, $V_{DD} = 3.3$ V
PFET	$L_{eff} = 0.36$ μm, $I_{DSAT} = 213$ μA/μm, $V_{DD} = 3.3$ V
Resistor	
*RP polysilicon resistor	220 Ω/square, $T_{CR} = -75$ ppm/OC
NS single-crystal resistor	8.1 Ω/square, $T_{CR} = 2050$ ppm/OC
RN single-crystal resistor	20.3 Ω/square, $T_{CR} = 1940$ ppm/OC
*RI single-crystal resistor	1750 Ω/square, $T_{CR} = 300$ ppm/OC
*FR polysilicon resistor	1500 Ω/square, $T_{CR} = -1360$ ppm/OC
Capacitor	
*Metal–metal	1.35 or 0.7 fF/μm^2 (max 5.5 V) (one only)
Polysilicon single crystal	1.5 fF/μm^2 (max 5.5 V)
Varactor	
Collector–base junction	1.7:1
Gated lateral PNP	Beta = 107@$V_{BE} = 0.72$ V
Schottky-barrier diode	$V_F \sim 300$ mV
PIN diode	$V_F \sim 800$ mV
Metal stack = 3–5 levels to LM	LM Rs = 15 mΩ/square LM to substrate = 6 μm (3 levels)–10 μm (5 levels)
*Metal stack = 3–6 levels to AM	AM Rs = 7 mΩ/square AM to substrate = 8 μm (3 levels)–14 μm (6 levels)

2. Overview of 0.5-μm Dual-Metal SiGe BiCMOS Offering (5DM)

A modular passives addition to SiGe 5HP

Same as SiGe 5HP with the following exceptions:

- MIM capacitor
- Far BEOL metal, e.g., inductors

New features added with SiGe 5DM:

- Dual metal inductor: high-Q > 28 or 4× inductance density
- Higher capacitance MIM: 1.35 and 2.7 fF/μm^2 single and dual MIM

Hyperabrupt junction varactor: 3.4:1 tuning range
MOS varactor
BEOL TaN nitride resistor (130 W/sq)

NPN	
High-performance SiGe HBT	f_T = 50 GHz, BV_{CEO} = 3.35 V, BV_{CBO} = 10.5 V
High-breakdown SiGe HBT	f_T = 27 GHz, BV_{CEO} = 5.5 V, BV_{CBO} = 14 V
FET	
NFET	L_{eff} = 0.39 μm, I_{DSAT} = 470 μA/μm, V_{DD} = 3.3 V
PFET	L_{eff} = 0.36 μm, I_{DSAT} = 213 μA/μm, V_{DD} = 3.3 V
Resistor	
*RP polysilicon resistor	220 Ω/square, T_{CR} = –75 ppm/OC
NS single-crystal resistor	8.1 Ω/square, T_{CR} = 2050 ppm/OC
RN single-crystal resistor	20.3 Ω/square, T_{CR} = 1940 ppm/OC
*RI single-crystal resistor	1750 Ω/square, T_{CR} = 300 ppm/OC
*FR polysilicon resistor	1500 Ω/square, T_{CR} = –1360 ppm/OC
*L1 thin-film BEOL resistor	130 Ω/square, T_{CR} = –726 ppm/OC
Capacitor	
*Single metal–metal	1.35 fF/μm^2 (max 5.5 V)
*Dual (stacked) metal–metal	2.7 fF/μm^2 (max 5.5 V)
Polysilicon single crystal	1.5 fF/μm^2 (max 5.5 V)
Varactor	
Collector–base junction	1.7:1
*Hyperabrupt junction	3.4:1 tuning range
MOS accumulation	2.8:1
Gated lateral PNP	Beta = 107@V_{BE} = 0.72 V
Schottky-barrier diode	V_F ~ 300 mV
PIN diode	~ 800 mV
Metal stack = 4–6 levels to MA	MA Rs = 7 mΩ/square, E1 Rs = 6 mΩ/square MA to substrate = 16 μm (4 levels)–20 μm (6 levels)

A high-voltage HBT modification to SiGe 5HP
SiGe NPN HBT (50 GHz ft)
Standard device, same as SiGe 5HP

High-breakdown device increased BV_{CBO} = 18 V from 10.4 V
MIM capacitor option at 14 V (0.3 fF/mm2)
AM metal option only

NPN	
High-performance SiGe HBT	f_T = 50 GHz, BV_{CEO} = 3.3 V, BV_{CBO} = 10.5 V
High-breakdown SiGe HBT	f_T = 25 GHz, BV_{CEO} = 6.4 V, BV_{CBO} = 18 V
FET	
NFET	L_{eff} = 0.39 μm, I_{DSAT} = 470 μA/μm, V_{DD} = 3.3 V
PFET	L_{eff} = 0.36 μm, I_{DSAT} = 213 μA/μm, V_{DD} = 3.3 V
Resistor	
*RP polysilicon resistor	220 Ω/square, T_{CR} = –75 ppm/OC
NS single-crystal resistor	8.1 Ω/square, T_{CR} = 2050 ppm/OC
RN single-crystal resistor	20.3 Ω/square, T_{CR} = 1940 ppm/OC
*RI single-crystal resistor	1750 Ω/square, T_{CR} = 300 ppm/OC
*FR polysilicon resistor	1500 Ω/square, T_{CR} = –1360 ppm/OC
Capacitor	
*Metal–metal	1.35, 0.7 (max 5.5 V) or 0.3 fF/μm² (max 14 V) (one only)
Polysilicon single crystal	1.5 fF/μm² (max 5.5 V)
Varactor	
Collector–base junction	1.7:1
Schottky-barrier diode	V_F ~ 300 mV
PIN diode	V_F ~ 800 mV
Metal stack = 4–6 levels to AM	AM Rs = 7 mΩ/square AM to substrate = 10 μm (4 levels)–14 μm (6 levels)

3. Overview of 0.35-μm SiGe BiCMOS (5HPE)

0.35-μm BiCMOS technology for RF and wired communications
SiGe NPN HBT (43 GHz ft)
Standard and high breakdown HBTs available
5-V CMOS FETs
0.40-μm L_{eff}, 12.1-nm gate oxide
Isolated NFET
Technology features

Shallow-trench isolation
Aluminum BEOL
Thick last-level dielectric and metal optimized for high-Q inductors
Suite of elements for analog designers
MOS caps, poly–poly cap, MOS/poly-poly stack cap
MIM capacitor (up to 2.7 fF/mm in stacked configuration)
Polysilicon and single-crystal resistors
Schottky diode, ESD protection
MOS varactor, varactor diode
LPNP 1.5 mm base @ 3.3 V and 2.5-mm base @ 6.5 V

NPN	
*High-performance SiGe HBT	f_T = 43 GHz, BV_{CEO} = 3.3 V, BV_{CBO} = 11 V
High-breakdown SiGe HBT	f_T = 19 GHz, BV_{CEO} = 9.6 V, BV_{CBO} = 23 V
FET	
NFET 5.0 V	L_{eff} = 0.4 μm, I_{DSAT} = 600 μA/μm, V_{DD} = 5.0 V
NFET 5.0 V in isolated p-well	L_{eff} = 0.4 μm, I_{DSAT} = 600 μA/μm, V_{DD} = 5.0 V
PFET 5.0 V	L_{eff} = 0.45 μm, I_{DSAT} = 265 μA/μm, V_{DD} = 5.0 V
PFET 3.3 V	L_{eff} = 0.35 μm, I_{DSAT} = 155 μA/μm, V_{DD} = 3.3 V
Resistor	
PC polysilicon resistor	220 Ω/square, T_{CR} = –270 ppm/OC
N^+ single-crystal resistor	64 Ω/square, T_{CR} = 1800 ppm/OC
P^+ single-crystal resistor	94 Ω/square, T_{CR} = 1500 ppm/OC
*PE polysilicon resistor	3000 Ω/square, T_{CR} = –1780 ppm/OC
Capacitor	
*Metal–metal	1.35 fF/μm^2 (max 5.5 V)
*Metal–metal (stacked)	2.70 fF/μm^2 (max 5.5 V)
Poly–single crystal	1.2 fF/μm^2 (max 5.5 V)
*Poly–poly	1.6 fF/μm^2 (max 5.5 V)
*Stacked Poly–Poly–single crystal	2.8 fF/μm^2 (max 5.5 V)
Varactor	
*Collector–base junction	1.7:1
High-breakdown collector–base junction	2.3:1

MOS accumulation	2:1
Lateral PNP	Beta = 25.5 @V_{BE} = 0.72
Schottky-barrier diode	V_F ~ 300 mV
*Metal stack = 3–6 levels to AM	AM Rs = 7 mΩ/square
	AM to substrate = 8 μm (3 levels)–14 μm (6 levels)

4. Overview of 0.25-μm SiGe BiCMOS (6HP)

0.25-μm BiCMOS technology for storage, RF, and wired communications
SiGe NPN HBT (47 GHz ft)
Lateral scaled for low power
Standard and high-breakdown HBTs available
ASIC-compatible CMOS 6SF FETs
0.18-μm L_{eff}, 5-nm gate-oxide, 2.5 V
Dual-oxide option: 3.3 V, 7-nm gate-oxide FETs
Technology features
Deep- and shallow-trench isolation
Aluminum BEOL, tight-pitch for densely wired CMOS
Thick last-level dielectric and metal optimized for high-Q inductors
Suite of elements for analog designers
MOS caps, MIM capacitor, inductors
polysilicon and single-crystal resistors
Collector–base junction varactor, HA varactor, MOS varactor
ESD protection

NPN	
High-performance SiGe HBT	f_T = 47 GHz, BV_{CEO} = 3.35 V, BV_{CEO} = 10.5 V
High-breakdown SiGe HBT	f_T = 27 GHz, BV_{CEO} = 5.7 V, BV_{CBO} = 14 V
FET	
NFET 2.5 V	L_{eff} = 0.190 μm, I_{DSAT} = 595 μA/μm, V_{DD} = 2.5 V
PFET 2.5 V	L_{eff} = 0.180 μm, I_{DSAT} = 280 μA/μm, V_{DD} = 2.5 V
*NFET 3.3 V	L_{eff} = 0.260 μm, I_{DSAT} = 580 μA/μm, V_{DD} = 3.3 V
*PFET 3.3 V	L_{eff} = 0.265 μm, I_{DSAT} = 285 μA/μm, V_{DD} = 3.3 V
Resistor	
Polysilicon resistor	210 Ω/square, T_{CR} = 200 ppm/OC
Polysilicon resistor	3600 Ω/square, T_{CR} = –2350 ppm/OC

N^+ single-crystal resistor	63 Ω/square, T_{CR} = 1500 ppm/OC
P^+ single-crystal resistor	100 Ω/square, T_{CR} = 1550 ppm/OC
Capacitor	
*Metal–metal	Nitride 1.35 fF/μm² or Oxide .7 fF/μm² (max 5.5 V)
*Metal–metal (stacked)	Nitride 2.7 fF/μm² , Oxide 1.4 fF/μm² (max 5.5 V)
Polysilicon single crystal	3.1 fF/μm² (max 3.6 V)
Varactor	
Collector–base junction	1.9:1
*Hyperabrupt junction	3.4:1 tuning range
MOS accumulation	2.5:1
*Metal stack = 4–6 levels to AM	AM Rs = 7 mΩ/square AM to substrate = 7.4 μm (4 levels)–10.1 μm (6 levels)

5. Overview of 0.25-μm Dual-Metal SiGe BiCMOS (6DM)

A modular addition of passive elements to SiGe 6HP
Features common with SiGe 6HP
0.25-μm BiCMOS technology for storage, RF and wired communications
SiGe NPN HBT (47 GHz ft)
Lateral scaled for low power
Standard and high-breakdown HBTs available
ASIC library compatibility CMOS 6SF FETs
0.18-μm L_{eff}, 5-nm gate-oxide, 2.5 V
Dual-oxide option: 3.3 V, 7-nm gate-oxide FETs
Technology features
Deep- and shallow-trench isolation
Aluminum BEOL, tight-pitch for densely wired CMOS
Suite of elements for analog designers
MOS caps, polysilicon and single-crystal resistors
ESD protection and collector–base junction diode varactor
New features added with 6DM:
Dual metal inductor: high-Q > 28 or 4× inductance density
Higher capacitance MIM: 1.35 and 2.7 fF/μm single and stacked
Hyperabrupt junction varactor: 3.4:1 tuning range
MOS varactor
BEOL TaN nitride resistor (130 W/sq)

NPN	
High-performance SiGe HBT	f_T = 50 GHz, BV_{CEO} = 3.35 V, BV_{CEO} = 10.5 V
High-breakdown SiGe HBT	f_T = 27 GHz, BV_{CEO} = 5.5 V, BV_{CBO} = 14 V
FET	
NFET 2.5 V	L_{eff} = 0.190 μm, I_{DSAT} = 585 μA/μm, V_{DD} = 2.5 V
PFET 2.5 V	L_{eff} = 0.180 μm, I_{DSAT} = 280 μA/μm, V_{DD} = 2.5 V
*NFET 3.3 V	L_{eff} = 0.260 μm, I_{DSAT} = 580 μA/μm, V_{DD} = 3.3 V
*PFET 3.3 V	L_{eff} = 0.265 μm, I_{DSAT} = 280 μA/μm, V_{DD} = 3.3 V
Resistor	
Polysilicon resistor	210 Ω/square, T_{CR} = ppm/OC
Polysilicon resistor	3600 Ω/square, T_{CR} = ppm/OC
N^+ single-crystal resistor	63 Ω/square, T_{CR} = ppm/OC
P^+ single-crystal resistor	100 Ω/square, T_{CR} = ppm/OC
*L1 thin-film BEOL resistor	130 Ω/square, T_{CR} = –726 ppm/OC
Capacitor	
*Metal–metal	1.35 fF/μm^2 (max 5.5 V)
*Dual (stacked) metal–metal	2.7 fF/μm^2 (max 5.5 V)
Polysilicon single crystal	3.1 fF/μm^2 (max 3.6 V)
Varactor	
Collector–base junction	1.9:1
*Hyperabrupt junction	3.4:1 tuning range
MOS accumulation	2.5:1
Metal stack = 5–7 levels to MA	MA Rs = 7 mΩ/square, E1 Rs = 6 mΩ/square MA to substrate = 14.4 μm (5 levels)–17.1 μm (7 levels)

6. Overview of 0.18-μm SiGe BiCMOS (7WL)

0.18-μm BiCMOS technology for RF communications
Advanced SiGe vertical-profile NPN HBT
High-performance: f_T = 60 GHz; BV_{CEO} = 3 V
Medium performance: f_T = 40; BV_{CEO} = 4 V
High-breakdown: f_T = 30 GHz; BV_{CEO} = 6 V
Technology features
Shallow-trench and medium-trench isolation
Copper first level metal

ASIC-compatible CMOS 7SF devices (6 types of FETs)

1.8 V CMOS FETs (T_{ox} = 35 A, L_{eff} = 0.11 ± 0.04 mm)

2.5 V/3.3 V CMOS FETs for I/O (w/dual-gate oxide; T_{ox} = 68 A)

Full suite of passive elements

1.9/3.8 fF/f"MathematicalPi 1">mm^2 MIM capacitor (dual MIM option), N-well capacitor

Diodes—MOS, base–collector, Schottky-barrier diode, P$^+$/N-well diode (for ESD), hyperabrupt

Inductors—ML, AM, dual metal (MA)

Single-crystal and polysilicon resistors, TaN BEOL resistor

LPNP, isolated 1.8 V/2.5 V/3.3 V NFET

NPN	
?*High-performance SiGe HBT	f_T = 60 GHz, BV_{CEO} = 3.0 V, BV_{CBO} = 10.8 V
Medium-performance SiGe HBT	f_T = 40 GHz, BV_{CEO} = 4.0 V, BV_{CBO} = 8.5 V
High-breakdown SiGe HBT	f_T = 30 GHz, BV_{CEO} = 6.0 V, BV_{CBO} = 16 V
FET	
NFET 1.8 V	L_{eff} = 0.11 μm, I_{DSAT} = 600 μA/μm, V_{DD} = 1.8 V
PFET 1.8 V	L_{eff} = 0.14 μm, I_{DSAT} = 260 μA/μm, V_{DD} = 1.8 V
*NFET 2.5/3.3 V	L_{eff} = 0.21/0.29 μm, I_{DSAT} = 425/550 μA/μm,
V_{DD} = 2.5/3.3 V	
*PFET 2.5/3.3 V	L_{eff} = 0.21/0.29 μm, I_{DSAT} = 185/229 μA/μm,
V_{DD} = 2.5/3.3 V	
NFET 1.8 V high gain	L_{eff} = 0.26 μm, I_{DSAT} = 517 μA/μm, V_{DD} = 1.8 V
PFET 1.8 V high gain	L_{eff} = 0.22 μm, I_{DSAT} = 177 μA/μm, V_{DD} = 1.8 V
Zero V_t FET 1.8 V	L_{eff} = 0.5 μm, I_{DSAT} = 450 μA/μm, V_{DD} = 3.3 V
*NFET 1.8 V isolated p-well	L_{eff} = 0.11 μm, I_{DSAT} = 600 μA/μm, V_{DD} = 1.8 V
*NFET 2.5/3.3 V isolated p-well	L_{eff} = 0.21/0.29 μm, I_{DSAT} = 425/550 μA/μm,
V_{DD} = 2.5/3.3 V	
Resistor	
Polysilicon resistor	260 Ω/square, T_{CR} = 99 ppm/OC
*Polysilicon resistor	1600 Ω/square, T_{CR} = –1105 ppm/OC
N$^+$ single–crystal resistor	72 Ω/square, T_{CR} = 1751 ppm/OC
P$^+$ single–crystal resistor	105 Ω/square, T_{CR} = 1401 ppm/OC
*K1 BEOL thin-film resistor	58 Ω/square, T_{CR} = –728 ppm/OC

Capacitor	
*Metal–metal	2.0 fF/μm^2 (max 5.5 V)
*Dual (stacked) metal–metal	4.0 fF/μm^2 (max 5.5 V)
NWell	9.0 fF/μm^2 (max 1.8 V)
Varactor	
Collector–base junction	1.3:1 tuning range
*Hyperabrupt junction	3.4:1 tuning range
MOS accumulation	2.5:1 tuning range (–0.5–1 V)
Schottky-barrier diode	$V_F \sim 340$ mV
*Lateral PNP	Beta = 14.6 @ $V_{BE} = 0.66$ V
Metal stack = 5–8 levels to MA	MA Rs = 7 mΩ/square, E1 Rs = 6 mΩ/square MA to substrate = 18.3 μm (5 levels)–20.3 μm (7 levels)

7. Overview of 0.18-μm High-Performance SiGe BiCMOS (7HP)

0.18-μm BiCMOS technology for RF communications
Advanced SiGe vertical profile NPN HBT with scaled ground rule
High-performance: WE = 0.2 mm; f_T/f_{MAX} = 120/100 GHz; BV_{CEO} = 1.8 V
High-breakdown: f_T/f_{MAX} = 30/50 GHz; BV_{CEO} = 4.5 V
ASIC-compatible CMOS 7SF devices (6 types of FETs)
1.8 V CMOS FETs w/optional Hi-Vt FETs (T_{ox} = 35 A, L_{eff} = 0.11 ± 0.04 mm)
2.5 V/3.3 V CMOS FETs for I/O (w/dual Gate oxide; T_{ox} = 68A)
Copper BEOL, tight-pitch, with thick aluminum last metal
CMOS 7SF Cu BEOL (M1–MT) with M2 to M4 as optional levels
Thick Al metals (LY and AM) with 4 mm tungsten studs
Full suite of passive elements
FEOL resistors ranging from 8.1 kW/E to 1.6 kW/E (NS, n and p diff, p-Poly, and RR)
BEOL TaN Resistor with low parasitic capacitance (142 W/E)
MIM (1.0 fF/μm^2) and FEOL MOS capacitors (2.6 fF/μm^2)
Inductors using thick analog metal

NPN	
High-performance SiGe HBT	f_T = 120 GHz, BV_{CEO} = 1.8 V, BV_{CEO} = 6.4 V
High-breakdown SiGe HBT	f_T = 30 GHz, BV_{CEO} = 4.2 V, BV_{CBO} = 12.5 V
FET	
NFET 1.8 V	L_{eff} = 0.11 μm, I_{DSAT} = 600 μA/μm, V_{DD} = 1.8 V

PFET 1.8 V	$L_{\text{eff}} = 0.14$ μm, $I_{\text{DSAT}} = 260$ μA/μm, $V_{DD} = 1.8$ V
NFET 1.8 V High-Vt	$L_{\text{eff}} = 0.11$ μm, $I_{\text{DSAT}} = 500$ μA/μm, $V_{DD} = 1.8$ V
PFET 1.8 V High-Vt	$L_{\text{eff}} = 0.14$ μm, $I_{\text{DSAT}} = 210$ μA/μm, $V_{DD} = 1.8$ V
*NFET 2.5/3.3 V	$L_{\text{eff}} = 0.29$ μm, $I_{\text{DSAT}} = 550$ μA/μm, $V_{DD} = 3.3$ V
*PFET 2.5/3.3 V	$L_{\text{eff}} = 0.29$ μm, $I_{\text{DSAT}} = 229$ μA/μm, $V_{DD} = 3.3$ V
*NFET 1.8 V isolated p-well **	$L_{\text{eff}} =$ NA μm, $I_{\text{DSAT}} =$ NA μA/μm, $V_{DD} = 1.8$ V
*NFET 2.5/3.3 V isolated p-well **	$L_{\text{eff}} =$ NA μm, $I_{\text{DSAT}} =$ NA μA/μm, $V_{DD} = 3.3$ V
Resistor	
Polysilicon resistor	270 Ω/square, $T_{\text{CR}} = 99$ ppm/OC
*Polysilicon resistor	1600 Ω/square, $T_{\text{CR}} = -1105$ ppm/OC
N^+ single-crystal resistor	72 Ω/square, $T_{\text{CR}} = 1751$ ppm/OC
NS single-crystal resistor	8.1 Ω/square, $T_{\text{CR}} = 1460$ ppm/OC
P^+ single-crystal resistor	105 Ω/square, $T_{\text{CR}} = 1401$ ppm/OC
*K1 BEOL thin-film resistor	142 Ω/square, $T_{\text{CR}} = -728$ ppm/OC
Capacitor	
*Metal–metal	1.0 fF/μm^2 (max 5.5 V)
Polysilicon single crystal	2.5 fF/μm^2 (max 5.5 V)
Varactor	
PN junction varactor	1.8:1
MOS accumulation	3.5:1 (-0.5 – 1 V)
*Metal stack = 4–7 levels to AM	AM Rs = 7 mΩ/square AM to substrate = 11.3 μm (4 levels)–13.3 μm (7 levels)

RF CMOS TECHNOLOGY EXAMPLES

1. Overview of 0.25-μm RF CMOS Offering (6RF)

Base technology features
- 0.24-μm photo
- 2.5-V and 3.3-V FETs
- 0 Vt nFET
- Polysilicon and silicon resistors

Features added for RF
- P-starting wafer
- 6.5-V FETs
- 1.4-fF/μm^2 (2.8 stacked) MIM

Thick-metal (Al) add-on module for high-Q inductors
Isolated NFET (triple well)
MOS varactor
RF characterized scalable ESD protection

FET	
NFET 2.5 V	$L_{\text{Drawn}} = 0.240$ μm, $I_{\text{DSAT}} = 595$ μA/μm, $V_{DD} = 2.5$ V
*NFET 2.5 V with isolated p-well	$L_{\text{Drawn}} = 0.240$ μm, $I_{\text{DSAT}} = 595$ μA/μm, $V_{DD} = 2.5$ V
NFET 2.5 V zero-Vt	$L_{\text{Drawn}} = 0.6$ μm, $L_{\text{eff}} = 0.568$ μm, $I_{\text{DSAT}} = 592$ μA/μm, $V_{DD} = 2.5$ V
PFET 2.5 V	$L_{\text{eff}} = 0.180$ μm, $I_{\text{DSAT}} = 280$ μA/μm, $V_{DD} = 2.5$ V
*NFET 3.3 V	$L_{\text{eff}} = 0.260$ μm, $I_{\text{DSAT}} = 580$ μA/μm, $V_{DD} = 3.3$ V
*PFET 3.3 V	$L_{\text{eff}} = 0.265$ μm, $I_{\text{DSAT}} = 285$ μA/μm, $V_{DD} = 3.3$ V
*NFET 6.5 V	$L_{\text{eff}} = 0.384$ μm, $I_{\text{DSAT}} = 560$ μA/μm, $V_{DD} = 6.5$ V
*PFET 6.5 V	$L_{\text{eff}} = 0.368$ μm, $I_{\text{DSAT}} = 280$ μA/μm, $V_{DD} = 6.5$ V
Resistor	
Polysilcon resistor	210 Ω/square, $T_{\text{CR}} = 300$ ppm/OC
Polysilcon resistor **	3200 Ω/square, $T_{\text{CR}} = -2200$ ppm/OC
N^+ single-crystal resistor	63 Ω/square, $T_{\text{CR}} = 1500$ ppm/OC
P^+ single-crystal resistor	100 Ω/square, $T_{\text{CR}} = 1500$ ppm/OC
Capacitor	
*Metal–metal	1.35 fF/μm^2 (max 5.5 V)
*Stacked metal–metal	2.70 fF/μm^2 (max 5.5 V)
Polysilicon single crystal	5.0 fF/μm^2
Varactor	
MOS accumulation (also DCAP)	2.5:1
*Metal stack = 3–6 levels to AM (ML option)	AM Rs = 7 mΩ/square AM to substrate = (3 levels)– μm (6 levels)

2. Overview of 0.18 μm RF CMOS Offering (7RF)

Base technology features

0.18 μm photo lithography
1.8 V FETs; 2.5/3.3 V I/O FETs
3.3 V HG FETs and 0 Vt NFET
MOS varactor
Polysilicon and silicon resistors

Features added for RF (adopted from SiGe 7WL)

- $2.0\ fF/\mu m^2$ ($4.0\ fF/\mu m^2$ dual) MIM
- Three last metal options for inductor Q: cost/performance trade off
- Thin last metal (low cost)
- Thick Al add on module for high Q inductors
- Thick Cu + Al add on module for higher Q inductors
- BEOL resistor
- High value polysilicon resistor
- Isolated NFET (triple well)
- Hyperabrupt junction varactor
- RF characterized scalable ESD protection

FET	
NFET 1.8 V	$L_{eff} = 0.11\ \mu m$, $I_{DSAT} = 600\ \mu A/\mu m$, $V_{DD} = 1.8$ V
PFET 1.8 V	$L_{eff} = 0.14\ \mu m$, $I_{DSAT} = 260\ \mu A/\mu m$, $V_{DD} = 1.8$ V
*NFET 2.5/3.3 V	$L_{eff} = 0.29\ \mu m$, $I_{DSAT} = 550\ \mu A/\mu m$, $V_{DD} = 3.3$ V
*PFET 2.5/3.3 V	$L_{eff} = 0.29\ \mu m$, $I_{DSAT} = 229\ \mu A/\mu m$, $V_{DD} = 3.3$ V
*NFET 1.8 V isolated p-well	L_{eff} = (NA) μm, I_{DSAT} = (NA) μA/μm, $V_{DD} = 1.8$ V
*NFET 2.5/3.3 V isolated p-well	L_{eff} = (NA) μm, I_{DSAT} = (NA) μA/μm, $V_{DD} = 3.3$ V
Resistor	
Polysilicon resistor	270 Ω/square, T_{CR} = 99 ppm/OC
*Polysilicon resistor	1600 Ω/square, T_{CR} = –1105 ppm/OC
N^+ single-crystal resistor	72 Ω/square, T_{CR} = 1751 ppm/OC
P^+ single-crystal resistor	105 Ω/square, T_{CR} = 1401 ppm/OC
*K1 BEOL thin-film resistor	58 Ω/square, T_{CR} = –728 ppm/OC
Capacitor	
*Metal–metal	$2.0\ fF/\mu m^2$ (max 5.5 V)
*Dual (stacked) metal–metal	$4.0\ fF/\mu m^2$ (max 5.5 V)
N-Well	$9.0\ fF/\mu m^2$ (max 1.8 V)
Varactor	
	No CB varactor in RF CMOS
*Hyperabrupt junction	3.4:1 tuning range
MOS accumulation	5:1 tuning range
Metal stack = 5–8 levels to MA	MA Rs = 7 mΩ/square, E1 Rs = 6 mΩ/square MA to substrate = 18.3 μm (5 levels)–20.3 μm (7 levels)

INDEX

ACPR, 302
Amplifier
  BBVGA, 284
  Input, 258
  Limiting, 264
  Low-noise, 279
  sense, 309
ASIC, 56
ASTC, 36

Balun, 71
BANANA, 30
Benchmarking, 173
BEOL, 62, 68, 149
BIST, 234
BSIM, 125, 142

C4, 300
Capacitors, 60, 153
  MIMCAP, 4, 61, 154
  MOSCAP, 60, 154
  PIPCAP, 61
CDF, 145
CDMA, 4, 271
CDR, 257
Clock Multiplier Unit, 257, 263
Charge pump, 238, 261
CISP, 151
CML, 306, 308
CMP, 64
Comparator, 312
Corners, 128
CVD, 27

Darlington, 73
Data converter
  ADC, 55
  DAC, 36
Deep Trench, DT, 64, 73, 108, 242
Demultiplexer, 246, 251
Design Entry, 164
Device simulation, 107
Distortion, 123
DMACS, 118
Downconverter, 282
DPSA, 28
DRAM, 64
DRC, 169, 170
Drift field, 3
Dual metal, 71
Duplexer filter, 274

Silicon Germanium: Technology, Modeling, and Design. By Singh, Harame, and Oprysko
ISBN 0-471-44653-X

Early voltage, 48
ECL, 132, 284, 305, 308
Eddy current, 151
Electromigration, 98
Epitaxial, EPI, 4, 25
ESD
  Design automation, 180
  Power clamps, 186
  Protection devices, 71
ETX, 33, 53, 79
Evaluation board, 286
Eye diagram, 248, 255, 256

Faraday shielding, 149
FET, 54
  MOSFET, 141
  BiFET, 38
  IGFET, 142
Fielday, 31

Gain stage, 238
Genetic algorithm, 143
GMSK modulation, 300
Green's function, 228
GSM, 4, 299
Guard ring, 209, 217, 221
Gummel, 26, 28
Gummel-Poon, 131

Halo
Harmonic balance, 173
HBT, 48, 131
HEMT, xii
HiCUM, 132
HiPOX, 22
Hot Carrier degradation, 94
HPSK, 301
HTO, 61

ICCR, 131
IIPx, 275, 279, 285, 292
Impact ionization
Inductor, 68, 148, 169
Inline subcircuit
Insertion loss, 286
Interconnect, 99, 194
  Extraction, 194
  Field solver solutions, 201
Inter-modulation distortion, 274

Jitter, 241, 269
  Deterministic, 268
Junction temperature, 89

Kerf, 128
Kroemer, xi, 1

LNA, 208, 273, 275, 279, 283
LO Leakage, 217, 273, 277, 292
LOCOS, 33
Low pass filter, 238
LTCC, 300
LTE, 24
LVS, 170

MAG, 119
MATLAB, 207
Memory, 304
  Cell design, 305
MEXTRAM, 132
Impedance mismatch, 286,
Mixed-Signal, 164
Mixer, 282
Modulator driver, 266
Monte Carlo, 106, 127
Multiplexer, 246, 251

NBTI, 96
Noise
  1/f (flicker), 6, 88, 119, 121, 145
  Characterization, 119
  Device, 85
  Figure, 87
  Receiver, 275
NSA, 50, 110
NTX, 25, 33

Output latch, 311

PAE, 121
Parasitics, 168
  Package, 253, 269
Pattern filling, 171
PCELL, 133, 144,167, 183
  Multifingered, 169
  Multiplicity, 169
PECVD, 61
Periodic steady state, 173
Phase detector, 238, 259

PLL, 236, 250
Poisson's equation, 107
Power amplifier, 79, 174, 274, 296
Power gain, 52
Proximity effect, 151

Q, 59, 68, 149, 153

Read address decoder, 306
Read Channel
  PRML, 13
Read word line driver, 308
Receiver
  Circuit design, 278
  Test-bed, 289
Register file, 304
Reliability, 92, 94, 97
Resistor, 59, 159
RTA, 80
RXB, 53

Safe Operating Area, 90, 298
SAW filter, 282
Schematic, 165
  Symbols, 166
SERDES, 11, 86, 235, 249
Shockley, 1
SIMS, 26, 35
Skew file, 128
SNR, 85, 119, 290
SoC, 15
Software baseband processor, 289
SOI, 71, 242
SONET, 8, 77, 234
SPC, 80
SPICE, 38, 127
SRAM, 82
STI, 64
Substrate, 217, 242
  Interconnect effects, 203
  Modeling, 228
  Isolation, 225
  Package, 253
SWCAD, 206
SXCUT, 172
System
  Performance tests, 287
  Test Results, 291

TCAD, 104, 218
Test-site, 56, 116, 125, 224, 236, 246, 312
Thermal effects, 89
Transmission line, 196
  Microstrip, 199
  CPW, 71
Transceiver, 244
Transformers, 68
Transmitter

UHV/CVD, 24, 25, 27
UMTS, 271
Unity current gain, 51

Varactor, 66, 156
VBIC, 112, 131
VCO, 179, 244, 245, 258, 262, 312
  Differential, 237
VLSI

W-CDMA, 4, 8, 271, 272
Write address decoder, 308

ABOUT THE AUTHORS

Raminderpal Singh was born in Essex, England, in 1970. He received a Bachelor of Engineering degree in 1981, at Imperial College, London University. He then spent a year as a venture capitalist with 3i plc (UK). In 1997, Singh received his Ph.D. in Electrical Engineering, focusing on *Efficient Substrate Modeling Techniques in Mixed-Signal IC Design,* from Newcastle University, United Kingdom. He then worked for Cadence Design Systems from 1997 to 2001, initially as the lead development engineer for the Substrate Coupling Analysis product, followed by lead technical roles in methodology development initiatives in SOC design and high-speed ASIC design. Since March 2001, Dr. Singh has been with IBM's RF/Mixed-Signal Design Kit Group in Burlington, Vermont, where he is currently a senior engineering manager leading a group of more than 25 staff members in various technical projects related to physical verification tools enablement, transmission line model development, substrate modeling, and RF/mixed-signal design methodologies. Dr. Singh has authored and co-authored numerous technical publications in the area of signal integrity, including editing a book, *Signal Integrity Effects in Custom IC and ASIC Designs* (IEEE Press, 2001). He regularly writes articles for leading trade magazines and is a recognized expert in the areas of signal integrity and analog IP integration into large ASIC and SoC designs. Dr. Singh is chair of the IP Implementation Working Group in the VSIA SOC standards body (http://vsi.org) and sits on VSIA Board of Directors. At VSIA, he led the development effort leading to the world's first Specifications document describing signal integrity issues for the import of analog and digital IP. Dr. Singh is a Senior Member of IEEE.

 Silicon Germanium: Technology, Modeling, and Design. By Singh, Harame, and Oprysko
ISBN 0-471-44653-X

David L. Harame was born in Pocatello, Idaho, in 1948. He received a Bachelor of Arts degree in Zoology from the University of California, Berkeley, in 1971, and a Master of Science degree in Zoology from Duke University, Durham, North Carolina, in 1973. He received a Master of Science degree in Electrical Engineering from San Jose State University, San Jose, California, in 1976, and a Master of Science degree in Materials Science and Ph.D. in Electrical Engineering from Stanford University, Stanford, California in 1984. In 1984, he joined IBM's Bipolar Technology group at the IBM T.J. Watson Research Center, in Yorktown Heights, New York, where he worked on the fabrication and modeling of silicon-based integrated circuits. His specific research interests there included silicon and SiGe-channel FET transistors, NPN and PNP SiGe-base bipolar transistors, complementary bipolar technology, and BiCMOS technology for digital and analog and mixed-signal applications. In 1993, he joined IBM's Semiconductor Research and Development Center in the Advanced Semiconductor Technology Center in Hopewell Junction, New York, where he was responsible for the development of SiGe technology for mixed-signal applications. He managed SiGe BiCMOS technology development at the ASTC through 1997. In 1998, he joined IBM's Manufacturing organization in Essex Junction, Vermont, where he managed an SiGe technology group and installed the 0.5 m SiGe BiCMOS process in the manufacturing line. In 1999, he rejoined the Semiconductor Research Corporation while remaining in Essex Junction and co-managed the qualification of a 0.25 m SiGe BiCMOS, as well as 0.18 m SiGe BiCMOS and numerous derivative SiGe BiCMOS technologies. In May 2002, he became director of the RF/Analog and Mixed Signal Process Development, Modeling, and Design Automation area. He is a Distinguished Engineer of the IBM Corporation and an IEEE Fellow, Executive Committee member of the Bipolar/BiCMOS Circuits and Technology Meeting (BCTM), and member of the Compact Model Council.

Modest M. Oprysko received his Ph.D. in Chemical Physics from Columbia University in 1983 and joined Gould Inc. Research Laboratory in Rolling Meadows, Illinois. At Gould, he invented a technique for repairing micron-scale defects in photo-masks that are used to define the patterns in integrated circuits. For years, that technology has been the industry standard for photo-mask repair. Dr. Oprysko joined IBM in 1986 at the Research Division in Yorktown Heights, New York, where he has held numerous technical and management positions. He worked on a broad range of technologies including opto-electonic packaging, fiber-optic data communications links, RF wireless communications subsystems for mobile computing applications, high-speed circuits in SiGe and CMOS technology and test. In June 2002, Dr. Oprysko assumed his current position of department group manager in Communications Technologies having responsibility for the work of more than 70 researchers in the areas of communications circuits and systems, optical communications and high-speed test, and digital communications engines. Over the course of his career, Dr. Oprysko has published more than 50 papers and patents. He has also served on many conference panel sessions and evaluation committees for the National Science Foundation. He has served on numerous technical conference committees having been one of the original organizers of the highly successful Optoelectronics Programs at the IEEE Electronic Components Technology Conference and the IEEE Radio and Wireless Communications Conference.